Nanoscale Sensors

Lecture Notes in Nanoscale Science and Technology

Volume 19

For further volumes:
http://www.springer.com/series/7544

Shibin Li • Jiang Wu • Zhiming M. Wang
Yadong Jiang
Editors

Nanoscale Sensors

Springer

Editors
Shibin Li
State Key Laboratory of Electronic Thin Film and Integrated Devices
School of Optoelectronic Information
University of Electronic Science and Technology of China
Chengdu, People's Republic of China

Jiang Wu
State Key Laboratory of Electronic Thin Film and Integrated Devices
University of Electronic Science and Technology of China
Chengdu, People's Republic of China

Zhiming M. Wang
State Key Laboratory of Electronic Thin Film and Integrated Devices
University of Electronic Science and Technology of China
Chengdu, People's Republic of China

Yadong Jiang
State Key Laboratory of Electronic Thin Film and Integrated Devices
School of Optoelectronic Information
University of Electronic Science and Technology of China
Chengdu, People's Republic of China

ISSN 2195-2159 ISSN 2195-2167 (electronic)
ISBN 978-3-319-02771-5 ISBN 978-3-319-02772-2 (eBook)
DOI 10.1007/978-3-319-02772-2
Springer Cham Heidelberg New York Dordrecht London

Library of Congress Control Number: 2013957068

Printed on acid-free paper

Springer is part of Springer Science+Business Media (www.springer.com)

Preface

Nanotechnology offers the promise of providing sensors capable of achieving many ideal characteristics, particularly those associated with speed, selectivity, sensitivity, and size reduction. Therefore, the development of nanoscale sensors is one of the most active areas in all of nanotechnology. Although the obvious advantages of miniaturization have driven a pronounced trend in sensor development for the past 50 years, new properties are realized at the nanoscale due to the inherent benefits of working directly at the molecular level where the sensing phenomena take place.

With immobilized bioreceptor probes selective for target analyte molecules, nanoscale biosensors have been extensively studied and used in different fields. As a result, the first group of the chapter contributions (r 1–4) describes general concepts of nanoscale biosensor design and application. Chapter 1 reviews the contribution of the evolving science of nanobiotechnology to precise sensing and quantification of water-borne contaminants. The use of portable nanobiosensors capable of instant field tests would help in screening sources of drinking water without the need for expensive and bulky analytical instruments. Future research directions for the fabrication of improved, fast-acting sensitive nanobiosensors are provided in this chapter as well. Chapter 2 discusses the general concept of surface-enhanced Raman spectroscopy (SERS), which is an increasingly exploited tool for the study of cell biology. Using SERS, noble-metal nanoparticles can be interrogated with wavelengths of light to which cells exhibit minimal autofluorescence, and return useful information about their immediate chemical environment. At the beginning of Chap. 3, nanowire-based field effect transistor biosensors are introduced. Afterwards, this chapter reviews the theoretical basis of BioFET-SIM models describing both single and multiple charges on the analyte. The use of the interface and its relative command line version are briefly shown. Finally, possible applications of the BioFET-SIM model are presented. In Chap. 4, the focus is on the basic photoelectrochemical sensor or biosensor, including basic principles and new progress using different semiconductor materials. A variety of detection mechanisms and the characteristics of photoelectrochemical sensors and biosensors are also introduced and their future developments are prospected and discussed.

The second part of this book (Chaps. 5 and 6) reviews the progress of hydrogen gas sensors and photodetectors based on ZnO nanostructures. Chapter 5 reviews the synthesis and fabrication of pure and doped ZnO nanoscaled materials for hydrogen sensing applications. Both gas-phase growth and solution growth methods are described. Afterwards, the fabrication of nanoscaled sensors based on ZnO nanostructures and their characteristics are discussed. The results include several methods that were employed to enhance nanosensor performance, such as increasing the surface/volume ratio, impurity doping, and surface functionalization. In Chap. 6, different types of photodetector device structures and architectures are presented. With a brief review of the growth/synthesis processes, this chapter also provides a synopsis of the most recent developments in the field of research and design of optoelectronic sensors and related devices based on nano- and microscale ZnO.

Another important part of this book (Chaps. 7 and 8) offers details of nanoscale sensors based on nanocarbon materials. Nanocarbons, such as carbon nanotubes (CNTs) and nanosheets of graphene or graphene oxide, as well as their hybrids with nanosized noble metals and semiconducting oxides, have emerged as new types of thin film gas sensors with advanced sensing performance. Chapter 7 aims to summarize progress in processing and characterization of nanosized hybrids based on nanocarbons and various nanoparticles for thin film gas sensors. The primary purpose of Chap. 8 is to assist and encourage researchers who are new to the field of carbon nanotubes and stretchable sensors in starting to explore and prototype nanotube-enabled devices and materials. While the primary focus is on how to use "off-the-shelf" components to prepare carbon nanotubes for use in a wide variety of applications, specific emphasis is placed on 2D patterning of nanotubes and stretchable electronics.

The last chapter (Chap. 9) presents the design, modeling, fabrication, and characterization of in-plane resonant nano-electro-mechanical (NEM) sensors based on the mass detection principle. In this chapter, the numerical modeling of IP-RSGFET sensors is also presented using a three-dimensional finite-element-method electro-mechanical simulation of IP-RNEM sensors combined with a NEM/MOS hybrid circuit simulation. The results of the modeling and experimental analysis provide a realistic guideline for further improvement and characterization of these ultrasensitive sensors.

We appreciate the creative work and contributions from all of the authors in this book. The book should be very useful to scientists and engineers who are already exploring the nanoscale sensors field, as well as to graduate and postgraduate students specializing in physics, chemistry, bioengineering, electronics, and materials science who are considering future projects and research opportunities. The editors also gratefully acknowledge the financial support of the National Natural Science Foundation of China through Grant NSFC- 61204098 and 61371046.

Chengdu, People's Republic of China

Shibin Li
Jiang Wu
Zhiming M. Wang
Yadong Jiang

Contents

Contributors

Mads Brandbyge Department of Micro & Nanotechnology, Technical University of Denmark, Kgs. Lyngby, Denmark, mads.brandbyge@nanotech.dtu.dk

Colin J. Campbell School of Chemistry, EaStCHEM, University of Edinburgh, Edinburgh, UK, colin.campbell@ed.ac.uk

Hongjun Chen ARC Centre of Excellence for Functional Nanomaterials, School of Chemical Engineering and Australian Institute for Bioengineering and Nanotechnology, The University of Queensland, Brisbane, QLD, Australia, h.chen11@uq.edu.au

Lee Chow Department of Physics, University of Central Florida, Orlando, FL, USA, lee.chow@ucf.edu

Daniel J. Cohen Joint Graduate Program in Bioengineering, University of California at Berkeley and San Francisco, Berkeley, CA, USA, daniel.cohen@berkeley.edu

Luca De Vico Department of Chemistry, University of Copenhagen, Copenhagen, Denmark, luca@chem.ku.dk

Ankush A. Gokhale Department of Chemical Engineering and Materials Science, Michigan State University, East Lansing, MI, USA, gokhale1@egr.msu.edu

Faezeh Arab Hassani School of Materials Science, Japan Advanced Institute of Science and Technology (JAIST), Nomi, Japan, ahfaezeh@jaist.ac.jp

Martin R. Hediger Department of Chemistry, University of Copenhagen, Copenhagen, Denmark, martin@chem.ku.dk

H. Huang Singapore Institute of Manufacturing Technology (SIMTech), Singapore, Singapore, hhuang@SIMTech.a-star.edu.sg

Adrian Mihai Ionescu Laboratory of Micro/Nanoelectronic Devices, Ecole Polytechnique Fédérale de Lausanne (EPFL), Lausanne, Switzerland, adrian.ionescu@epfl.ch

Jan H. Jensen Department of Chemistry, University of Copenhagen, Copenhagen, Denmark, jhjensen@chem.ku.dk

L. B. Kong School of Materials Science and Engineering, Nanyang Technological University, Singapore, Singapore, ELBKONG@ntu.edu.sg

Ilsoon Lee Department of Chemical Engineering and Materials Science, Michigan State University, East Lansing, MI, USA, leeil@egr.msu.edu

Jue Lu Technova Corporation, Lansing, MI, USA, juelu@technovacorp.com

Oleg Lupan Department of Physics, University of Central Florida, Orlando, FL, USA, lupan@physics.ucf.edu

Michel M. Maharbiz Department of Electrical Engineering, University of California, Berkeley, CA, USA, maharbiz@eecs.berkeley.edu

Karen L. Martinez Bio-Nanotechnology and Nanomedicine Laboratory, Department of Chemistry & Nano-Science Center, University of Copenhagen, Copenhagen, Denmark, martinez@nano.ku.dk

Hiroshi Mizuta School of Electronics and Computer Science, University of Southampton, Southampton, UK, hm2@ecs.soton.ac.uk

Jesper Nygård Nano-Science Center & Center for Quantum Devices, Niels Bohr Institute, University of Copenhagen, Copenhagen, Denmark, nygard@nbi.dk

Yuri M. Strzhemechny Texas Christian University, Fort Worth, TX, USA, Y.Strzhemechny@tcu.edu

Patrick I. T. Thomson School of Chemistry, EaStCHEM, University of Edinburgh, Edinburgh, UK, patrick.thomson@gmail.com

Yoshishige Tsuchiya School of Electronics and Computer Science, University of Southampton, Southampton, UK, yt2@ecs.soton.ac.uk

Lianzhou Wang ARC Centre of Excellence for Functional Nanomaterials, School of Chemical Engineering and Australian Institute for Bioengineering and Nanotechnology, The University of Queensland, Brisbane, QLD, Australia, l.wang@uq.edu.au

W. G. Zhu School of Electrical and Electronic Engineering, Nanyang Technological University, Singapore, Singapore, EWZHU@ntu.edu.sg

Chapter 1
Recent Progress in the Development of Novel Nanostructured Biosensors for Detection of Waterborne Contaminants

Ankush A. Gokhale, Jue Lu, and Ilsoon Lee

Abstract Rapid industrialization and exploitation of natural resources to accommodate for the demands of the increasing human population have contributed to large-scale contamination of the ecosystem. The presence of environmental contaminants and toxins in the ecosystem can have a deleterious effect on human health. To counter this, there is a need to come up with effective strategies to detect and quantify the presence of these pollutants in the environment. In this review, the contribution of the evolving science of nanobiotechnology for precise sensing and quantification of waterborne contaminants will be presented. The use of portable nanobiosensors capable of instant field tests would help in screening sources of drinking water, thus eliminating the need for expensive analytical instruments. Recent advances in nanotechnology tools have enabled the fabrication of integrated nanostructured bioelectronic interfaces that are capable of sensing minute concentrations of specific analytes. The sensing principles and the quantification capability of the sensor depend on the type of nanomaterial used, its morphology, as well as the microenvironment surrounding the biological component. Nano-architectures that enable increased interaction between the desired analyte and the biological component followed by effective signal transduction to the electronic component help to improve the sensitivity and the response of the sensor. In this review, recent progress in the development of nanomaterial incorporated transducer components and biorecognition elements will be discussed. Finally, this review will also provide future research directions for the fabrication of improved, fast-acting sensitive nanobiosensors.

A.A. Gokhale • I. Lee (✉)
Department of Chemical Engineering and Materials Science, Michigan State University, East Lansing, MI 48824-1226, USA
e-mail: gokhale1@egr.msu.edu; leeil@egr.msu.edu

J. Lu
Technova Corporation, 1926 Turner Street, Lansing, MI 48906, USA
e-mail: juelu@technovacorp.com

S. Li et al. (eds.), *Nanoscale Sensors*, Lecture Notes in Nanoscale Science and Technology 19, DOI 10.1007/978-3-319-02772-2_1,

1.1 Introduction

The alarming impact of environmental pollution on human health has sparked serious introspection in recent years. Anthropogenic activities such as agriculture and consumption of fossil fuels have produced unprecedented deterioration of air, soil, and aquatic environments [1, 2]. The presence of pollutants and toxins in the ecosystem has contributed to a marked increase in the level of greenhouse gases and produced dramatic inconsistencies in the global weather patterns. To counter this, there is a need to come up with effective strategies to detect and quantify the presence of these pollutants in the environment. Environmental monitoring and control has thus emerged as a major preventive measure against the rise of these pollutants [3, 4]. These techniques enable regulators to continuously monitor the prevalence of contaminants in risk-prone areas and take immediate corrective action if their levels rise above the permissible limits. The collection of environmentally sensitive parameters can also provide crucial historical data to the future generations of environmental scientists [5]. However, rising costs of environmental monitoring have been a major dampener for the growth of this sector. According to Looney and Falta [6], manual collection and off-site analysis of thousands of samples collected daily can be an extremely labor-intensive process with costs ranging from $100 to 1,000 per sample.

Contamination of sources of drinking water is fast emerging as a major threat thanks to wide-scale use of pesticides and nitrogen-based fertilizers. Application of pesticides to improve agricultural productivity is so rampant that around three billion kilograms of pesticides are consumed globally every year [7]. Blatant application of pesticides and fertilizers along with other agricultural activities has put an enormous strain on the world's aquatic drinking resources since most of toxins released from these materials invariably enter the water cycle. Along with that, organic pollutants and heavy metal particulates released from manufacturing units can also end up in the drinking water supplies. In order to protect the general public from major health hazards due to exposure to waterborne toxins and contaminants, the US Environmental Protection Agency (EPA) has imposed strict drinking water standards for public distribution networks to comply with.

1.2 Biosensors for Environmental Monitoring

Decision making for effective environmental monitoring requires all round efforts by regulators, scientists, and local stakeholders. Danielsen and co-workers [8] advocated that the contribution of the local stakeholders is extremely important. While regulators and scientists can come up with administrative tools and advanced protocols, respectively, the final implementation of the policy rests with the local population. Each region has a unique set of flora and fauna. The local populace is best suited to identify the debilitating environmental factors that can disturb the local

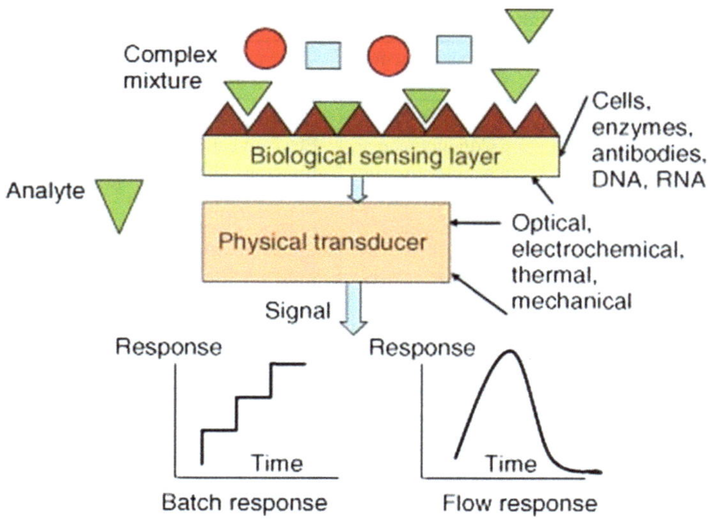

Fig. 1.1 Schematic diagram of a typical biosensor. Reproduced with permission from [11]. The quantification of the biological signal is made possible through the use of transducers. Transducers play an important role in the functioning of the biosensor because of their ability to smartly exploit the behavior of recognition element under certain conditions and convert it into a measurable output. Table 1.1 given below classifies the different types of biosensors on the basis of their transducer elements along with their general applications

ecological balance. Vigilance exercised by local groups can help in effective environmental monitoring and could also speed up the process to take corrective action in case of an imbalance. Methods for detection and quantification of environmental toxins include techniques such as ion chromatography (IC), complexation electrospray mass spectrometry (cESI-MS), and Raman spectroscopy [9]. However, these methods require extensive instrumentation, need professional assistance, and are not particularly attractive for instant field tests. Local environmental monitoring groups would therefore require access to analytical laboratories equipped with technical personnel trained in toxin analysis. Biosensors on the other hand have demonstrated a great potential for the development of simple portable systems capable of detection and quantification even at minute concentrations. Biosensors as the name suggests include the successful incorporation of a biorecognition element mounted on a transducer assembly capable of intercepting and transmitting the output signal given out by the biological receptor. Figure 1.1 shows a schematic diagram of a simple biosensor with biological and transducer elements. Biosensors have advantages such as high selectivity, minimum sample preparation, and successful application in field tests which makes it possible to overlook their relatively low accuracy and reproducibility as compared to other conventional analytical techniques [10].

Scognamiglio and co-workers [13] cite a report published by Global Industry Analysts Inc. that indicates that the global biosensor market stood at US $8.2 billion in 2009. The report further adds that at an average annual growth of about 6.3%,

Table 1.1 Classification of biosensors based on their transducer elements

Transducer system	Measurement mode	Typical applications
Electrochemical		
Conductometric	Conductance	Enzyme-catalyzed reactions
Enzyme electrode	Amperometric (current)	Enzyme substrates and immunological systems (antibody–antigen)
Field-effect transistors (FET)	Potentiometric (voltage)	Ions, gases, enzyme substrates, and immunological analytes
Ion-selective electrodes (ISE)	Potentiometric (voltage)	Ions in biological media, enzyme electrodes, immunoelectrodes
Gas-sensing electrodes	Potentiometric (voltage)	Gases, enzymes, organelle, cell or tissue electrodes, enzyme immunoelectrodes
Impedimetric	Impedance	Enzyme immunosensors
Piezoelectric crystals, surface acoustic devices	Mass change	Volatile gases, vapors, and immunological analytes
Optoelectronic, fiber optics, and waveguide devices	Optical	pH, enzyme substrates, immunological analytes
Thermistors, diodes	Thermometry/ calorimetric (heat)	Enzyme, organelle, whole cell or tissue sensors for substrates, gases, pollutants, antibiotics, vitamins, immunological analytes

Reproduced with permission from [12]

the biosensor market would be dominated by devices for biomedical and life sciences followed by environmental monitoring and remediation applications [13]. It was predicted over a decade ago that process industries and environmental monitoring applications would constitute to about one-fifth of the biosensor market [14]. Though that target has not been achieved, mainly due to technical and commercialization hurdles, further impetus can be provided for the growth of the biosensor market by improving the performance of the biological component and the transducer technology. Thanks to the great advancement in the field of proteomics in the past few decades, enzymology and whole-cell biocatalysis are now mature technologies and are credited to developing highly substrate-specific biological mechanisms [15]. With the successful incorporation of nanoscale materials, transducer technology has also shown promising outcomes.

The aim of this paper is to review the recent progress made in the sensor industry by the incorporation of nanomaterials and biological components to build novel synergistic devices capable of detecting minute concentrations of waterborne contaminants. An overview of the different types of pollutants/contaminants that carry a risk of endangering drinking water supplies would be briefly covered in the next section. In order to understand the recent advances in the field of nanobiosensing for contaminant detection, the sensors have been classified into three different types on the basis of the transducing element used. Finally innovative solutions based on current research or emerging trends in the nanobiosensor industry are also reported.

1.3 Occurrence of Contaminants in Drinking Water Supplies

A number of pollutants or trace contaminants can be found in drinking water supplies as a result of industrial effluents, agricultural practices, and domestic sewage. Given the immense research carried out in this area, it would be hard to summarize all the contaminants and their complex interactions with the rest of the environment. Detailed studies regarding contaminants and their ill-effects can be found elsewhere [16–18]. In this review, we present a broad classification of these contaminants into three groups:

1. *Organic contaminants*: Alkanes, aromatic compounds, and polycyclic aromatic hydrocarbons collectively abbreviated as PAHs pose serious environmental risks [16]. They are mostly formed during incomplete combustion or pyrolysis. These contaminants are extremely toxic and can cause cancer or related diseases. Phenols, perfluorates, phthalates, polybrominated diphenyl ethers (PBDEs), and triazoles can cause significant damage to human health [17]. Some of these contaminants are extremely water soluble and can cause disruption of the endocrine functioning in the human body.
2. *Agricultural chemicals*: A number of harmful components are added to soil in the form of agricultural pesticides and fertilizers. These chemicals can leach out of the soil and enter groundwater supplies. The extensive use of Chilean saltpeter as a fertilizer in the USA is responsible for the dual addition of nitrate and perchlorate to the environment. Excessive nitrate consumption can lead to methemoglobinemia (type of cellular anoxia), whereas perchlorate can cause cancers and thyroid malfunction. Pesticides are composed of dangerous components, some of which can directly affect the human nervous system. Table 1.2 lists the ill-effects of certain types of pesticide residue on the human body. Table 1.3 lists EPA guidelines for regulating the amount of pesticide residues in drinking water supplies.

Table 1.2 Impairment of body functioning due to pesticide residue

Pesticide	Type of pesticide	Mode of action
Carbaryl	Insecticide	Inhibits acetylcholine esterase
Malathion	Insecticide	Inhibits acetylcholine esterase
Chlorpyrifos	Insecticide	Inhibits acetylcholine esterase
Diazinon	Insecticide	Inhibits acetylcholine esterase
Endosulfan	Insecticide	Nervous system stimulant producing convulsions
Acetochlor	Herbicide	Inhibits cell division
Metolachlor	Herbicide	Inhibits cell division
Glyphosate	Herbicide	Inhibits amino acid synthesis
2,4-D	Herbicide	Auxin mimic
Atrazine	Herbicide	Inhibits photosystem II

Reproduced with permission from [19]

Table 1.3 EPA drinking water standards for a select group of pesticide components

Pesticide	Maximum concentrations observed in water bodies (p.p.b.)	Maximum contaminant level or lifetime health advisory (p.p.b.)	Criteria for continuous concentration (p.p.b.)[a]
Carbaryl	2,500	No standard	No standard
Malathion	583	100	0.1
Chlorpyrifos	2	2	0.041
Diazinon	33,000	1	0.17
Endosulfan	9	No standard	0.056
Acetochlor	21	No standard	No standard
Metolachlor	124	70	No standard
Glyphosate	5,200	700	No standard
2,4-D	692	70	No standard
Atrazine	172	3	No standard

Reproduced with permission from [19]

[a]Defined by the EPA as "the highest concentration of a material in surface water to which an aquatic community can be exposed indefinitely without resulting in an unacceptable effect"

Table 1.4 EPA regulatory limits for inorganic/particulate contaminants in drinking water supplies

Contaminant	Max. contaminant level goal (MCGL) (mg/l)	Max. contaminant level (mg/l)
Antimony	0.006	0.006
Arsenic	–[a]	0.010 (as of 01/23/06)
Asbestos (fiber >10 μm)	7 million fibers per liter	7 million fibers per liter
Barium	2	2
Beryllium	0.004	0.004
Cadmium	0.005	0.005
Chromium (total)	0.1	0.1
Copper	1.3	Action level = 1.3[b]
Cyanide (as free cyanide)	0.2	0.2
Fluoride	4.0	4.0
Lead	zero	Action level = 1.3[b]
Mercury (inorganic)	0.002	0.002
Nitrate (measured as nitrogen)	10	10
Nitrite (measured as nitrogen)	1	1
Selenium	0.05	0.05
Thallium	0.0005	0.002

Reproduced with permission from [18]

[a]Therefore, there is no MCLG for arsenic since MCLGs were not established before the 1986 Amendments to the Safe Drinking Water Act

[b]If more than 10% of tap water samples exceed the action level, water systems must take additional steps

3. *Particulates and inorganics*: Heavy metal particulates have a high residence time in the environment because they tend to accumulate in the bodies of organisms. Heavy metals impair the normal functioning of the body by inhibiting the action of enzymes pertinent to metabolism. Table 1.4 lists EPA standards for inorganic/particulate contaminants in drinking water supplies.

1.4 Nanostructured Biosensors for Environmental Monitoring

The growth of nanotechnology-enabled goods has been phenomenal in recent years. The market potential for these goods is already estimated to be US $40 billion and is supposed to register significant gains in the coming years to reach as much as US $1 trillion by 2015–2020 [20]. Environmental monitoring is one of the seven core areas predicted to show appreciable growth in nanotechnology-based applications. Market research companies have predicted huge employment opportunities in nanotechnology-driven industries in the coming years. Based on such reports, Liota and Tzitzios [21] concluded that the jobs in manufacturing sector will undergo a major reorganization with about ten million people (about 11% of the global workforce in manufacturing) will be employed in nanotechnology-driven industries by 2014. The versatility of use of nanomaterials in biosensing applications has been aptly described by Wang [22]. Tailor-made nanoparticles and colloidal quantum dots have been exploited for their biotagging applications [23–25]. Similarly, 1-D nanostructures (nanowires and nanotubes) serve as excellent bioelectronic interfaces even in poorly conducting medium [26, 27]. This is because unlike flat 2-D films which are prone to lateral current shunting, 1-D nanostructures are the smallest dimension that allow efficient transport of electrons enabling detection of single molecules [28]. In the following sections, nanobiosensors for the purpose of detection of waterborne contaminants have been classified on the basis of the transducer element.

1.4.1 Electrochemical Nanostructured Biosensors

Electrochemical biosensors employ a bioelectrochemical reaction to produce a quantifiable signal. Cases where the measurable signal is in the form of current (amperometry) or potential (potentiometry) are very common. Similarly, other cases include change in conductive properties (conductometry), resistance and reactance (electrochemical impedance spectroscopy), and field effect (transistor technology) [29]. Most electrochemical biosensors more particularly amperometric sensors make use of oxidoreductase, a class of enzymes capable of catalyzing the transfer of electrons from the oxidant molecule to the reductant molecule [30]. To improve the performance of electrochemical biosensors, there is a need to employ electroactive agents that are capable of amplifying the shuttling of electrons between the redox site of the enzyme and the electrode surface. These agents commonly known as mediators therefore find large-scale application in the use of electrochemical biosensors. Electrochemical biosensors are particularly attractive because of their simple operation, very small reaction volumes, and little or no sample preparation [31]. The fusion of biorecognition and catalytic abilities of enzymes on one hand and the electronic properties of conductive nanomaterials

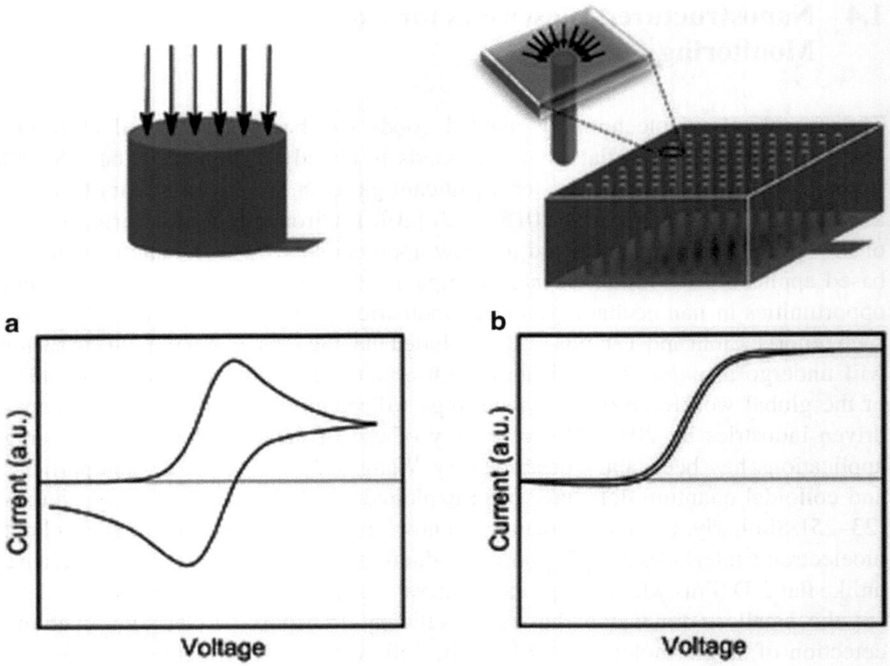

Fig. 1.2 Typical cyclic voltammograms of (**a**) macroelectrode (planar diffusion) and (**b**) nanoelectrode (radial diffusion). Reproduced with permission from [45]

on the other hand can help fabricate novel biosensors with improved accuracy and sensitivity [32]. Several examples featuring the incorporation of nanomaterials such as metal nanoparticles [33–35], layer by layer (LbL) techniques [36–39], carbon nanostructures [40–42], and clay platelets [43, 44] are available in literature. Wei and co-workers [45] in an exhaustive review regarding the different types of electrochemical nanostructured biosensors report a number of benefits arising from the incorporation of nanoscale materials into the biosensor architecture. The success of the improved sensor response lies in the high surface area of the nanomaterials. Electrobiochemical reactions typically generate very small currents. However, the large surface area of the nanoelectrode enables measurement of this current even in poorly conducting medium due to the reduction in the ohmic drop across the electrode surface. It has been also predicted that nanoelectrodes show radial diffusion as opposed to planar diffusion resulting in higher mass transfer rates and higher signal-to-noise ratio (S/N) [45, 46]. Figure 1.2 shows typical cyclic voltammograms of macroelectrodes and nanoelectrodes. Planar diffusion in case of macroelectrode leads to the formation of peaks, whereas radial diffusion as a result of nanoelectrodes leads to the formation of a sigmoid curve.

1.4.1.1 Amperometric Nanostructured Biosensors

Amperometric biosensors are devices where current between the working electrode and the reference electrode is monitored as a function of time. The applied potential remains constant throughout the process. Biochemical reactions taking place at the surface of the electrode generate a steady-state or a limiting current that is proportional to the concentration of the electroactive species [12].

As mentioned earlier, phenol is one of the water-soluble contaminants that can pose high risk to human health. Phenol has been found in domestic water supply in the USA at a concentration of 1 ppb [47], whereas as much as 40 ppm has been observed in river water in the vicinity of petrol processing plants [48]. Environmental monitoring for phenolic compounds can be accomplished by using amperometric nanobiosensors and has been reported extensively in literature [49–51]. Zhang and co-workers [52] replaced carbon nanotube (CNT) with carbon nanofiber (CNF) to build an electroactive composite with polyaniline. The resultant composite (PANI-CNT) enjoys several advantages such as easy surface chemistry and high mechanical strength as compared to PANI-CNT composite. The addition of ionic liquid, a solvent that can potentially serve as a supporting electrolyte in biosensing applications, promotes adhesion between the composite and surface of the support. The one-step in situ electropolymerization of aniline on the surface of CNF helps in the growth of nanofibrillar structures. The immobilization of tyrosinase, an enzyme capable of converting monophenols into quinone derivatives, is accomplished by cross-linking the enzyme with the nanofiber using glutaraldehyde. The resultant phenol biosensor is highly sensitive and has a detection limit of the order of 0.1–0.5 nM for different types of phenolic derivatives.

Liu and co-workers [53] used biocompatible colloidal gold particles to improve the performance of carbon-paste phenol biosensor. The addition of gold colloidal particles helped to improve the performance of the sensor making it fast, accurate, and highly sensitive about 4.25 times better than the sensor without the gold particles. The improved performance is attributed to the biocompatibility that the gold colloidal particles bring to the carbon paste. The immobilization of tyrosinase on anionic colloidal gold ensures the formation of a microenvironment that closely resembles that of the native state. Firstly, the addition of gold particles helps to improve the conductivity of the carbon-paste electrode, thus improving the electrical connection with the enzyme and the electrode surface. Secondly, on account of the curvature of the particles, the addition of colloidal gold is hypothesized to reduce the restrictive effect caused due to immobilization, allowing the enzyme molecules to freely orient themselves.

In another variation, Song and co-workers [54] replaced carbon paste with graphene oxide to construct a gold nanoparticle-based tyrosinase biosensor. Graphene oxide is a flat 2-D nanomaterial that can be easily dispersed in water making it a promising candidate for biological applications. GO can be easily functionalized, thus opening doors to a variety of surface chemistry techniques for effective immobilization of biomolecules. In this example, tyrosinase was covalently bound using a supramolecular assembly of 1-pyrenebutanoic acid succinimidyl ester (PASE) on GO sheets. The high enzyme loading on account of large surface area of GO matrix coupled with the bioaffinity due to the incorporation of gold nanoparticles provides superior sensing capabilities to this electrode. The authors used a screen-printed electrode (SPE) for the deposition of the different biosensor components. There has been a growing trend towards using SPE technology for disposable biosensor applications because of their cheap costs, easy portability, and low sample volume requirement. The ability to blend together nanomaterials with low-cost SPE technology could open doors to fabrication of economical nanobiosensors capable of environmental monitoring field tests.

Li and co-workers [55] tested the efficacy of using zinc oxide (ZnO) nanoparticle matrix for immobilization of tyrosinase. ZnO has an isoelectric point close to 9.5. The high isoelectric point contributes to the creation of a microenvironment that remains stable over a wide range of pH. This facilitates improved enzyme stability which in turn allows for the operation of the biosensor at neutral pH. Chitosan, a promising biocompatible polymer, was used in the fabrication of the ZnO tyrosinase biosensor on account of its bioaffinity and adhesion properties. Another observation made by Li and co-workers indicates that the no mediator is necessary in presence of ZnO nanoparticles. ZnO nanoparticles can promote direct electron transfer between the redox site of tyrosinase and the electrode surface.

Dubacheva and co-workers [56] used a polyelectrolyte LbL approach to incorporate tyrosinase on a graphite electrode. A cationic polyelectrolyte poly (diallyldimethylammonium chloride) (PDDA) was coupled with negatively charged tyrosinase to deposit a certain number of bilayers. The stabilization of the tyrosinase via cross-linking with glutaraldehyde promoted the fabrication of a highly sensitive (detection limit 10 nM) phenol biosensor. The authors however note that increasing number of bilayers do not improve the performance of the electrode. As more bilayers are deposited, the underlying layers may lack accessibility with the substrate. Arecchi and co-workers [57] used a nanofibrous nylon membrane for the immobilization of tyrosinase. The nanofibrous polyamide nylon membrane was prepared by electrospinning and mounted on a glassy carbon electrode (GCE). Thereafter the enzyme was loaded using the drop-cast method followed by drying. The authors report that the use of the nanofibrous nylon membrane has a significant effect on the selectivity of the enzyme. The nanobiosensor shows different levels of sensitivity towards various types of phenolic derivatives that decreases in the order: pyrocatechol > caffeic acid > epicatechin. The authors hypothesized that the use of nanofibrous nylon membrane improved the selectivity of the enzyme towards certain phenolic derivatives such as pyrocatechol as compared to caffeic acid. This hypothesis was verified by testing in presence of free enzyme with and without the nylon coating on the GCE. The presence of nylon membrane showed improved signal in case of pyrocatechol as compared to caffeic acid. The authors also studied the effect of covalent immobilization on the tyrosinase selectivity. Experiments with the immobilized enzyme and those with free enzymes confirmed the same level of selectivity towards pyrocatechol w.r.t caffeic acid, indicating that the active sites of the enzymes are not significantly affected during the immobilization process.

Korkut and co-workers [58] developed multiwalled carbon nanotube-poly(pyrrole)-horseradish peroxidase nanobiocomposite to serve as a sensing element for the detection of phenolic compounds. Tyrosinase has been long used for identification of phenols and its derivatives. However, the use of this enzyme is restricted to phenolic compounds with at least one ortho position free [59]. Horseradish peroxidase (HRP) on the other hand can respond to a wide range of phenolic derivatives. The quantification of phenolics is made possible through the double-displacement mechanism commonly known as "ping-pong" mechanism where the presence of two substrates, namely, peroxide and the phenolic derivative, under consideration is required for accurate quantification [60]. The nanostructured biosensor was subjected to testing with 18 different phenolic derivatives. With the exception of *o*-cresol, in all the other cases, the sensor displayed an extremely fast response time (about 2 s) with different levels of sensitivities for each substrate. The incorporation of CNTs helped in achieving two objectives. Firstly, it resulted in improved electrical conductivity of the electrode. Secondly as reported in literature elsewhere [61], it is also believed to function as an electron mediator between the potential range -1.5 to $+1.5$ V because of absence of significant oxidation or reduction peak in that range.

Amperometric biosensors have been designed to detect and quantify nitrate residues in drinking water [62, 63]. Can and co-workers [64] used a carbon nanotube/polypyrrole nanocomposite for the fabrication of nitrate biosensor. Nitrate reductase, an enzyme capable of converting nitrate to nitrite, was co-immobilized with the nanocomposite using a one-step electrochemical polymerization approach. The concentration of pyrrole (monomer) in the polymerization medium is important. High concentration of monomer was found to result in the development of thicker films that limited the ability to transfer electrons leading to a poor sensor response. Nicotinamide adenine dinucleotide (NADH) is commonly used cofactor for the efficient conversion of nitrate to nitrite. However, disadvantages associated with the use of NADH such as low current response, harmful by-products, and high cost necessitate the addition of redox mediators. The authors report the use of potassium ferricyanide $K_3Fe(CN)_6$ for the first time as a redox mediator. Compared with other redox mediators, potassium ferricyanide displayed a much higher sensitivity as well as a reasonable response time. The authors also reported that the addition of lithium chloride to the buffer solution helps to scale down the potential at which the mediator functions efficiently. Lower potential helps in maintaining the integrity of the immobilized nitrate reductase.

Moretto and co-workers [65] used an ultrathin film composite membrane composed of 1-methyl-3-(pyrrol-1-ylmethyl)pyridinium tetrafluoroborate (MPP), to compartmentalize the analyte and the aqueous nitrate reductase solution. An electrocatalyst in the form of a redox mediator can also be added to the enzyme solution in appropriate concentration. The advantage of using a thin film membrane allows for the permeation of analyte but restricts the leaching of the aqueous enzyme/electrocatalyst solution. Just like in the previous example, the thickness of the film is of significant importance. Higher film thickness can restrict the motion of analyte into the other compartment. Very thin films on the other hand can result in leaching of the electrocatalyst and/or the enzyme solution. A thickness of about 1 μm was reported to be the most optimum. Da Silva and co-workers [66] reported a novel method to fabricate an amperometric nanostructured biosensor to immobilize nitrate reductase. The enzyme was immobilized by incorporating it within a hydrophilic matrix consisting of laponite colloidal particles. The enzyme is contained within the interlamellar region of the clay, thus maintaining its integrity. The clay matrix was then used as a scaffold to deposit an electropolymerized form of pyrrole–viologen derivative to ensure electrical connectivity. The cationic exchange capacity of laponite particles presents an opportunity to integrate the water-soluble viologen within the bioinorganic coating. According to the authors, in the absence of the laponite particles, the biosensor acquired a more hydrophobic configuration due to the presence of alkyl chains between the viologen groups and the polymer backbone. A kinetics study confirmed the efficacy of using a clay scaffold for the fabrication of the biosensor. The low value of the apparent Michaelis–Menten constant K_m for the clay-modified biosensor indicated increased affinity between the substrate and the enzyme.

As shown earlier in Sect. 1.1, nitrite is also a potent water-soluble contaminant that can adversely affect human health. Nitrite is believed to be a precursor for the formation of several carcinogens. Hence detection and quantification of nitrite as a part of the environmental pollutant monitoring process is essential. Salimi and co-workers [67] developed a catalase-based GCE modified with carbon nanotubes for detection of different analytes including nitrite. Using the Laviron method which helps in estimating the electron transfer rate coefficient, the authors reported the improved shuttling of electrons between the catalase and the electrode surface on account of the incorporation of carbon nanotubes. The authors also noted the significance of pH in the measurement of analytes. The electrode showed lower electrocatalytic activity at a $pH > 5$. Under weakly acidic conditions, the availability of protons starts to diminish starving the electrochemical reaction. Therefore, a highly acidic condition ($pH = 2$) was chosen.

Besides nitrates and nitrites, contamination due to organophosphate residues from pesticides is also of grave concern [68–70]. Du and co-workers [71] used a sol–gel process for the development of a silicate template to immobilize acetyl cholinesterases (AChE). Gold nanoparticles were added for improving the conductivity and bioaffinity. The deposition gold nanoparticle incorporated silica sol–gel (Au SiSG) on GCE resulted in a porous weblike structure, thus increasing the effective surface area for the immobilization of AChE. Acetylthiocholine chloride (ATCl) is added for the buffer solution for evaluating the performance of the nanobiosensor. ATCl undergoes hydrolysis to produce thiocholine. The oxidation of thiocholine is monitored to get a quantifiable current. The presence of Au nanoparticles in silica sol–gel facilitates measurement of enzyme activity at low potentials. The state in which the nanoparticles are present in the silica gel can affect the strength of the output current. If present on the surface, Au nanoparticles can cause a large background current to flow through the system. Au nanoparticles entrapped in the sol–gel is hence a more favored configuration. Similarly, a very high loading of gold nanoparticles can also result in poor electrical wiring because of increased resistance

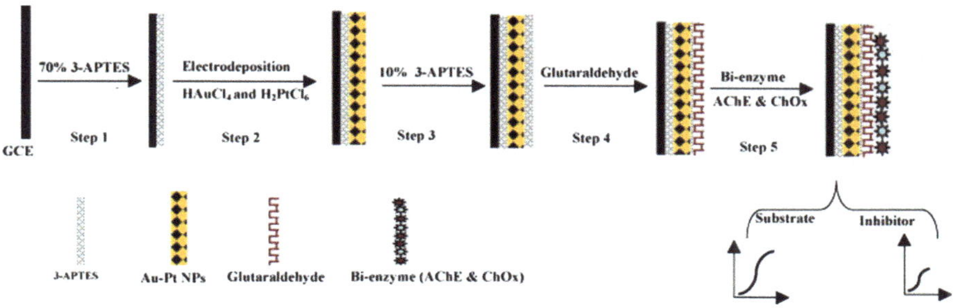

Fig. 1.3 Fabrication of bienzyme, bimetallic nanoparticle sensor for detection of organophosphates. Reproduced with permission from [73]

and double-layer capacitance. The silica sol–gel with its numerous hydroxyl groups can form a tight cross-linked hydrogen-bonded matrix that prevents the leaching of the immobilized enzyme, thus facilitating improved performance as compared to conventional biosensors. In another example, Du and co-workers [72] used a thin film copolymer consisting of polypyrrole and polyaniline that was integrated with CNTs for the immobilization of AChE. The two polymers were grown separately such that polyaniline was deposited on a nanocomposite comprising of polypyrrole + CNT. The polypyrrole-CNT composite exhibits porous morphology. However, if subjected to more than eight cyclic voltammetry (CV) scans, it resulted into a strong attachment of polypyrrole to the GCE electrode with fewer pores. The compact nature of the polymer matrix also restricted the transfer of electrons and hence such configuration should be generally avoided. The subsequent growth of polyaniline film resulted in a uniform homogenous film that is ideal for enzyme immobilization. The lower value of the electron transfer resistance R_{ct} estimated by using EIS revealed that that the incorporation of copolymer-CNT nanocomposite facilitated fast shuttling of electrons.

Upadhyay and co-workers [73] used a novel bienzyme, bimetallic nanoparticle system for effective detection of organophosphates. Figure 1.3 shows the assembly of the enzyme electrode using electrodeposition followed by cross-linking with glutaraldehyde. The authors used a combination of AChE and choline oxidase (ChOx) immobilized on a bimetallic support consisting of gold–platinum nanoparticles. During the enzymatic reaction, AChE catalyzed the conversion of acetylcholine to acetate and choline. The choline so formed was then subjected to oxidation in the presence of ChOx to produce H_2O_2. The detection of H_2O_2 can be achieved by using the electrochemical sensors. Platinum (Pt) is an excellent electrode material to facilitate oxidation/reduction of H_2O_2. The incorporation of Pt nanoparticles in the fabrication of GCE improved the apparent surface area and decreased the overpotential of H_2O_2. According to the authors, the addition of conductive metals such as gold nanoparticles further strengthened the electrocatalytic ability of the electrode. High enzyme loading for both ChOx and AChE was used so that all the reactants are converted to their respective products and there is no interference due to unreacted reactants.

Because of the extreme toxicity displayed by cyanide, the World Health Organization (WHO) has capped the maximum limit of cyanide present in water supplies to 1.9 μM. Shan and co-workers [74] have developed a electrochemical biosensor to detect the presence of cyanide in aqueous medium. The nanostructured device was designed by immobilizing the enzyme polyphenol oxidase in Zn–Al nanoclay matrix. Cyanide has an inhibitory effect on polyphenol oxidase and this forms the basis for the construction of cyanide sensing biosensors. However, as per the WHO standards for detection of cyanide, the biosensor needs to be equipped to handle extremely low concentrations as well. The mere inhibition of enzyme is not sufficient to produce a quantifiable signal. The Zn–Al nanoclays have higher positive charge density as compared to other types of clay such as laponite. The higher charge density facilitates greater binding of the negatively charged analytes just improving the chances of getting a quantifiable current output. The authors also noted the importance of the thickness of the coating in the functioning of the biosensor. Thicker films of clay/enzyme may block or slow down the flow of electrons between the enzyme and the electrode surface. Therefore thinner films are preferred. In another interesting application, Wang and co-workers [75] used a template-assisted electrochemical deposition to grow a hydroxyapatite nanoarray-based cyanide biosensor. The hydroxyapatite nanowires modified with chitosan can serve as excellent scaffolds for the immobilization of horseradish peroxidase (HRP). Cyanide can inhibit the activity of HRP. In presence of the cyano group, the attachment of H_2O_2 to the active center of the enzyme is obstructed leading to subdued H_2O_2 reduction current. Though cyanide is capable of blocking the active site of the enzyme, this effect is reversible. Washing the electrode with ultrapure water several times can help to relieve the inhibitory effect facilitating repetitive use of the sensor.

1.4.1.2 Potentiometric Nanostructured Biosensors

Potentiometric devices help to estimate the buildup of charge potential at the working electrode w.r.t the reference electrode in absence of any appreciable current flowing between them [29]. Common examples include glass pH electrodes and ion-selective electrodes (ISE). Potentiometric biosensors such as ISE show a large dependence on the buffer capacity of the sample and its ionic strength [76]. The addition of a biological element to the device helps to catalyze the production of an ion that the sensor under consideration is designed to detect [31].

There are several examples in literature that illustrate the use of this technique for the construction of biosensors [77–79]. Sohail and Adeloju [80] carried out potentiometric detection of nitrate using a nitrate reductase/polypyrrole thin film biosensor. The sensor was constructed with and without NADH. Longer polymerization time facilitated the formation of thicker films with higher enzyme loading. The addition of an electrolyte (such as KCl) helps in reducing the potential necessary for the growth of the enzyme–polymer–NADH composite and is a critical step in formation of thin film biosensor. NADH promotes amplified potentiometric response due to improved electron shuttling. In another study by the same authors [81], redox mediators such as thionin acetate (ThAc), safranin (Saf), and azure A (AzA) were used to replace NADH. Redox mediators are commonly used in amperometric detection. However the enzymatic conversion of nitrate to nitrite aided by redox mediators also causes a change in the H^+ ions making the process amenable to potentiometric detection. High overpotential and generation of by-products can be avoided by use of mediators leading to better sensitivity and selectivity. Besides nitrate reductase, organophosphorus hydrolase (OPH) is another enzyme whose catalytic action generates protons. Hence, OPH-based biosensors fabricated using potentiometric techniques have been used for detection of certain types of pesticide residues. Unlike the amperometric-based systems which rely on production of para-nitrophenol, potentiometric systems depend on change in pH as a result of release of H^+ due to the enzymatic action of OPH. All pesticides in the presence of OPH show pH change (of different intensity), facilitating simultaneous detection of different types of pesticides [82]. Mulchandani and co-workers [83] developed an OPH-based biosensor by depositing a thin layer of enzyme on a pH electrode. The enzyme was cross-linked with bovine serum albumin (BSA) and glutaraldehyde. As mentioned earlier, potentiometric sensors are very much susceptible to the concentration of the buffer. The authors observed an inverse relation between the potential change and the buffer concentration because high concentration buffers counteracted the pH change better than low concentration buffers. The use of sodium chloride as an electrolyte helps to stabilize the buffer and generate reproducible data. Schöning and co-workers [84] developed a dual amperometric–potentiometric thin film biosensor for the detection of organophosphates. The combination of amperometric and potentiometric transducer elements helps to conveniently distinguish between different types of organophosphates. Paraoxon and parathion were detected using the amperometric transducer, whereas paraoxon, dichlorvos, parathion, and diazinon were sensed using the potentiometric transducer.

1.4.1.3 Impedimetric Nanostructured Biosensors

The change in the resistance of the solution as a result of the biocatalytic activity can be exploited in the development of impedimetric nanostructured biosensors. The depletion of the analytes under consideration followed by formation of new products as a result of enzymatic biocatalysis can alter the solution conductivity. Conductometric measurements help the user to estimate the solution conductivity of a thin layer of electrolyte just adjacent to the electrode surface. Conductometric biosensors are generally considered to be a subset of impedimetric sensors. Impedimetric biosensors are easy to fabricate primarily because most sensors utilize the standard thin film deposition techniques [85]. These sensors also offer the attractive option to continuously monitor the conductivity/resistance data. This not only helps to record the progress of the reaction but also offers an opportunity to modify the process parameters if the need arises. Electrochemical impedance spectroscopy (EIS) is an important tool that helps to analyze the changes in the conductance of the electrode as a result of biocatalytic activity, protein immobilization or antigen–antibody binding interactions [29]. However, these methods are also prone to inaccuracies. The double-layer capacity and the electrode polarization during the enzymatic reaction and the low specificity of the electrode are some major obstructions that limit the widespread use of these sensors [86].

Whole-cell biosensor is a special type of sensor which consists of living cells that act as biological recognition elements. Unlike enzyme incorporated biosensors which require extensive enzyme purification, most whole-cell biosensors are simple to use and inexpensive to fabricate. When assembled on interdigitated electrodes, they offer distinct advantages for impedimetric measurements. A whole-cell biosensor consisting of platinum interdigitated electrodes and the microalgae (biological element) is reported to be used for monitoring the aquatic environments for traces of heavy metal particulates [87]. Interdigitated electrodes enjoy several advantages. They have low cell constant which allows accurate measurements even in case of low conductivity solutions, high interelectrode capacitance, and short penetration depth for electric fields [88]. Biocompatible materials such polysilicon have been proposed for the fabrication of interdigitated electrodes because of the robust immobilization of proteins, enzymes, or cellular organisms following the one-step silanization reaction. Other materials can be made biocompatible by suitable pretreatment [88]. Guedri and Durrieu [87] fabricated a Pt interdigitated nanobiosensor consisting of nanoscale self-assembled monolayers of alkanethiols. These monolayers are utilized for the immobilization of microalgae *Chlorella vulgaris*. Most of the other immobilization techniques either can result in harsh modifications of the surface moieties of the microalgae resulting in a drop in the catalytic activity or can create physical barriers (as in entrapment or encapsulation) that prevent interaction between the substrate and the algae. However, the use of self-assembled monolayers helps to overcome these problems. In case of immobilization techniques based on attachment of the algae using BSA reticulated by glutaraldehyde, the possibility of the substrate getting entrapped in BSA gels is high. Such techniques therefore produce sensors which are less sensitive at lower concentrations of the substrate. In another example involving whole-cell nanobiosensors, Hnaien and co-workers [89] reported the fabrication of a whole-cell nanotube-incorporated bacterial biosensor capable of detecting trichloroethylene (TCE), a toxic carcinogen commonly found in groundwater. A 3-D self-assembly of alkanethiol monolayer/carbon nanotube integrated with *Pseudomonas* antibodies was utilized for the immobilization of a bacterial strain *Pseudomonas putida F1* (PpF1). The deposition process includes self-assembly of an alkanethiol monolayer on gold interdigitated microelectrodes followed by covalent assembly of carbon nanotubes and the antibody. PpF1 is incorporated in the assembly through the antigen–antibody interaction mechanism. PpF1 consists of toluene dioxygenase (TOD), an enzyme that can degrade TCE into glyoxylate and formate. The ionic species produced during the reaction can cause a change in the conductance of the solution. This can be detected by using a two-electrode system consisting of a reference and the working electrode. The reference electrode construction involves replacing PpF1 with PpF4 which contains a defective mutant gene that is incapable of catalyzing the conversion of TCE into the above products. The use of carbon nanotubes is said to improve the sensitivity of the electrode by a factor of 80. Hnaien and co-workers have also reported another design for a nanostructured biosensor using the same alkanethiol/carbon nanotube/antibody assembly but by changing the detection mode to EIS [90]. Another example of a thin film conductometric interdigitated electrode was reported by Wang and co-workers [91]. The authors report the co-immobilization of nitrate reductase and redox mediator (methyl viologen) in a conductive Nafion film for the detection of nitrate ions. The presence of nitrate ions is evaluated as a function of the conductance of the solution.

1.4.1.4 Nanostructured Biosensors Based on Field-Effect Transistor Technology

Field-effect transistor (FET)-based biosensors have been gaining popularity in recent years. Their ability to operate under weak signal or high impedance conditions makes them a viable choice in electrobiochemical reactions [29]. Singh and co-workers [92] produced a miniaturized potentiometric biosensor by effective integration with the FET technology. They immobilized OPH at the gate of a pH-sensitive field-effect transistor (ISFET). In order to maximize the loading of OPH at the gate area, the authors used a silica sol–gel process to produce a porous structure. The silica gel with its Si–O–Si network is structurally and electrically compatible with the underlying SiO_2 layer. The formation of alcohol during the silica gel process can potentially denature the protein/enzyme when a one-step co-immobilization process is used. Instead the authors recommend an additional high temperature curing step to stabilize the structure and to drive away the volatile alcohol. Immobilization of biomolecules can be continued as a successive step. The authors noted that increasing the number of sol–gel coats helps in improving the enzyme loading. However, coatings beyond four layers resulted in an additional diffusion barrier that restricts the OPH activity. Schoning and co-workers [82] suggested the use of capacitive electrolyte–insulator–silicon (EIS) field-effect transducer in place of ISFET to minimize biosensor instability and high costs of fabrication. The overall setup of this sensor is similar to ISFET minus the photolithography steps. Easy mounting, lower costs, and long-term stability are some attractive features of using EIS electrodes over ISFET. The authors reported the use of thin film techniques to grow three different types of pH-sensitive materials (Ta_2O_5, Al_2O_3 and Si_3N_4) on the electrode surface. The enzyme OPH was immobilized using various immobilization strategies. The EIS biosensor displayed superior response time and good reusability making them attractive candidates for online monitoring systems.

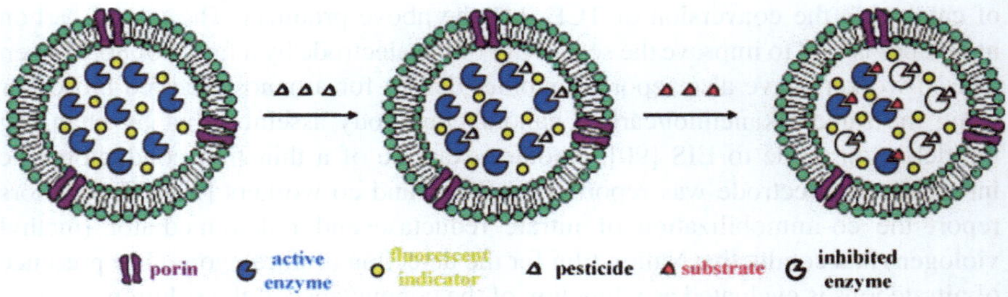

Fig. 1.4 Schematic diagram of liposome-based AChE biosensor. The incorporation of porins within the liposome membrane helps in substrate partitioning. Reproduced with permission from [98]

1.4.2 Optical Nanostructured Biosensors

Optical nanostructured biosensors have been used on a large scale for detection of water-soluble pollutants by examining properties such as change in absorption, luminescence, and florescence. Optical biosensors are reported to have higher signal-to-noise ratio [93]. This significantly contributes to improved accuracy. The use of nanoparticles and quantum dots in optical biosensors to modify planar electrode surface has facilitated amplification of signal and obviated the need for labels [94]. The ability to analyze a broad spectrum of samples using different wavelengths of light with little or no interference of individual components adds to the robustness of the design [95].

Various colorimetric and spectroscopic approaches have been provided in literature regarding quick detection of toxic contaminant residues. The Hestrin's method [96] (for unreacted acetylcholine (Ach) and corresponding esters) and Ellman's method [97] (for identification of thiols) are some of the techniques used in the past for estimation of organophosphorus pesticide residues. In recent years, a number of nanostructured materials have been incorporated to improve the optical biosensor technology. Vamvakaki and Chaniotakis [98] used a novel liposome-based AChE biosensor for detection of pesticide residues. Figure 1.4 shows the general working mechanism of the liposome-based biosensor. The liposome is a robust nanostructure capable of maintaining the integrity of the immobilized biomolecule. In the past, liposomes with key biorecognition elements have been applied to bioelectronic interfaces to act as bilayer lipid membranes (BLMs) and to mimic cellular membrane functions [99, 100]. In the above example, a fluorescence detecting scheme with pyranine serving as the indicator dye was utilized for quantification of the pesticide residues. Pyranine is pH sensitive with its pK_a very close to the optimum pH for AChE. The creation of porins that serve as selective channels for the permeation of substrate molecules of certain size protects the enzymes contained within the liposome. The interaction between the enzyme and the substrate results in the formation of acetic acid. The presence of acetic acid can reduce the local pH of the solution. This results in lower fluorescence intensity for the pyranine dye which can be correlated to the original organophosphate pesticide residue. Liposome encased nanobiosensors are highly stable. These biosensors are reported to be stable for about 50 days when stored at 4 °C. The free enzyme in the same time interval under similar conditions lost close to two-third of its original activity.

Haron and co-workers [101] used a polyelectrolyte-based nanobiosensor for determination of pesticide residues using the same pH-based approach. A planar waveguide was constructed by sandwiching a 190 nm silicon nitride (Si_3N_4) layer between micron-sized silicon dioxide (SiO_2) cladding layers. Nanofilms consisting of alternate deposition of positively charged polyelectrolytes and negatively charged species (such as an organic indicator like cyclotetrachromotropylene (CTCT) or enzymes such as AChE) were fabricated on top of Si_3N_4 layer. The large difference between the refractive indices between the Si_3N_4 layer and the cladding layers composed of SiO_2 presents a large evanescent field volume resulting in improved intensity of the output signal. The introduction of the pesticide residue results in inhibition of the enzyme producing lower amounts of products. The output signal intensity of the planar waveguide can be correlated to initial concentration of the residues. Simonian and co-workers [102] used a gold nanoparticle-based biosensor for effective detection of the organophosphates (OP). Unlike most approaches where a change in the local pH as a result of OP hydrolysis triggers a reduction in the intensity of the dye, the authors used a competitive inhibitor of the OPH enzyme to accomplish the same task. When this inhibitor which is in the form of a fluorophore is displaced by the OP substrate, it triggers change in the fluorescence. The fluorophore when introduced in the solution containing OPH conjugated gold nanoparticles gets weakly bound to the active site of the enzyme. The distance between the fluorophore and the gold nanoparticle is of critical interest. Higher distances (>40 nm) can reduce the intensity of the fluorescence. Smaller distances typically between 10 and 40 nm enhance the intensity. The addition of analyte of interest can induce a detachment of the fluorophore from the active site followed by attachment of OP in its place. The motion of the fluorophore away from the gold nanoparticle (distance more than 40 nm) can trigger a change in the intensity of fluorescence that can be used to track the concentration of the analyte. The performance of gold nanoparticle-based OPH biosensor strongly depends on the binding capabilities of the OPs (denoted by the constant K_M) and that of the fluorophore (K_I). It does not depend on the hydrolyzing ability of the OPH or the local pH of the buffer solution. In yet another example involving gold nanoparticles where competitive substrate interaction plays an important role, Liao and co-workers [103] developed an AChE-based liquid crystal (LC) biosensor. The enzyme-assisted growth of gold nanoparticles has potential applications in the sensing industry because of enhancement of optical signal associated with the increase in the size of these nanoparticles. The larger-sized gold nanoparticles can also disturb the orientation of the LC. In presence of ATCl, AChE catalyzes the production of thiocholine. When introduced to a solution containing gold precursors, thiocholine induces the reduction of $AuCl_4^-$ to neutral Au nanoparticles. The presence of large Au nanoparticles can substantially alter the LC configuration, resulting in a change in the optical signal. However, when an inhibitor in the form

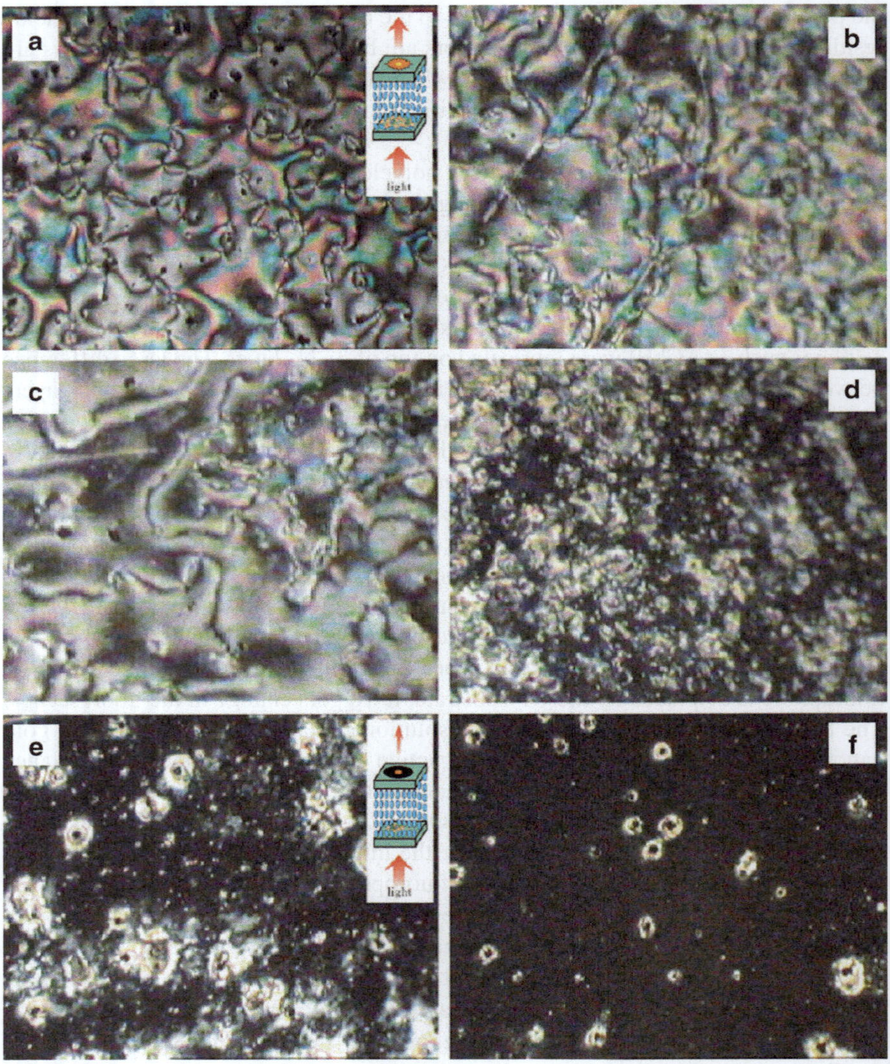

Fig. 1.5 Optical images of LC biosensor at different concentrations of OPs. (**a**) 0, (**b**) 0.3, (**c**) 3, (**d**) 30, (**e**) 300, and (**f**) 3,000 nM. Inset pictures in (**a**) and (**e**) show change in optical intensity as a result of the inhibitor effect of OPs. Reprinted with permission from [103]. Copyright (2012) American Chemical Society

of OP residue is present in the solution, it restricts the formation of thiocholine. Thus, the biometallic transformation of gold precursors to gold nanoparticles gets inhibited causing little or no change in the LC configuration. Figure 1.5 shows the optical images obtained at different concentrations of OPs. The inhibitory effect of OPs is not seen at lower concentrations. As the concentration increases, there is a decrease in the optical signal.

Zheng and co-workers [104] used the versatile technique of polyelectrolyte LbL assembly to fabricate a OP biosensor by depositing nanofilms of AChE coupled with photoluminescent (PL) CdTe quantum dots (QD). Two types of polyelectrolytes were used in this system: the cationic poly(allylamine hydrochloride) (PAH) and the anionic sodium polystyrene sulfonate (PSS). Initially a certain number of PAH and QD layers denoted as PAH/QD were deposited on glass or quartz substrates. The PAH/AChE bilayers were deposited at the top. The three bilayers of PAH/PSS sandwiched between the bottommost PAH/QD layers and the PAH/AChE layers. The presence of PAH/PSS layers obstructs the passage of large molecules such as enzymes to the bottom layers. Increasing number of PAH/QD layers facilitates uniform surface coverage leading to higher absorbance and PL intensity. After addition of acetylthiocholine, the PL intensity decreases. This is because AChE catalyzes the production of thiocholine that can quench the photoluminescence. However, in presence of OPs the enzymatic activity of AChE is inhibited leading to lower production of thiocholine. This has significant effect on the PL intensity. The quenching rate is monitored before and after the addition of the OPs to estimate the concentration of the pesticide residues. However, the fluorescence following the quenching action of thiocholine cannot be completely recovered. This makes it difficult for the sensor to be reused. Secondly, the nanobiosensor shows different levels of photoluminescence towards different types of OPs. Therefore when subjected to a mixture of OPs, the accuracy of the sensor is a suspect. Kim and co-workers [105] designed a self-assembled peptide hydrogel matrix incorporated with QDs and enzymes to serve as an optical biosensor. Though not specifically designed to detect contaminants, this method proposed by the authors can be easily extended to other types of biorecognition elements by simple modifications. The following schematic diagram (Fig. 1.6) shows the application of peptide hydrogels for the enzymatic detection of analytes. The peptide hydrogels are formed by self-assembly of *N*-fluorenylmethoxycarbonyl diphenylalanine (Fmoc-FF) building blocks.

Abdullah and co-workers [106] chose a biocompatible chitosan matrix for the immobilization of tyrosinase. The phenolic residues in the water samples can be estimated by adding a colored dye 3-methyl-2-benzothiazolinone hydrazone (MBTH). MBTH reacts with the end products of enzymatic reaction (*o*-quinone) to produce a maroon color that is characterized spectrophotometrically. In another example involving use of chitosan matrix, Abdullah and co-workers [107] used hybrid nafion/sol–gel silicate for MBTH immobilization. Sol–gel technology allows for low-temperature fabrication of porous glasses with the attractive option of mixing organic reagent dyes and biomolecules during the fabrication process. However, sol–gel can undergo cracking with passage of time making it a major impediment in the successful use of this technology. In order to prevent cracking, a hybrid material containing nafion integrated with sol–gel was proposed. The tyrosinase was encapsulated in a chitosan layer and deposited as a separate layer. The new design which allowed the integration of nafion gave the hybrid material a slightly hydrophobic character. This ensured the successful immobilization of MBTH with low risk of the dye leaching out. The composition of

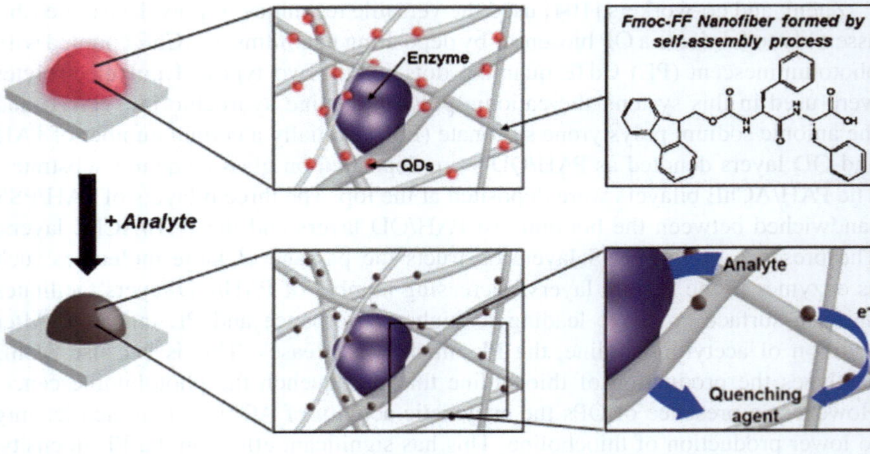

Fig. 1.6 Schematic diagram of PL quenching of peptide hydrogel. PL quenching is achieved as a result of enzymatic action on the analytes. Reproduced with permission from [105]

nafion in the hybrid composite is of significant interest. Increased concentration of nafion in the composite led to decreased porosity. This can adversely affect the immobilization of MBTH or the enzyme lowering the output signal. The leaching of MBTH from the sol–gel silicate film was quantified to study the effect of the hybrid nafion composite. Leaching of MBTH was much higher in absence of the nafion/sol–gel silica thus validating the importance of using the composite. Park and co-workers [108] developed a microdevice capable of detecting phenolic residues in water supplies. This device consisted of tyrosinase–quantum dot conjugates encased in poly(2-hydroxyethyl methacrylate) (PHEMA) hydrogel microparticles. Most biosensor applications suffer from poor stability due to immobilization on the enzyme on hard solid surfaces. In order to avoid this, hydrogel microparticles were chosen as immobilization supports. The hydrogel particles encase the enzyme molecules protecting them from sudden changes in the microenvironment. This encapsulation is achieved by photopolymerization. The biosensor also makes use of the microfluidics technology to entrap the enzyme-encased hydrogel particles. The incorporation of quantum dots into the biosensor design allows for the use of optical analytical techniques for evaluating the concentration of phenolic compounds in the water samples. The enzymatic catalysis facilitates the production of quinone intermediates from phenolic compounds. The quinone intermediates are known to cause a quenching effect that reduces the fluorescence intensity of the quantum dots. In another variation involving the same quenching approach, Dong and co-workers [109] used cysteine protected gold nanoparticles–tyrosinase conjugates for estimation of phenolic compounds. The enzyme tyrosinase shows strong affinity for the cysteine-coated gold nanoparticles because of the supramolecular chemistry that exists between them. The local cationic amino moieties existing on the tyrosinase molecule can electrostatically bind to the anionic cysteine covered nanoparticles

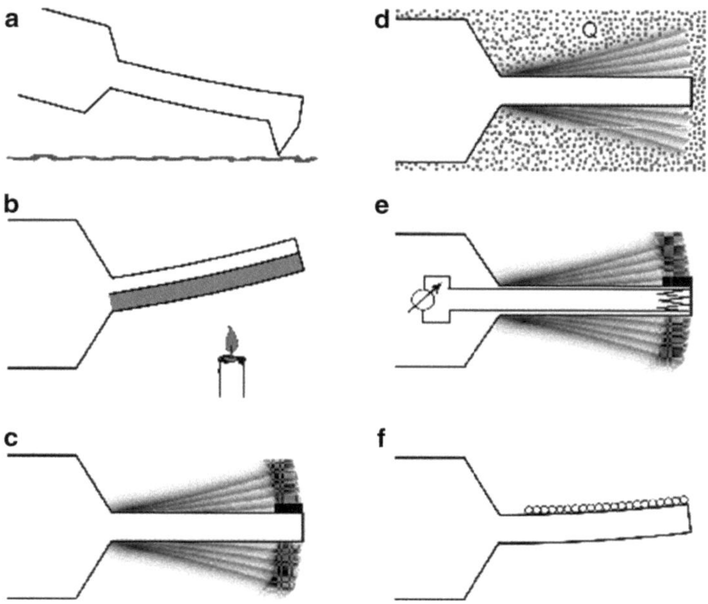

Fig. 1.7 Schematic diagram showing different configurations of cantilever-based sensors. (**a**) Force sensor used for AFM imaging. (**b**) Bimetallic thermal sensor. (**c**) Mass-loading sensor. (**d**) Medium viscoelasticity sensor. (**e**) Thermogavimetric sensor. (**f**) Stress sensor. Reproduced with permission from [112]

preventing leaching of the enzyme during the washing steps. An indirect method of estimation of the phenolic residues is made possible by utilizing the near-infrared (NIR) photoluminescence properties of cysteine-coated gold nanoparticles. The preparation of this sensor is easy, the device itself is made by nontoxic components, and its performance is comparable to the quantum dot-based biosensors.

1.4.3 Nanomechanical Biosensors

Nanomechanical biosensors utilize nano- or microcantilevers as transducer elements to detect the change in the process conditions (mostly mass) as a result of the interaction of the analyte with the biomolecule. Cantilever biosensors typically convert the catalytic ability of the biomolecule into nanomechanical motion, and this motion can be detected by using optical, piezo-resistive, or other modes of detection systems [110]. Microcantilevers undergo deflection due to accumulation of mass as a result of molecules getting adsorbed on one side of the lever. This induces changes in the resonance frequency producing differential surface stress across the cantilever [111]. The cantilever transducer principle has been exploited to develop a variety of sensors. Figure 1.7 provides a few examples of sensors based on cantilever principles.

These devices are generally made of polymer or silicon/silicon-derived materials with nanoscale thickness. The silicon layers are usually coated with another thin film of freshly deposited gold. The gold film plays a very important role in the operation of the nanomechanical biosensors. Firstly, gold is considered to have a benign effect on biomolecule immobilization. Secondly, the kinetics of deposition and its subsequent orientation depends on the binding sites available on the gold surface. Mertens and co-workers [113] therefore studied the impact of the morphology of the deposited gold layer on the molecular adsorption process. The authors report that thin gold films with low deposition rates give rise to well-defined grain structures that favor improved performance in terms of sensitivity and reproducibility. Thicker films with high deposition rates lead to the formation of nanoislands that can hinder molecular adsorption. Nanomechanical biosensors featuring micro- or nanocantilevers require careful handling. There is a need to guard the sensor against inaccuracies as a result of improper bending due to surface contamination [114]. Similarly, efforts need to be directed towards developing new universal immobilization techniques to ensure reproducibility of results [114]. However, the ability to fabricate large arrays of these sensors followed by simultaneous detection of several samples at the same time (increased throughput) makes these devices particularly attractive for commercialization [110].

Yan and co-workers [115] developed an AChE-coated nanomechanical biosensor for the detection of organophosphate residues. To get the maximum biosensor performance, it is necessary to ensure full surface coverage of molecular recognition elements (in this case AChE enzyme). The addition of pesticide residues causes an irreversible binding of the organophosphates with the enzymes. The enzymes undergo a conformational change as a result of this binding. This conformational change in the structure of AChE produces small surface stresses of the order of nano-Newtons per unit length. The magnitude of bending of the cantilever as a result of these surface stresses can then be related to the concentration of organophosphates introduced in the cell. However, the authors point out that biosensor so developed was heavily dependent on the ability of the target molecule (analyte) to inhibit the AChE activity. Different pesticide precursors have different abilities to inhibit the enzyme activity. This necessitates the calibration of the sensor against each type of pesticide residue. The authors believe that decreasing the thickness of the silicon microcantilever can improve the deflection several folds enabling the possibility of using these devices in real time field tests.

In a study based on similar principles, Chen and co-workers [116] used an acoustic resonator mass-loading sensor coated with AChE for accurate quantification of organophosphates. Using a mass sensitive quartz crystal microbalance (QCM), the authors recorded the change in mass as result of the binding of the pesticide residue to the catalytic center of AChE. The increase in mass causes a corresponding decrease in the resonant frequency of the crystal. However, the use of conventional quartz crystals may not solve the problem of sensitivity because they generally tend to provide fundamental resonant frequencies of order of megahertz. For the biosensor to operate at very low concentrations of organophosphates, much higher frequencies of the order of gigahertz are required. This can be achieved by using

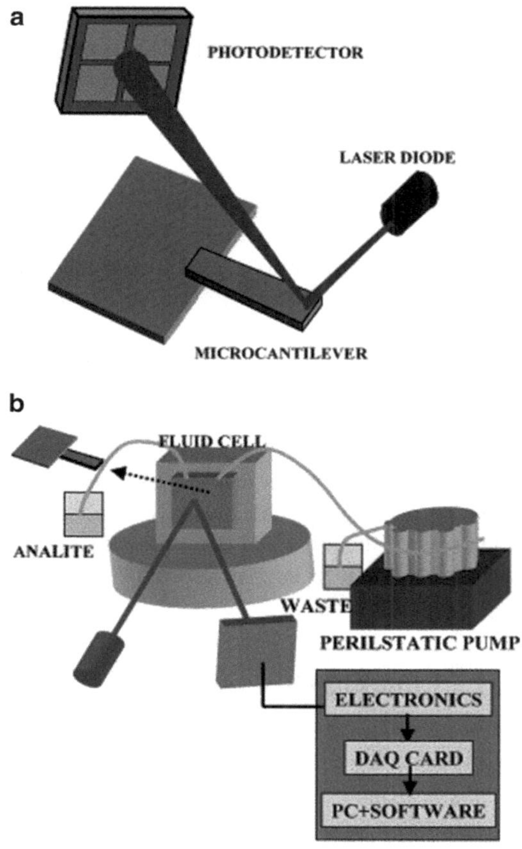

Fig. 1.8 Nanobiosensor for detection of DDT residues. (**a**) Optical beam deflection method used for quantification of cantilever motion. (**b**) Experimental setup. Reproduced with permission from [117]

a thin film bulk acoustic resonator (TFBAR). It consists of a piezoelectric film centered between two electrodes and a Bragg acoustic reflector. A microcantilever subjected to polyelectrolyte/enzyme LbL deposition was reported by Karnati and co-workers [72]. The surface of the cantilever was coated with nanoscale precursor films of polyethyleneimine (PEI) and PSS. The precursor films are followed by deposition of OPH/PSS multilayers. The authors report an order of magnitude improvement in the detection limit as compared to potentiometric and optical biosensors. Alvarez and co-workers [117] developed a nanomechanical biosensor to estimate the concentration of dichlorodiphenyltrichloroethane (DDT), a widely used insecticide. Optical methods are utilized to estimate the cantilever motion. Figure 1.8 shows the schematic diagram of the setup used for analysis.

The sensor consists of a bimetallic strip where one side is made of silicon nitride (inert layer) and the other side is made of nanothin gold films (active side). Exclusion of silicon nitride layers from the subsequent reactions can be achieved by the use of thiol chemistry which imparts specific functionality to the gold surface. The highly active thiol-terminated gold surface is further exposed to cystamine–

glutaraldehyde making it conducive for the immobilization of proteins via covalent binding. A synthetic hapten of DDT mixed with bovine serum albumin is deposited on the gold layer. This is followed by exposure to a monoclonal antibody. The microcantilever bending before and after the addition of the antibody is monitored to estimate the deflection due to the addition of the antibody. Moreover, the authors found that it is possible to regenerate the microcantilever sensor by simply adding 100 mM hydrochloric acid (HCl). The addition of dilute acid helps in breaking the hapten/antibody complex. The regenerated sensor can be reused again for the same hapten/antibody pair to generate reproducible results with little or no degradation. However, the dilute HCl regeneration method appears to be specific for hapten/antibody pair under consideration. Different pairs may show differential regenerative abilities. In yet another example involving antibody binding, Bache and co-workers [118] tested the performance of a pesticide detecting microcantilever sensor coated with a conjugate of ovalbumin and dichlorobenzamide popularly known as BAM. The subsequent addition of BAM antibody resulted in the formation of a complex whose presence can be quantified by the bending action of the cantilever. A piezoelectric layer embedded in the cantilever is used to quantify the extent of bending and determine the concentration of the residue. Fluorescent-marked antibodies can give a visual confirmation of the antibody binding.

1.5 Emerging Trends

Nanotechnology is a developing science. Over the past decade, novel methods to process nanomaterials and integrate them with biological components have energized the field of nanobiosensors. The use of silicon nanofabrication technology for development of enzyme-based biosensors is a promising route to produce cheaper devices. Screen printing, a technique that helps to form patterns on ceramic or plastic surfaces by passing inks through masks and photolithography, is one of the popular methods to produce inexpensive silicon-based devices [16].

Nabok and co-workers [119] developed silicon-based multienzyme planar waveguide device capable of detecting traces of waterborne pollutants such as heavy metal ions and pesticides. The method for fabrication of the enzyme sensor is similar to the one explained earlier [101]. As per the authors, the use of chromophores and the application of the principle of attenuated total reflection (ATR) in both the examples can help to improve the sensitivity of the nanobiosensor by 1,000 times as compared to conventional UV–vis absorption spectroscopy.

Another area that is showing significant progress is aptamer-based nanobiosensors. Aptamers are nucleic acids (DNA or RNA) capable of exhibiting immense binding power with biological components or small molecules [120]. Aptamers are easy to functionalize and are more stable under harsh conditions as compared to antibodies [121]. In a recent example, He and co-workers developed an acetamiprid biosensor by utilizing the binding affinity of nucleic acids (aptamers). Further advances in the systematic evolution of ligands by exponential enrichment (SELEX) technology are expected to help in developing numerous aptamers with improved binding specificity. Surface plasmon resonance (SPR) spectroscopy is yet another promising technique. It is displayed by certain nanoparticles or thin metal films upon continuous irradiation. Zayats and co-workers [122] exploited the photochemical charging behavior shown by gold nanoparticles in an array consisting of Au nanoparticles and cadmium sulfide (CdS) nanoparticles and studied the response of the photo-induced current using SPR spectroscopy. Thiocholine, a product of the enzymatic hydrolysis, donates electrons to produce a photocurrent. Presence of pesticide residues restricts the formation of thiocholine blocking the generation of the photocurrent. SPR sensors are extremely sensitive. They have the capability of detecting toxic components up to parts per trillion (ppt) levels. Similarly, as mentioned in the earlier sections, the combination of different types of transducer elements such as the dual amperometric/potentiometric biosensor [84] and hybrid optical/nanomechanical biosensor [117] can be shown with great promise in achieving accurate and quick detection of harmful waterborne contaminants.

1.6 Conclusion

The biosensor industry has come a long way since Clark and co-workers [123] developed the first glucose biosensor. The interesting combination of nano-enabled techniques and biotechnology is now set to herald a new future for the biosensor industry. The introduction of nanomaterials in biosensor applications has improved the spatial resolution, reduced the sample volume, amplified output signal, and enhanced the sensory response to the order of milliseconds [123].

Nanotechnology-based techniques are a huge success in the semiconductor and electronics industry. Mass production of electronic chips and devices at an extremely cheap price has become a reality because of recent advances in silicon-based manufacturing techniques. The commercial success of silicon-based nanofabrication methods can be easily extended to biosensing applications [16]. Similarly, new technological advances such as replacing electrochemical cells with screen-printed electrodes and using wireless technology to remotely monitor the levels of waterborne contaminants have the potential of making nanostructured biosensors cheaper and easier to use. With the advent of the industrial age, the destruction of the delicate balance among the different components of the Earth's ecosystem is proceeding at an alarming rate. Environmental monitoring is today emerging as a leading preventive measure in the fight against pollution. Given their high accuracy and substrate specificity, nanobiosensors are uniquely positioned to deliver key outcomes. However, a general sense of pessimism that has long clouded the use of nanostructured biological sensors stems from the lack of understanding of how these systems would behave under different operating conditions. This can be overcome by increasing public awareness and investing significantly in research and development efforts to fabricate newer and safer nanobiosystems.

The US federal government has made substantial investment in nanotech goods by setting up the National Nanotechnology Initiative (NNI). Through this Initiative, as of 2005, investments worth US $ 6.5 billion have been channelized into developing cutting-edge technologies [21]. This enthusiasm needs to spread globally especially to the developing countries which are most likely to feel the brunt of environmental degradation in the near future. However care should be taken to ensure that the use of nanobiosensors or the production of thereof for the purpose of environmental monitoring should not contribute to the very problem that we seek to solve. Innovative solutions such as simultaneous incorporation of multiple biorecognition elements to detect threats from a wide range of waterborne environmental pollutants by immobilizing on nano-/microarrays to amplify the transducer response would help transform these systems into a commercial reality.

Acknowledgments The authors greatly appreciate the contributions and support from students and our collaborators. Financial support in part from the National Science Foundation (0609164, 0832730, and 0928835), the Department of Defense Strategic Environmental Research and Development Program (DOD SERDP W912HQ-12-C-0020), the USDA National Institute of Food and Agriculture (USDA-SBIR, 2011-33610-30822), the Michigan University Research Corridor, the Michigan Initiative for Innovation and Entrepreneurship, and the MSU Foundation are greatly appreciated.

References

1. Moss, B.: Water pollution by agriculture. Phil. Trans. Biol. Sci. **363**, 659–666 (2008)
2. Bose, B.: Global warming: energy, environmental pollution, and the impact of power electronics. IEEE Ind. Electron. Mag. **4**, 6–17 (2010)
3. Lindenmayer, D.B., Likens, G.E.: The science and application of ecological monitoring. Biol. Conservat. **143**, 1317–1328 (2010)
4. Lindenmayer, D.B., Likens, G.E.: Adaptive monitoring: a new paradigm for long-term research and monitoring. Trends Ecol. Evol. **24**, 482–486 (2009)
5. Lovett, G.M., Burns, D.A., Driscoll, C.T., Jenkins, J.C., Mitchell, M.J., Rustad, L., Shanley, J.B., Likens, G.E., Haeuber, R.: Who needs environmental monitoring? Front. Ecol. Environ. **5**, 253–260 (2007)
6. Looney, B.B., Falta, R.W.: Vadose Zone Science and Technology Solutions. Battelle Press Columbus, Ohio (2000)
7. Peshin, R., Dhawan, A., Pimentel, D.: Environmental and Economic Costs of the Application of Pesticides Primarily in the United States, in: Integrated Pest Management: Innovation-Development Process, pp. 89–111. Springer, Netherlands (2009)
8. Danielsen, F., Burgess, N.D., Jensen, P.M., Pirhofer-Walzl, K.: Environmental monitoring: the scale and speed of implementation varies according to the degree of people's involvement. J. Appl. Ecol. **47**, 1166–1168 (2010)
9. Urbansky, E.T.: Perchlorate as an environmental contaminant. Environ. Sci. Pollut. Res. **9**, 187–192 (2002)
10. Rodriguez-Mozaz, S., de Alda, M.J.L., Barcelo, D.: Biosensors as useful tools for environmental analysis and monitoring. Anal. Bioanal. Chem. **386**, 1025–1041 (2006)
11. Bassi, A.S., Flickinger, M.C.: Biosensors, Environmental, in: Encyclopedia of Industrial Biotechnology. John Wiley & Sons Inc., Hoboken, NJ (2009)

12. Sethi, R.S.: Transducer aspects of biosensors. Biosens. Bioelectron. **9**, 243–264 (1994). Reprinted from Gec Journal Research Vol 9, Pg 81, 1991
13. Scognamiglio, V., Pezzotti, G., Pezzotti, I., Cano, J., Buonasera, K., Giannini, D., Giardi, M.T.: Biosensors for effective environmental and agrifood protection and commercialization: from research to market. Microchim. Acta **170**, 215–225 (2010)
14. Bogue, R.: Developments in biosensors-where are tomorrow's markets? Sensor Rev. **25**, 180–184 (2005)
15. Gokhale, A.A., Lee, I.: Cellulase immobilized nanostructured supports for efficient saccharification of cellulosic substrates. Top. Catal. **55**, 1231–1246 (2012)
16. Rodriguez-Mozaz, S., Marco, M.P., Lopez de Alda, M., Barcelo, D.: Biosensors for environmental applications: future development trends. Pure Appl. Chem. **74**(4), 723–752 (2004)
17. Murray, K.E., Thomas, S.M., Bodour, A.A.: Prioritizing research for trace pollutants and emerging contaminants in the freshwater environment. Environ. Pollut. **158**, 3462–3471 (2010)
18. Ho, C.K., Robinson, A., Miller, D.R., Davis, M.J.: Overview of sensors and needs for environmental monitoring. Sensors **5**, 4–37 (2005)
19. Relyea, R.A.: A cocktail of contaminants: how mixtures of pesticides at low concentrations affect aquatic communities. Oecologia **159**, 363–376 (2009)
20. Tegart, G.: Nanotechnology: the technology for the twenty-first century. Foresight **6**, 364–370 (2004)
21. Liota, T., Tzitzios, V.: Investing in nanotechnology. Nanotech. L & Bus **3**, 521 (2006)
22. Wang, J.: Nanomaterial-based electrochemical biosensors. Analyst **130**, 421–426 (2005)
23. Salata, O.V.: Applications of nanoparticles in biology and medicine. J. Nanobiotechnol. **2**, 3 (2004)
24. Wu, X., Liu, H., Liu, J., Haley, K.N., Treadway, J.A., Larson, J.P., Ge, N., Peale, F., Bruchez, M.P.: Immunofluorescent labeling of cancer marker Her2 and other cellular targets with semiconductor quantum dots. Nat. Biotechnol. **21**, 41–46 (2002)
25. Jaiswal, J.K., Mattoussi, H., Mauro, J.M., Simon, S.M.: Long-term multiple color imaging of live cells using quantum dot bioconjugates. Nat. Biotechnol. **21**, 47–51 (2002)
26. Misra, N., Martinez, J.A., Huang, S.-C.J., Wang, Y., Stroeve, P., Grigoropoulos, C.P., Noy, A.: Bioelectronic silicon nanowire devices using functional membrane proteins. Proc. Natl. Acad. Sci. U. S. A. **106**, 13780–13784 (2009)
27. Laocharoensuk, R., Bulbarello, A., Mannino, S., Wang, J.: Adaptive nanowire-nanotube bioelectronic system for on-demand bioelectrocatalytic transformations. Chem. Comm. **32**, 3362–3364 (2007)
28. Wanekaya, A.K., Chen, W., Myung, N.V., Mulchandani, A.: Nanowire-based electrochemical biosensors. Electroanalysis **18**, 533–550 (2006)
29. Grieshaber, D., MacKenzie, R., Voros, J., Reimhult, E.: Electrochemical biosensors - sensor principles and architectures. Sensors **8**, 1400–1458 (2008)
30. Frew, J.E., Hill, H.A.O.: Electrochemical biosensors. Anal. Chem. **59**, 933–944 (1987)
31. Ronkainen, N.J., Halsall, H.B., Heineman, W.R.: Electrochemical biosensors. Chem. Soc. Rev. **39**, 1747–1763 (2010)
32. Li, H., Liu, S., Dai, Z., Bao, J., Yang, X.: Applications of nanomaterials in electrochemical enzyme biosensors. Sensors **9**, 8547–8561 (2009)
33. Min, I.H., Choi, L., Ahn, K.S., Kim, B.K., Lee, B.Y., Kim, K.S., Choi, H.N., Lee, W.Y.: Electrochemical determination of carbohydrate-binding proteins using carbohydrate-stabilized gold nanoparticles and silver enhancement. Biosens. Bioelectron. **26**, 1326–1331 (2010)
34. Jena, B.K., Raj, C.R.: Electrochemical biosensor based on integrated assembly of dehydrogenase enzymes and gold nanoparticles. Anal. Chem. **78**, 6332–6339 (2006)
35. Wang, H., Wang, X., Zhang, X., Qin, X., Zhao, Z., Miao, Z., Huang, N., Chen, Q.: A novel glucose biosensor based on the immobilization of glucose oxidase onto gold nanoparticles-modified Pb nanowires. Biosens. Bioelectron. **25**, 142–146 (2009)
36. Zhao, W., Xu, J.J., Shi, C.G., Chen, H.Y.: Multilayer membranes via layer-by-layer deposition of organic polymer protected Prussian blue nanoparticles and glucose oxidase for glucose biosensing. Langmuir **21**, 9630–9634 (2005)

37. Lvov, Y., Ariga, K., Ichinose, I., Kunitake, T.: Assembly of multicomponent protein films by means of electrostatic layer-by-layer adsorption. J. Am. Chem. Soc. **117**, 6117–6123 (1995)
38. Hassler, B.L., Kohli, N., Zeikus, J.G., Lee, I., Worden, R.M.: Renewable dehydrogenase-based interfaces for bioelectronic applications. Langmuir **23**, 7127–7133 (2007)
39. Hassler, B.L., Amundsen, T.J., Zeikus, J.G., Lee, I., Worden, R.M.: Versatile bioelectronic interfaces on flexible non-conductive substrates. Biosens. Bioelectron. **23**, 1481–1487 (2008)
40. Liu, Y., Wang, M., Zhao, F., Xu, Z., Dong, S.: The direct electron transfer of glucose oxidase and glucose biosensor based on carbon nanotubes/chitosan matrix. Biosens. Bioelectron. **21**, 984–988 (2005)
41. Hrapovic, S., Liu, Y., Male, K.B., Luong, J.H.T.: Electrochemical biosensing platforms using platinum nanoparticles and carbon nanotubes. Anal. Chem. (Washington, DC, U. S.) **76**, 1083–1088 (2003)
42. Wang, J.: Carbon nanotube based electrochemical biosensors: a review. Electroanalysis **17**, 7–14 (2004)
43. Zhao, X., Mai, Z., Kang, X., Zou, X.: Direct electrochemistry and electrocatalysis of horseradish peroxidase based on clayâ€"chitosan-gold nanoparticle nanocomposite. Biosens. Bioelectron. **23**, 1032–1038 (2008)
44. Li, Z., Hu, N.: Direct electrochemistry of heme proteins in their layer-by-layer films with clay nanoparticles. J. Electroanal. Chem. **558**, 155–165 (2003)
45. Wei, D., Bailey, M.J.A., Andrew, P., Ryhanen, T.: Electrochemical biosensors at the nanoscale. Lab Chip **9**, 2123–2131 (2009)
46. Zoski, C.G.: Ultramicroelectrodes: design, fabrication, and characterization. Electroanalysis **14**, 1041–1051 (2002)
47. Michalowicz, J., Duda, W.: Phenols-sources and toxicity. Pol. J. Environ. Stud. **16**, 347 (2007)
48. Bruce, R.M., Santodonato, J., Neal, M.W.: Summary review of the health effects associated with phenol. Toxicol. Ind. Health **3**, 535–568 (1987)
49. Stanca, S.E., Popescu, I.C.: Phenols monitoring and Hill coefficient evaluation using tyrosinase-based amperometric biosensors. Bioelectrochemistry **64**, 47–52 (2004)
50. Kulys, J., Vidziunaite, R.: Amperometric biosensors based on recombinant laccases for phenols determination. Biosens. Bioelectron. **18**, 319–325 (2003)
51. Chang, S.C., Rawson, K., McNeil, C.J.: Disposable tyrosinase-peroxidase bi-enzyme sensor for amperometric detection of phenols. Biosens. Bioelectron. **17**, 1015–1023 (2002)
52. Zhang, J., Lei, J., Liu, Y., Zhao, J., Ju, H.: Highly sensitive amperometric biosensors for phenols based on polyanilineâ€"ionic liquidâ€"carbon nanofiber composite. Biosens. Bioelectron. **24**, 1858–1863 (2009)
53. Liu, S., Yu, J., Ju, H.: Renewable phenol biosensor based on a tyrosinase-colloidal gold modified carbon paste electrode. J. Electroanal. Chem. **540**, 61–67 (2003)
54. Song, W., Li, D.W., Li, Y.T., Li, Y., Long, Y.T.: Disposable biosensor based on graphene oxide conjugated with tyrosinase assembled gold nanoparticles. Biosens. Bioelectron. **26**, 3181–3186 (2011)
55. Li, Y.F., Liu, Z.M., Liu, Y.L., Yang, Y.H., Shen, G.L., Yu, R.Q.: A mediator-free phenol biosensor based on immobilizing tyrosinase to ZnO nanoparticles. Anal. Biochem. **349**, 33–40 (2006)
56. Dubacheva, G.V., Porus, M.V., Sigolaeva, L.V., Pergushov, D.V., Tur, D.R., Papkov, V.S., Zezin, A.B., Yaroslavov, A.A., Eremenko, A.V., Kurochkin, I.N.: Nanostructured polyelectrolyte films for engineering highly sensitive tyrosinase biosensors: Specifics of enzyme-polyelectrolyte structures. Nanotech. Russia **3**, 221–227 (2008)
57. Arecchi, A., Scampicchio, M., Drusch, S., Mannino, S.: Nanofibrous membrane based tyrosinase-biosensor for the detection of phenolic compounds. Anal. Chim. Acta. **659**, 133–136 (2010)
58. Korkut, S., Keskinler, B., Erhan, E.: An amperometric biosensor based on multiwalled carbon nanotube-poly(pyrrole)-horseradish peroxidase nanobiocomposite film for determination of phenol derivatives. Talanta **76**, 1147–1152 (2008)

59. Serra, B., Benito, B., Agüí, L., Reviejo, A.J., Pingarrón, J.M.: Graphite-teflon-peroxidase composite electrochemical biosensors. A tool for the wide detection of phenolic compounds. Electroanalysis **13**, 693–700 (2001)
60. Ryan, O., Smyth, M.R., Fagain, C.O.: Horseradish peroxidase: the analyst's friend. Essays Biochem. **28**, 129 (1994)
61. Sotiropoulou, S., Gavalas, V., Vamvakaki, V., Chaniotakis, N.A.: Novel carbon materials in biosensor systems. Biosens. Bioelectron. **18**, 211–215 (2003)
62. Kirstein, D., Kirstein, L., Scheller, F., Borcherding, H., Ronnenberg, J., Diekmann, S., Steinrucke, P.: Amperometric nitrate biosensors on the basis of Pseudomonas stutzeri nitrate reductase. J. Electroanal. Chem. **474**, 43–51 (1999)
63. Glazier, S.A., Campbell, E.R., Campbell, W.H.: Construction and characterization of nitrate reductase-based amperometric electrode and nitrate assay of fertilizers and drinking water. Anal. Chem. **70**, 1511–1515 (1998)
64. Can, F., Ozoner, S.K., Ergenekon, P., Erhan, E.: Amperometric nitrate biosensor based on Carbon nanotube/Polypyrrole/Nitrate reductase biofilm electrode. Mater. Sci. Eng. C **32**, 18–23 (2012)
65. Moretto, L.M., Ugo, P., Zanata, M., Guerriero, P., Martin, C.R.: Nitrate biosensor based on the ultrathin-film composite membrane concept. Anal. Chem. (Washington, DC, U. S.) **70**, 2163–2166 (1998)
66. Da Silva, S., Shan, D., Cosnier, S.: Amperometric detection of nitrite, iodate and periodate at glassy carbon electrode modified with catalase and multi-wall carbon nanotubes. Sens. Actuators B **103**, 397–402 (2004)
67. Salimi, A., Noorbakhsh, A., Ghadermarzi, M.: Amperometric detection of nitrite, iodate and periodate at glassy carbon electrode modified with catalase and multi-wall carbon nanotubes. Sens. Actuators B **123**, 530–537 (2007)
68. Kohli, N., Lee, I., Richardson, R.J., Worden, R.M.: Theoretical and experimental study of bi-enzyme electrodes with substrate recycling. J. Electroanal. Chem. **641**, 104–110 (2010)
69. Kohli, N., Srivastava, D., Sun, J., Richardson, R.J., Lee, I., Worden, R.M.: Nanostructured biosensor for measuring neuropathy target esterase activity. Anal. Chem. **79**, 5196–5203 (2007)
70. Srivastava, D., Kohli, N., Richardson, R.J., Worden, R.M., Lee, I.: Neuropathy target esterase biosensor. In: Somerset, V.S. (ed.) Intelligent and Biosensors. In-Tech, New York (2010)
71. Du, D., Chen, S., Cai, J., Zhang, A.: Talanta **74**, 766–772 (2008)
72. Karnati, C., Du, H.W., Ji, H.F., Xu, X.H., Lvov, Y., Mulchandani, A., Mulchandani, P., Chen, W.: Organophosphorus hydrolase multilayer modified microcantilevers for organophosphorus detection. Biosens. Bioelectron. **22**, 2636–2642 (2007)
73. Upadhyay, S., Rao, G.R., Sharma, M.K., Bhattacharya, B.K., Rao, V.K., Vijayaraghavan, R.: Immobilization of acetylcholineesterase choline oxidase on a gold-platinum bimetallic nanoparticles modified glassy carbon electrode for the sensitive detection of organophosphate pesticides, carbamates and nerve agents. Biosens. Bioelectron. **25**, 832–838 (2009)
74. Shan, D., Mousty, C., Cosnier, S.: Subnanomolar cyanide detection at polyphenol oxidase/clay biosensors. Anal. Chem. **76**, 178–183 (2004)
75. Wang, S., Lei, Y., Zhang, Y., Tang, J., Shen, G., Yu, R.: Hydroxyapatite nanoarray-based cyanide biosensor. Anal. Biochem. **398**, 191–197 (2010)
76. Thevenot, D.R., Toth, K., Durst, R.A., Wilson, G.S.: Electrochemical biosensors: recommended definitions and classification - (technical report). Pure Appl. Chem. **71**, 2333–2348 (1999)
77. Ghindilis, A.L., Morzunova, T.G., Barmin, A.V., Kurochkin, I.N.: Potentiometric biosensors for cholinesterase inhibitor analysis based on mediatorless bioelectrocatalysis. Biosens. Bioelectron. **11**, 873–880 (1996)
78. Reybier, K., Zairi, S., Jaffrezic-Renault, N., Fahys, B.: The use of polyethyleneimine for fabrication of potentiometric cholinesterase biosensors. Talanta **56**, 1015–1020 (2002)
79. Keusgen, M., Kloock, J.P., Knobbe, D.T., Junger, M., Krest, I., Goldbach, M., Klein, W., Schoning, M.J.: Direct determination of cyanides by potentiometric biosensors. Sens. Actuators B **103**, 380–385 (2004)

80. Sohail, M., Adeloju, S.B.: Electroimmobilization of nitrate reductase and nicotinamide adenine dinucleotide into polypyrrole films for potentiometric detection of nitrate. Sens. Actuators B Chem. **133**, 333–339 (2008)
81. Sohail, M., Adeloju, S.B.: Fabrication of redox-mediator supported potentiometric nitrate biosensor with nitrate reductase. Electroanalysis **21**, 1411–1418 (2009)
82. Schoning, M.J., Arzdorf, M., Mulchandani, P., Chen, W., Mulchandani, A.: A capacitive field-effect sensor for the direct determination of organophosphorus pesticides. Sens. Actuators B Chem. **91**, 92–97 (2003)
83. Mulchandani, P., Mulchandani, A., Kaneva, I., Chen, W.: Biosensor for direct determination of organophosphate nerve agents. 1. Potentiometric enzyme electrode. Biosens. Bioelectron. **14**, 77–85 (1999)
84. Schoning, M.J., Krause, R., Block, K., Musahmeh, M., Mulchandani, A., Wang, J.: A dual amperometric/potentiometric FIA-based biosensor for the distinctive detection of organophosphorus pesticides. Sens. Actuators B Chem. **95**, 291–296 (2003)
85. Berezhetskyy, A. Development of conductometric biosensors based on alkaline phosphatases for the water quality control. arXiv preprint arXiv:0809.3578 (2008)
86. Jaffrezic-Renault, N., Dzyadevych, S.V.: Conductometric microbiosensors for environmental monitoring. Sensors **8**, 2569–2588 (2008)
87. Guedri, H., Durrieu, C.: A self-assembled monolayers based conductometric algal whole cell biosensor for water monitoring. Microchim. Acta **163**, 179–184 (2008)
88. de la Rica, R., Fernandez-Sanchez, C., Baldi, A.: Polysilicon interdigitated electrodes as impedimetric sensors. Electrochem. Commun. **8**, 1239–1244 (2006)
89. Hnaien, M., Lagarde, F., Bausells, J., Errachid, A., Jaffrezic-Renault, N.: A new bacterial biosensor for trichloroethylene detection based on a three-dimensional carbon nanotubes bioarchitecture. Anal. Bioanal. Chem. **400**, 1083–1092 (2011)
90. Hnaien, M., Bourigua, S., Bessueille, F., Bausells, J., Errachid, A., Lagarde, F., Jaffrezic-Renault, N.: Impedimetric microbial biosensor based on single wall carbon nanotube modified microelectrodes for trichloroethylene detection. Electrochim. Acta **56**, 10353–10358 (2011)
91. Wang, X.J., Dzyadevych, S.V., Chovelon, J.M., Renault, N.J., Chen, L., Xia, S.Q., Zhao, J.F.: Development of a conductometric nitrate biosensor based on Methyl viologen/Nafion (R) composite film. Electrochem. Commun. **8**, 201–205 (2006)
92. Singh, A.K., Flounders, A.W., Volponi, J.V., Ashley, C.S., Wally, K., Schoeniger, J.S.: Development of sensors for direct detection of organophosphates. Part I: immobilization, characterization and stabilization of acetylcholinesterase and organophosphate hydrolase on silica supports. Biosens. Bioelectron. **14**, 703–713 (1999)
93. Ramsden, J.J.: Optical biosensors. J. Mol. Recognit. **10**, 109–120 (1997)
94. Ligler, F.S.: Perspective on optical biosensors and integrated sensor systems. Anal. Chem. **81**, 519–526 (2008)
95. Velasco-Garcia, M.N.: Optical biosensors for probing at the cellular level: a review of recent progress and future prospects. Semin. Cell Dev. Biol. **20**, 27–33 (2009)
96. Hestrin, S.: The reaction of acetylcholine and other carboxylic acid derivatives with hydroxylamine, and its analytical application. J. Biol. Chem. **180**, 249–261 (1949)
97. George, L., Ellman, K., Valentino Jr., A.: A new and rapid colorimetric determination of acetylcholinesterase activity. Biochem. Pharmacol. **7**, 88–95 (1961)
98. Vamvakaki, V., Chaniotakis, N.A.: Pesticide detection with a liposome-based nano-biosensor. Biosens. Bioelectron. **22**, 2848–2853 (2007)
99. Kohli, N., Vaidya, S., Ofoli, R.Y., Worden, R.M., Lee, I.: Arrays of lipid bilayers and liposomes on patterned polyelectrolyte templates. J. Colloid Interface Sci. **301**, 461–469 (2006)
100. Kohli, N., Hassler, B.L., Parthasarathy, L., Richardson, R.J., Ofoli, R.Y., Worden, R.M., Lee, I.: Tethered lipid bilayers on electrolessly deposited gold for bioelectronic applications. Biomacromolecules **7**, 3327–3335 (2006)
101. Haron, S., Nabok, A.V., Ray, A.K.: Optical biosensing transducer based on silicon waveguide structure coated with polyelectrolyte nano layers. Proc. SPIE **2003**, 101 (2003)

102. Simonian, A.L., Good, T.A., Wang, S.S., Wild, J.R.: Nanoparticle-based optical biosensors for the direct detection of organophosphate chemical warfare agents and pesticides. Anal. Chim. Acta. **534**, 69–77 (2005)
103. Liao, S.Z., Qiao, Y.A., Han, W.T., Xie, Z.X., Wu, Z.Y., Shen, G.L., Yu, R.Q.: Acetylcholinesterase liquid crystal biosensor based on modulated growth of gold nanoparticles for amplified detection of acetylcholine and inhibitor. Anal. Chem. **84**, 45–49 (2012)
104. Zheng, Z.Z., Zhou, Y.L., Li, X.Y., Liu, S.Q., Tang, Z.Y.: Highly-sensitive organophosphorous pesticide biosensors based on nanostructured films of acetylcholinesterase and CdTe quantum dots. Biosens. Bioelectron. **26**, 3081–3085 (2011)
105. Kim, J.H., Lim, S.Y., Nam, D.H., Ryu, J., Ku, S.H., Park, C.B.: Self-assembled, photoluminescent peptide hydrogel as a versatile platform for enzyme-based optical biosensors. Biosens. Bioelectron. **26**, 1860–1865 (2011)
106. Abdullah, J., Ahmad, M., Karuppiah, N., Heng, L.Y., Sidek, H.: Immobilization of tyrosinase in chitosan film for an optical detection of phenol. Sens. Actuators B Chem. **114**, 604–609 (2006)
107. Abdullah, J., Ahmad, M., Heng, L.Y., Karuppiah, N., Sidek, H.: Chitosan-based tyrosinase optical phenol biosensor employing hybrid nafion/sol-gel silicate for MBTH immobilization. Talanta **70**, 527–532 (2006)
108. Park, S.A., Jang, E., Koh, W.G., Kim, B.: Development of analytic microdevices for the detection of phenol using polymer hydrogel particles containing enzymeâ€“QD conjugates. Talanta **84**, 1000–1003 (2011)
109. Dong, W., Dong, C., Shuang, S., Choi, M.M.F.: Near-infrared luminescence quenching method for the detection of phenolic compounds using N-acetyl-L-cysteine-protected gold nanoparticles-tyrosinase hybrid material. Biosens. Bioelectron. **25**, 1043–1048 (2010)
110. Carrascosa, L.G., Moreno, M., Alvarez, M., Lechuga, L.M.: Nanomechanical biosensors: a new sensing tool. Trends Anal. Chem. **25**, 196–206 (2006)
111. Adrian, J., Fernandez, F., Muriano, A., Obregon, R., Ramon-Azcon, J., Tort, N., Marco, M.P.: Biosensors for pharmaceuticals and emerging contaminants based on novel micro and nanotechnology approaches, vol 5J. In: The Handbook of Environmental Chemistry. Springer, Berlin, Heidelberg, pp. 47–68 (2009)
112. Raiteri, R., Grattarola, M., Berger, R.: Micromechanics senses biomolecules. Mater. Today **5**, 22–29 (2002)
113. Mertens, J., Calleja, M., Ramos, D., Taryn, A., Tamayo, J. Role of the gold film nanostructure on the nanomechanical response of microcantilever sensors. Journal of Appl. Phys. **101** (034904):1 (2007)
114. Datar, R., Kim, S., Jeon, S., Hesketh, P., Manalis, S., Boisen, A., Thundat, T.: Detection of organophosphates using an acetyl cholinesterase (AChE) coated microcantilever. MRS Bull. **34**, 449–454 (2009)
115. Yan, X.D., Tang, Y.J., Ji, H.F., Lvov, Y., Thundat, T.: Highly sensitive detection of organophosphorus pesticides by acetylcholinesterase-coated thin film bulk acoustic resonator mass-loading sensor. Instrum. Sci. Technol. **32**, 175–183 (2004)
116. Chen, D., Wang, J., Xu, Y., Li, D., Zhang, L., Li, Z.: Highly sensitive detection of organophosphorus pesticides by acetylcholinesterase-coated thin film bulk acoustic resonator mass-loading sensor. Biosens. Bioelectron. **41**, 163–167 (2012)
117. Alvarez, M., Calle, A., Tamayo, J., Lechuga, L.M., Abad, A., Montoya, A.: Development of nanomechanical biosensors for detection of the pesticide DDT. Biosens. Bioelectron. **18**, 649–653 (2003)
118. Bache, M., Taboryski, R., Schmid, S., Aamand, J., Jakobsen, M.H.: Investigations on antibody binding to a microcantilever coated with a BAM pesticide residue. Nanoscale Res. Lett. **6**, 386 (2011)
119. Nabok, A., Tsargorodskaya, A., Haron, S., Travis, J.: Planar waveguide enzyme sensor array for water pollution monitoring. IET **2006**, 119–126 (2006)
120. Chiu, T.C., Huang, C.C.: Aptamer-functionalized nano-biosensors. Sensors **9**, 10356–10388 (2009)

121. Sassolas, A., Prieto-Simon, B., Marty, J.L.: Biosensors for pesticide detection: new trends. Am. J. Anal. Chem. **3**, 210–232 (2012)
122. Zayats, M., Kharitonov, A.B., Pogorelova, S.P., Lioubashevski, O., Katz, E., Willner, I.: Probing photoelectrochemical processes in Au-CdS nanoparticle arrays by surface plasmon resonance:application for the detection of acetylcholine esterase inhibitors. J. Am. Chem. Soc. **125**, 16006–16014 (2003)
123. Clark, L.C., Lyons, C.: Electrode systems for continuous monitoring in cardiovascular surgery. Ann. N. Y. Acad. Sci. **102**, 29–45 (2006)
124. Demidov, W.: Nanobiosensors and molecular diagnostics: a promising partnership. Expert. Rev. Mol. Diagn. **4**, 267–268 (2004)

Chapter 2
Nanosensors for Intracellular Raman Studies

Patrick I.T. Thomson and Colin J. Campbell

Abstract Raman spectroscopy is an increasingly exploited tool for the study of cell biology, and the focus of this chapter is surface-enhanced Raman spectroscopy (SERS). Using SERS, noble metal nanoparticles can be interrogated with wavelengths of light to which cells exhibit minimal autofluorescence and return useful information about their immediate chemical environment. Using established gold–thiol surface chemistry, gold nanoparticles can be functionalised with suitable reporter molecules used to make nanosensors, which can be chosen to be sensitive to intracellular variables such as pH, protein concentration/activity or redox potential. This chapter presents the general concept of SERS for use in the monitoring of intracellular variables, and we will describe the preparation, use and functions of the nanosensors.

2.1 Background: Intracellular Nanosensors

The chemical environment inside cells is of great interest, and many methods are known which can destructively assay properties such as redox potential [1] or the concentration of arbitrary metabolites [2]. However, if such information is desired about the progress of individual cells over time, or living tissues, then non-destructive methods of obtaining information are needed. Fluorescent sensors are widely used [3, 4], with fluorescent proteins and fluorescent nanoparticles being particularly popular because they can be engineered to respond to an environmental stimulus and report it as a measurable change in fluorescence (which is a well-established modality in cellular imaging). Nanoparticles are well suited since they are readily taken up by cells, and techniques to engineer the architecture and

P.I.T. Thomson • C.J. Campbell (✉)
EaStCHEM School of Chemistry, University of Edinburgh, Edinburgh EH9 3JJ, UK
e-mail: patrick.thomson@gmail.com; colin.campbell@ed.ac.uk

S. Li et al. (eds.), *Nanoscale Sensors*, Lecture Notes in Nanoscale Science and Technology 19, DOI 10.1007/978-3-319-02772-2_2,

composition of nanoparticles are well established. Fluorescent proteins exhibit a complementary set of benefits; cells can be engineered to stably express them and genetic engineering techniques can be used to modify their function. However, fluorescence-based methods have limitations, such as background fluorescence of the cell and medium limiting the lower limits of detection, and the requirement of a photoactivatable electronic transition which is sensitive to environmental conditions. Raman spectroscopy and imaging offers an alternative to fluorescence, and in this chapter, we provide some examples where the features of Raman spectroscopy can be used to the benefit of the experimentalist.

2.2 Background: Raman and SERS

When photons are scattered by an atom or molecule, most of the light is reemitted unchanged (elastic or Rayleigh scattering). These resultant photons have the same kinetic energy as the incident ones, and are responsible for such phenomena as the colour of the sky [5]. However, approximately one in ten million reemitted photons have a different (usually lower) kinetic energy to the one absorbed (inelastic or Stokes Raman scattering)—indicating that the target molecule has undergone a transition in vibrational energy levels (Fig. 2.1) [6]. The frequency shift in these photons can be directly correlated to vibrational energy levels and thus gives us information about the vibrational modes of the target molecule in a similar fashion to infrared spectroscopy. The phenomenon was first observed and described by Sir Venkata Rāman, for which he was knighted and subsequently awarded the Nobel Prize in Physics in 1930 [7].

Raman and infrared spectroscopy can both probe the vibrational modes of target molecules, but the two techniques have the potential to return complementary information due the differing selection rules that govern which vibrational modes are IR

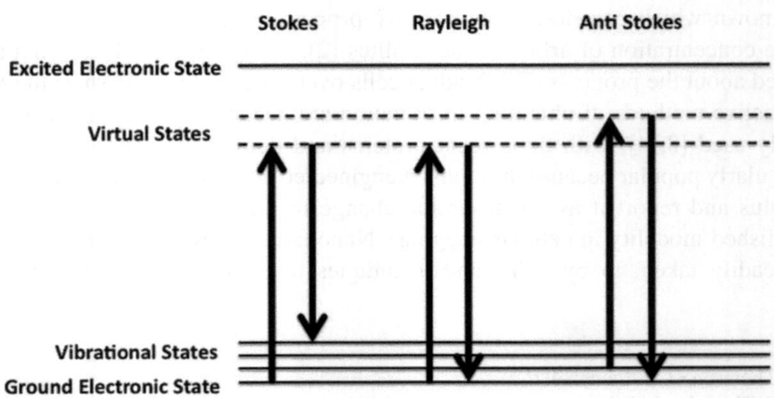

Fig. 2.1 Scattering modes for the interaction of photons with matter

active and which are Raman active. The main drawback of Raman spectroscopy is its low sensitivity—and the complex techniques thus required to separate out the Raman-scattered photons from the much more intense Rayleigh background [6].

The utility of Raman spectroscopy received a boost in 1974, when a phenomenon called surface-enhanced Raman spectroscopy (SERS) was first observed by Fleischmann [8, 9]. Specifically, he observed an anomalous enhancement of the Raman spectrum of pyridine adsorbed onto the surface of a roughened silver electrode. Theories to explain the phenomenon were proposed concurrently by Van Duyne [10] and Creighton [11], with the predominant theory today being Van Duyne's proposed involvement of surface plasmon resonance. Incident light on a metallic surface can excite localised surface plasmons, which increases the local electric field and can enhance the Raman scattering of adsorbed molecules by a factor ranging from 4 to 14 orders of magnitude [12–15]. Silver and gold surfaces and nanoparticles are particularly useful substrates for SERS studies of biological systems since they are relatively inert and the wavelength required to excite their surface plasmons falls in the visible/near-IR range, to which biological systems are permeable [16]. For spherical homogeneous nanoparticles, the SERS signal is greatest when they are clustered together; the nanoparticle–nanoparticle junction experiences a localised enhancement in plasmon intensity, and this dominates signals from single nanoparticles. Various strategies have been developed in order to use single nanoparticles as effective SERS sensors without relying on aggregation, such as dielectric-cored gold nanoshells [17], gold nanorods or gold nanostars [18]. Exploiting these features, the SERS enhancement in some cases can permit the detection of single molecules [13, 19].

2.3 SERS Measurements in Cells

As a result of the high sensitivity, SERS is a very useful technique for measuring the Raman spectra of small numbers of molecules [19] or nanoparticles, and as a result it has been used to explore the environment inside individual cells. Even a single gold nanoparticle can return useful spectral information about analytes close to its surface, and this has been used to probe the internal environments of both bacterial [20] and eukaryotic [21, 22] cells directly. To use SERS in this way, it is necessary to know where in the cell the nanoparticle locates and whether its presence has any detrimental effect on the cell (e.g., does it cause stress or induce cell death?).

Gold nanoparticles delivered to cells in this manner do not typically affect cell viability, and we have demonstrated that they do not induce cell death (either apoptotic or necrotic) or cause oxidative stress. Cellular uptake of gold nanoshells [22] is dependent on a variety of factors. We tested various culture conditions in order to probe different endocytosis mechanisms (e.g., by depleting clathrin, cholesterol or ATP from the cells), but we found that the greatest level of cellular delivery is when using cell culture media containing no calf serum (Fig. 2.2). We attribute this finding to the fact that serum proteins are known to coat nanoparticles,

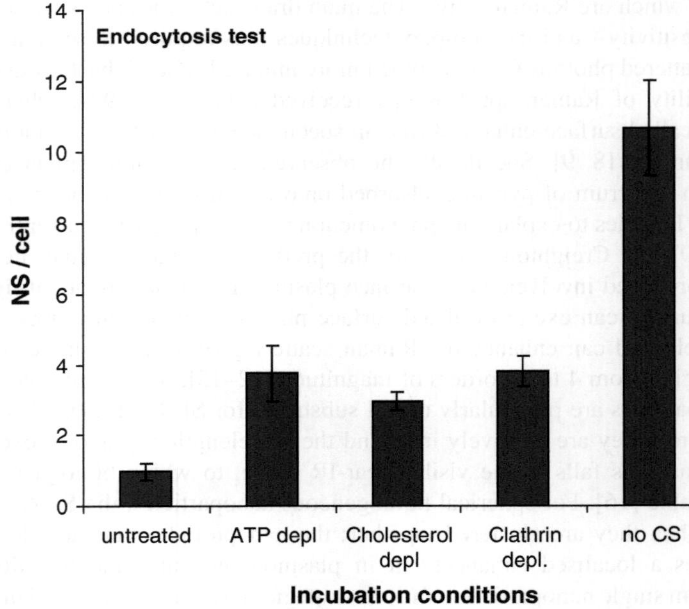

Fig. 2.2 Estimated amount of NS per cell according to particles found in 100 TEM sections. NS incubations for this test were conducted for 5 h with 10 fm NS in 10 % calf serum (CS)-containing media, except the condition with no CS on the far *right*. The *bars* show the amount of NS (monomers, dimers) found in 100 cell sections

thus changing their physicochemical surface characteristics. This change in surface characteristics seems to be sufficient to change the way in which some cell types react to the nanoshells.

An important consequence of removing serum proteins from the culture medium is that the nanoshells which are taken up tend to be more likely to be free in the cytosol, whereas those taken up in the presence of serum are more likely to be found in endosomal structures. Figure 2.3 shows TEM images of gold nanoshells in different locations within mouse fibroblast cells. In Fig. 2.3a, b, nanoshell dimers are encapsulated in endosomes as a result of incubation in serum-containing culture media. Similarly Fig. 2.3c shows a nanoshell which had been incubated in serum-containing media coated in a biomolecular corona. Only in serum-free conditions (Fig. 2.3d) can the nanoshells be seen free in the cytosol.

When we used gold nanoshells [17] to measure SERS spectra inside cells, we found that the spectra contained peaks characteristic of a variety of biomolecular vibrational modes, mostly originating from proteins. Examples of these spectra can be seen in Fig. 2.4. Specifically, Fig. 2.4 shows two important features. Firstly, while spectra b–e show the heterogeneity between SERS spectra measured at different spatial locations within the cell, they do not show any vibrations characteristic of nucleic acids—this suggests that the nanoshells do not access the nucleus. Secondly,

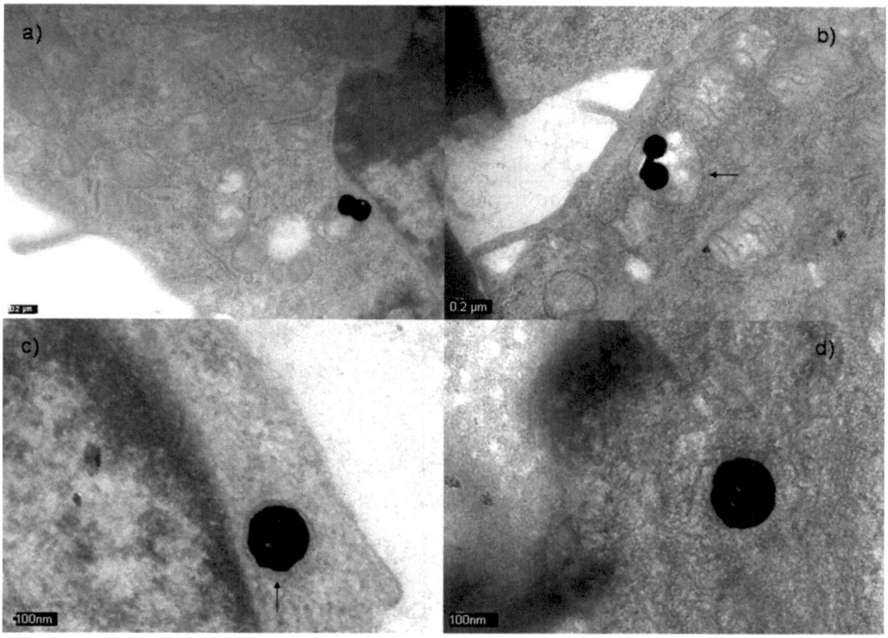

Fig. 2.3 Set of TEM images of NS inside cells. Images (**a**, **b**, **c**) are all of NS which were incubated in media containing 10 % CS and can be found either (**a**) free suspended in the cytosol (scalebar 0.2 µm) or (**b**) taken up in oversized vesicles or as in (**c**) surrounded by what may be a dense coating of protein. (**d**) Representative NS which has been taken up during incubation without any CS in the growth media. None of the NS found in these conditions were surrounded by any kind of coating or vesicles inside the cells

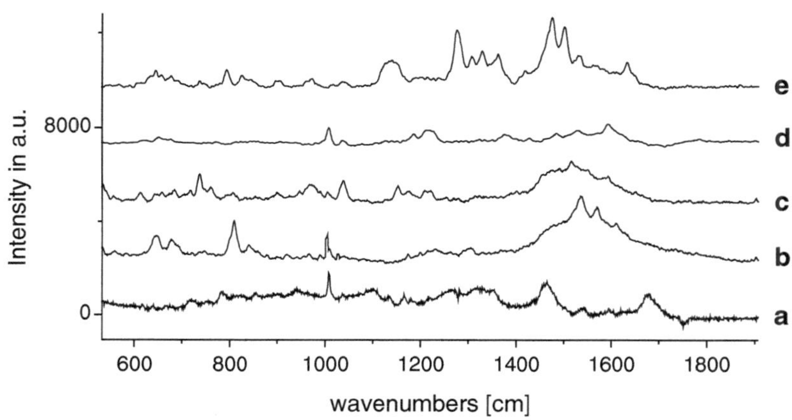

Fig. 2.4 (**a**) Averaged Raman spectrum of 50 different NIH/3T3 fibroblast cells measured at a 785.32 nm excitation wavelength, 50 mW laser power at the sample for 180 s. (**b–e**) SER spectra taken by excitation of NS in different cells at the same wavelength with 5 s acquisition time and 3 mW laser power at the sample

Table 2.1 Vibrational modes and assignments in the Raman spectra of cellular contents as shown in Fig. 2.4. SERS-enhanced signals from samples b–e are correlated with the assigned peaks in the whole-cell Raman spectrum

		NS in different cells (cm^{-1})			
Whole-cell resonance (cm^{-1})	Assignment of Raman features	(b)	(c)	(d)	(e)
640	C–C twist in tyrosine	646		642	644
675	C–S stretch in cysteine	678	684		
720	C–N stretch in lipid/adenine		736		737
783	DNA: O–P–O backbone stretch in thymine/cytosine				
825	DNA: O–P–O backbone stretch/out of plane ring breath in tyrosine				825
856	In plane ring breathing mode in tyrosine/C–C	848		850	
893	C–C skeletal stretch in protein		900	905	896
941	C–C skeletal stretch in protein		967		972
1,008	Symmetric ring breathing mode of phenylalanine	1,005	1,005	1,007	1,428
1,036	C–H in plane bending mode of phenylalanine		1,037	1,038	1,036
1,100	DNA: O–P–O backbone stretching				
1,109	DNA: O–P–O backbone stretching				
	C–N stretch in polypeptide chains		1,133	1,138	1,139
1,166	C–C stretching in proteins	1,174	1,174		
1,217	Amide III: beta-sheet	1,216	1,209	1,214	1,214
1,266	Amide III: beta-sheet/adenine/cytosine				1,275
1,329	Guanine			1,323	1,329
1,354	Polynucleotide chain (DNA bases)				
	Possible porphyrin stretches			1,375	1,366
1,465	$(\nu\,(C=C)+\nu(C-C)+\nu(C=O-H))$, Chromophore			1,485	1,475
	Aromatic ring stretches		1,516		1,511
1,543	Lipid stretches	1,535		1,528	1,534
	Tyrosine stretch	1,570		1,593	1,570
1,595	Ring mode (adenine/guanine)				
1,619	C═C bending in phenylalanine and tyrosine	1,616			
1,677	Amide I: alpha-helix				

a comparison of the bulk Raman signal from cells (Fig. 2.4a) with SERS spectra (Fig. 2.4b–e) shows the intensity of the SERS effect. The bulk Raman measurement is only possible by using high laser powers, long acquisition times and averaging over several cells; SERS measurements are made using low laser power and short acquisition times and have subcellular resolution. Table 2.1 shows the peak positions and assignments of vibrational modes typically recorded when measuring Raman spectra in cells.

2.4 SERS Nanosensors

For constructing SERS nanosensors, gold and silver are the most popular nanoparticle substrates by far [18]. The use of these metals, particularly gold, means that surface functionalisation of the nanoparticles can be relatively simple. Sulphur and gold exhibit a specific interaction in which organic thiols and disulfides spontaneously ligate onto a gold surface and form a monolayer, under ambient conditions from dilute aqueous or organic solutions [23–25]. In this way, monolayer surfaces of organic molecules can be constructed to put them in contact with both a gold nanoparticle surface and the surrounding environment, so that each probe molecule is presented with the same chemical environment. Thus, all the necessary requirements for a nanosensor are fulfilled: the molecules are SERS active, they are stably attached to the particles, and the sensors report on their environment. Delivering the assembled nanosensors into a cell is usually extremely simple; incubating eukaryotic cells with a high-femtomolar concentration of nanosensors in serum-free media is usually sufficient to cause uptake directly and without affecting cell viability [22]. Additionally, nanosensor distribution in cells is found to be directly proportional to the extracellular concentration, giving an easy way to tune intracellular sensor density. This is in sharp contrast to methods using modified fluorescent proteins, which require the genetic engineering of specific cell lines.

In general, a sensor is a system which detects a particular stimulus by changing its properties in a measurable manner; SERS sensors typically consist of a metal nanoparticle substrate, functionalised with a reporter molecule adsorbed on the surface. The functionalised nanoparticle can act as both a delivery mechanism and as a SERS-active surface for monitoring, and the reporter molecule typically has a structure sensitive to a particular stimulus. When the structure of the reporter molecule changes, so does its Raman spectrum, and so comparison of different regions of the Raman spectrum will give a direct read-out of the relative proportions of a reporter in its distinct conformational states—and this can be calibrated with, e.g., buffers with known analyte concentration or of known pH or redox potential (Fig. 2.5).

This type of sensor is distinct from the majority of fluorescent reporters which exhibit an irreversible change in fluorescence as a result of a covalent reaction with an analyte. For example, DCFH-DA is a fluorescent dye commonly used for the detection of reactive oxygen species (ROS). In the presence of ROS, it switches from a nonfluorescent dihydrodichlorofluorescein to the fluorescent form, dichlorofluorescein. While this gives important information on the formation of ROS, the reaction is irreversible and as a result the reporter cannot report on a change back to low ROS concentrations. Similarly, "sandwich assays" which use two probe molecules in order to measure analyte concentrations (one of which is labelled with a fluorescent reporter) are irreversible and cannot measure dynamic changes in concentration of an analyte.

Sandwich assays are typically used to measure concentrations of molecules such as proteins, and we recently developed a SERS nanosensor for the detection of

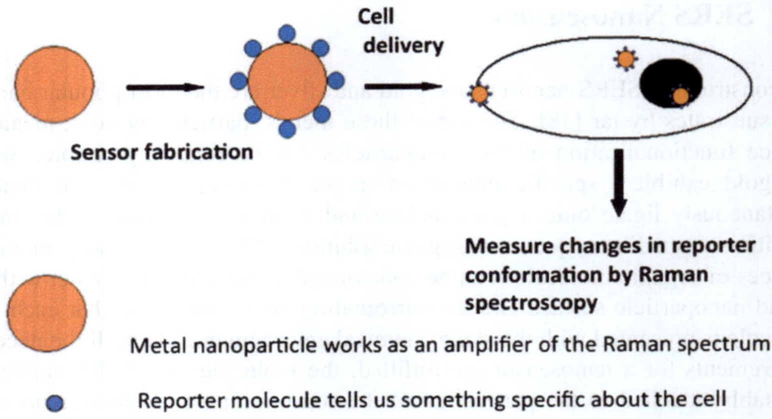

Fig. 2.5 Schematic for intracellular SERS sensing

the protein thrombin. Thrombin is an important target molecule clinically since it is an important regulator of the cascade of reactions which lead to blood clot formation. Blood clots cause an estimated 25,000 deaths in the UK per year and recent clinical guidelines have suggested that evaluation of the risk of clot formation could significantly reduce this toll [26, 27]. A simple, sensitive test of thrombin levels could be a valuable tool in such evaluations.

2.5 Nanosensors: Protein Quantification

In a recent paper, we considered that an aptamer, specific for thrombin (referred to herein as TBA), would be a probe molecule well suited to SERS sensors [28]. Aptamers are probe molecules made from nucleic acid oligomers, which have been selected (via a process called SELEX) for their ability to bind non-nucleic acid target molecules. Since aptamers often undergo a conformational change on binding their cognate target and since the probe and target are physicochemically distinct (nucleic acid and protein respectively), we predicted that we would see a change in the SERS spectrum as a result of thrombin binding to TBA immobilised on the surface of a gold nanoshell (shown schematically in Fig. 2.6).

The thrombin nanosensor was assembled from gold nanoshells which were surface functionalised with TBA, then the remaining surface blocked with mercaptohexanol (MCH). MCH blocking allows better sensitivity in the sensor since it inhibits non-specific binding to the gold surface and therefore inhibits non-specific signals such as those seen in Fig. 2.4. The Raman spectra from the nanosensors show very low intensity (Fig. 2.7), with a prominent and stable feature at 1,089 cm^{-1} assignable to the DNA O–P–O backbone stretch, and this does not change when the nanosensor is incubated with either bovine serum albumin (BSA),

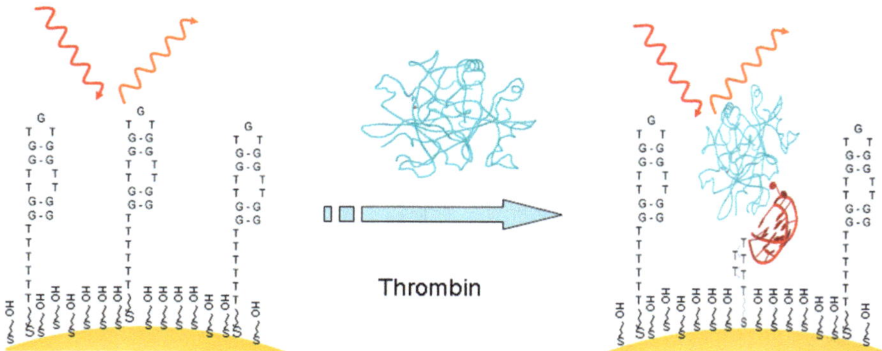

Fig. 2.6 The assembled mixed monolayer of prehybridised TBA oligonucleotide and MCH and the model for the detection

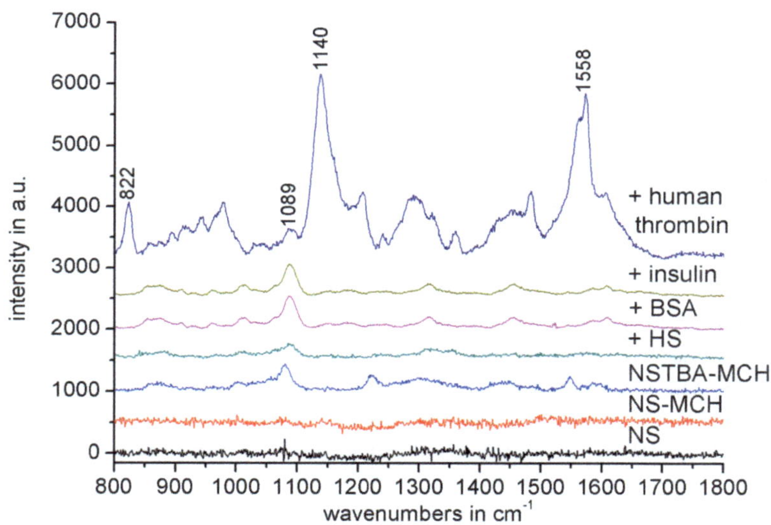

Fig. 2.7 SERS spectra from thrombin nanosensor. The *top trace* shows the spectrum measured as the result of a specific interaction between thrombin and TBA—dominant peaks can be assigned to DNA vibrational modes (see text). Controls with BSA (10 mg/ml), insulin 600 nM and 1 % human serum show no evidence of protein binding. Lower three spectra show spectra of NS, NS plus blocking agent (mercaptohexanol) and the fully assembled NSTBA sensor. Spectra are averages of ten acquisitions

human serum (HS) or insulin—this shows that there is very little signal change associated with non-specific binding of non-cognate proteins to the nanosensor. On incubation with its cognate target (thrombin), there is a large change in the nanosensor SERS spectrum. The major spectral changes include increases in signal at 822, 1,140 and 1,558 cm^{-1}. These can be assigned to the combined C2′-endo and C3′-endo modes of the 2′-deoxyribose sugars, the C–O–C stretch at 1,140 cm^{-1}

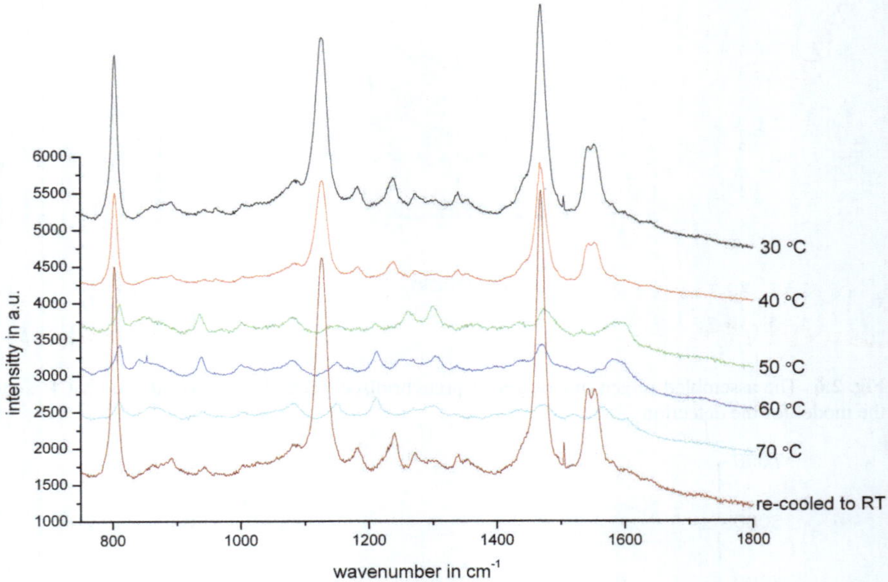

Fig. 2.8 SERS spectra of a G-quadruplex forming oligonucleotide. Heating and cooling cycles show that the secondary structure can be reversibly disrupted allowing a structural switch

and guanine ring modes at 1,558 cm^{-1}. At 1,480 cm^{-1} the low intensity feature can be assigned to a guanine ring mode [29, 30]. Further signal increases can be assigned to vibrational modes of protein, such as the amide III backbone vibration at 1,220 cm^{-1}, CH_2 stretching mode at 1,440–1,460 cm^{-1} or tyrosine aromatic ring vibrations at 1,610 cm^{-1} [31, 32]. When we used an oligonucleotide with no specific affinity for thrombin, we saw no change in its SERS signal—this confirms that the spectral change correlates with a specific interaction between TBA and thrombin.

Our investigations showed that the majority of the spectral change could be attributed to the formation of a G-quadruplex motif in the aptamer structure, and to prove this, we functionalised NS with an oligonucleotide known to form a G-quadruplex structure. Not only was its spectrum very similar to that of the TBA–thrombin complex, but controlled heating demonstrated that the secondary structure was responsible for the strong spectral signals and that the structure reformed on cooling (Fig. 2.8). We also showed that the sensor response to thrombin scaled with concentration and that we could detect low, clinically relevant concentrations (Fig. 2.9).

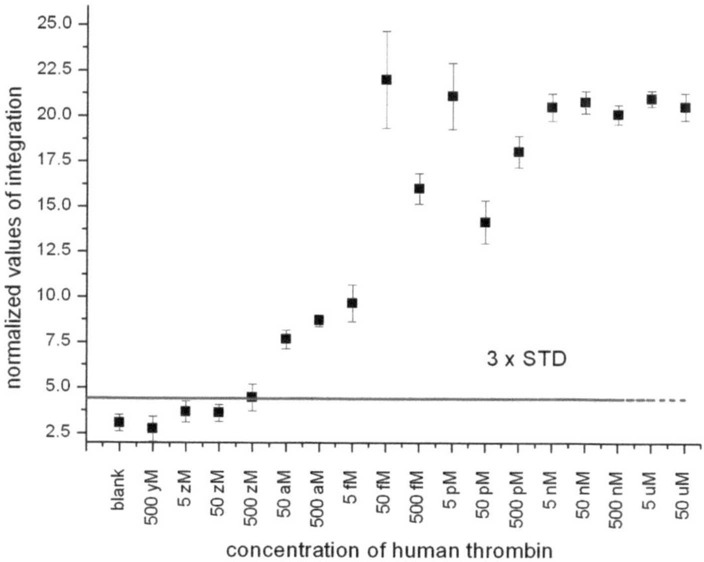

Fig. 2.9 Calibration curve for thrombin detection. *Error bars* represent single standard deviations

2.6 Nanosensors: Peptide Structure and Protease Activity

In a subsequent publication we adapted the concept of using SERS to investigate biomolecular structure in order to study the potential dependence of the structure of a peptide implicated in Goodpasture's disease and furthermore to determine how the peptide structure altered its interaction with a protease (cathepsin D) [33]. Goodpasture's disease is a rare autoimmune condition of the kidneys and lungs in which there is an immune response to a misfolded domain of the alpha-3 chain of type IV collagen ($\alpha3(IV)NC1_{67-85}$). It is thought that in healthy individuals $\alpha3(IV)NC1_{67-85}$ is completely broken down (destructive processing) [34] and as a result is never normally presented to T cells. Destructive processing of the antigen is thought to require reduction of a disulfide bridge that holds it in a hairpin-like conformation, which is then followed by proteolytic cleavage by an aspartate protease, cathepsin D [35, 36]. The importance of the oxidation state of the cysteine thiols in the peptide suggests that the redox environment during $\alpha3(IV)NC1_{67-85}$ processing may be crucial in determining whether processing proceeds as described with destruction of $\alpha3(IV)NC1_{67-85}$ or some other route in which $\alpha3(IV)NC1_{67-85}$ is preserved and autoreactive T cells are activated. It is therefore important to determine the standard redox potential of the disulfide bond of the peptide and investigate how redox potential influences proteolysis.

Since we had shown the utility of SERS in determining the conformation of an oligonucleotide, we speculated that SERS could similarly be used to determine

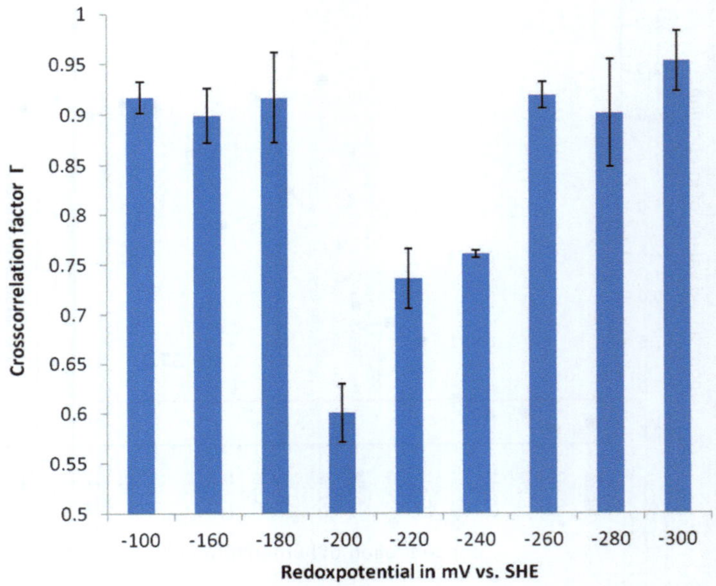

Fig. 2.10 Graph showing diagonal cross-correlation factors Γ (correlating 25 spectra) versus redox potential from -100 to -300 mV. Values close to 1 describe conformational stability and decreasing Γ values depict a dynamic conformational change. The dip at -200 mV suggests the dynamic breaking and formation of the internal disulfide bond of AS35. The *error bars* describe a single standard deviation over three independent experiments

the conformation (and therefore the oxidation state) of a peptide. In order to measure SERS spectra from the peptide, we made a nanosensor from the peptide (attached via a free cysteine at its c-terminus) to gold nanoshells. We controlled the solution redox potential (E) using defined pH and concentrations of glutathione and glutathione disulfide in accordance with the Nernst equation for glutathione (where E^{θ} is the standard redox potential for glutathione/glutathione disulfide):

$$E = E^{\theta} + \frac{RT}{nF}\ln\frac{[GSSG]\,[H^{+}]^{2}}{[GSH]^{2}}$$

and measured the cross-correlation factor between consecutive spectra (Γ), which we suggest is a reporter of population heterogeneity

At reducing potentials the peptide molecules should be a homogeneous population in a hairpin structure, at oxidising potentials the peptide molecules should be a homogeneous population in a linear structure and at intermediate potentials the molecules should be a heterogeneous mixture. This is indeed what we see (Fig. 2.10) with a pronounced dip in Γ at around -200 mV versus NHE.

This conformational change also correlated well with the digestion of the peptide by the protease cathepsin D. By controlling redox potential and measuring

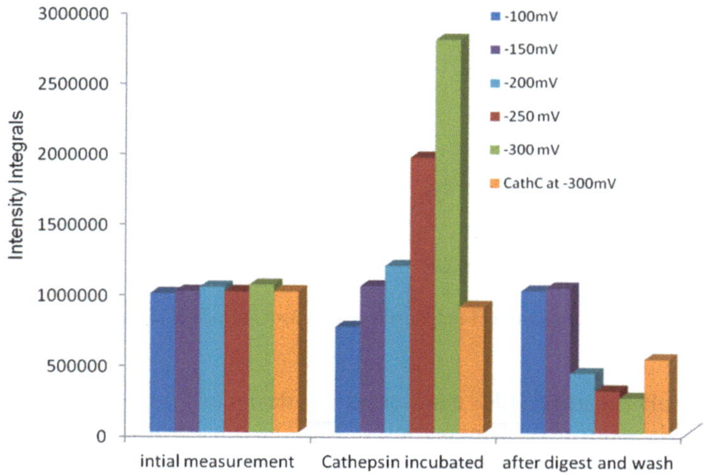

Fig. 2.11 Integrals from spectra recorded during digestion experiments at redox potentials of −100, −250 and −300 mV and control experiments with less active protease cathepsin C and pepstatin A inhibited cathepsin D at −300 mV, compared before, during and after the incubation with cathepsin

SERS before, during and after addition of the protease, we could measure both binding of the protease during digestion and the loss of specific peptide signal as a result of digestion (Fig. 2.11). Significantly, the addition of pepstatin A (an inhibitor of cathepsin D) stopped any digestion of the peptide. Taken together these findings suggest that dysregulation of redox potential can lead to defective antigen processing. When the antigen is improperly processed, this can lead to the autoimmune reaction that causes degeneration of kidney and lung tissue.

2.7 pH-Sensitive Intracellular SERS Nanosensors

When the functionalised nanoparticles are delivered inside cells, they act as nanosensors and can measure physiologically interesting events in single live cells in real time (Fig. 2.5). The first and most well-explored class of intracellular SERS nanosensors have been those sensitive to changes in pH. For the measurement of pH, reporter molecules should incorporate an acid or basic group which is sensitive to pH changes and a thiol group for ligation to a metal nanoparticle. The protonation state of the probe molecule can then be used to calculate the pH via the Henderson–Hasselbalch equation. The range of this type of sensor is determined by the pK_a of the reporter molecule attached to the particle, and so careful selection of the probe molecule can give sensors sensitive to different ranges of pH and therefore be useful in different applications (Fig. 2.12).

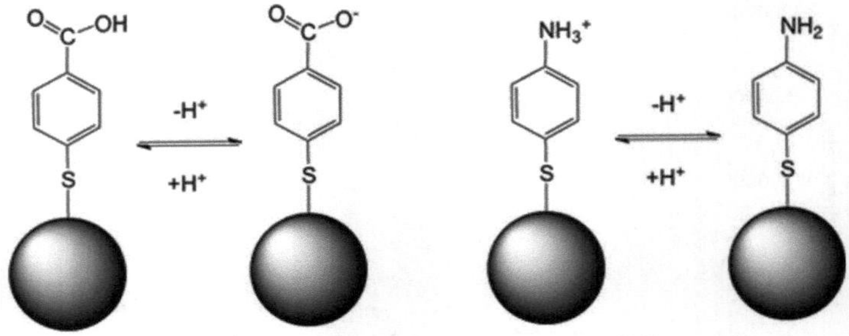

Fig. 2.12 Two different types of pH-responsive nanosensors sensitive to base (*left*) and acid (*right*)

The first pH-sensitive SERS nanosensor was designed for the measurement of pH in living cells using silver nanoparticles [37]. The sensor used 4-mercaptobenzoic acid as reporter, which ligated to the metal surface via the thiol group, and was responsive to pH ranging from 6 to 8 pH units due to the carboxylic acid. However, the SERS enhancement was found to originate solely from clusters of nanoparticles, which also exhibit a large variability in the electrical double layer close to the nanoparticle–nanoparticle junctions and thus the local pH. Unfortunately these nanoparticle–nanoparticle junctions are also SERS "hotspots" and the measurements, though accurate, encompass individual reporter molecules over widely varying local pH values. For these reasons, the nanosensor suffered from a wide variability in the reported pH (± 1 pH unit). Use of gold nanoshells [22, 38] overcame the problem with aggregation, as these engineered particles comprising a dielectric silica core coated in a film of gold gave a further enhancement of the Raman signal from the surface monolayer of 4-mercaptobenzoic acid. In this way, single nanoparticles still act as effective nanosensors and the "hotspots" created by nanoparticle–nanoparticle junctions no longer overwhelm the signal from the rest of the surface reporter molecules. The pH resolution could be improved to ± 0.1 pH units over the range of 5.8–7.6 pH units, but with some sensitivity between 4 and 9 pH units.

As well as carboxylic acid probes for basic pH, there have been amine-based nanosensors developed which are sensitive in acidic pH [39]. In this case, the reporter molecule 2-thioaniline was sensitive to pH ranging from 3 to 8, although the use of silver nanoparticles led to the same aggregation-induced error as before, limiting the sensor to ± 1 pH unit in living cells. pH-sensitive nanosensors have also been applied to probing the dynamic environment inside cells as the pH changes over time [40], in one case studying the resilience of cancer cells to photodynamic therapy [41]. Endosomes have been specifically targeted for pH sensing [42, 43] by using a nanosensor functionalised with 4-mercaptopyridine for acidic sensing, alongside 2,4-ε-dinitrophenol-L-lysine to target the FcεRI receptor-mediated endocytic pathway.

2.8 Redox-Sensitive Intracellular Nanosensors

In cells, the redox potential is a tightly regulated and compartmentalised intracellular variable. The intracellular redox potential is controlled by a variety of mechanisms, such as couples between oxidised and reduced forms of glutathione, NAD^+/NADH and $NADP^+$/NADPH, or reactive oxygen species (ROS) [44, 45]. Different organelles can have a substantially different redox potential depending on function, for example, more oxidising in the endoplasmic reticulum and more reducing in the mitochondria, to enable the distinct chemistries required in these compartments [46]. When the redox potential of cells is perturbed towards substantially more oxidising potentials, the cells are said to be in a state of oxidative stress, which has been implicated in many diseases and dysfunctions such as inflammation, cancers and neurodegeneration [46, 47]. Redox potential has also been theorised to underpin the entire cascade of programmed cell death (apoptosis), and understanding its regulation or dysregulation could also have implications for cancer treatment [48]. Being able to monitor the intracellular redox potential in single living cells, then, is a desirable outcome.

Microinjected protein crystals of glutathione reductase have been used to monitor the redox potential of the cytoplasm of human fibroblast cells [49], determining it to be approximately -270 mV versus NHE, although the read-out was via relatively insensitive optical microscopy, and the cells were only reported to have survived the injection procedure for a few hours. A modified redox-sensitive green fluorescent protein (roGFP) has also been used to give a read-out of intracellular redox potential [47], but this method has limitations intrinsic to the protein-based nature of the sensor: they require specific per-cell-type genetic engineering, are constrained to a narrow window of potentials [50] and do not survive the relatively oxidising potentials of some subcellular compartments [51]. RoGFP is sensitive to the glutathione redox couple [52], but specific modifications are required in order to detect other species [53]. Additionally, background fluorescence of the cell and medium places a limit on the sensitivity of the technique.

While we previously demonstrated that SERS could be used to investigate the biological significance of redox potential, in the study of a peptide implicated in a human autoimmune disease [33], we have also recently developed a series of SERS nanosensors that can quantitatively report on local redox potential [54].

In this case the reporter is a small redox-active molecule and can exist in either an oxidised or reduced state. The thiol–gold interaction was used to assemble the nanosensors, using hydroquinone and naphthoquinone reporters (Fig. 2.13). Specifically, the naphthoquinone reporter has a distinct series of spectra at a physiologically useful range of redox potentials which exceeds that of fluorescent roGFP. These sensors have been delivered to cells, and we have demonstrated that they can reversibly report on redox potential changes (i.e., both oxidative and reductive changes). We used these sensors to measure changes in intracellular redox

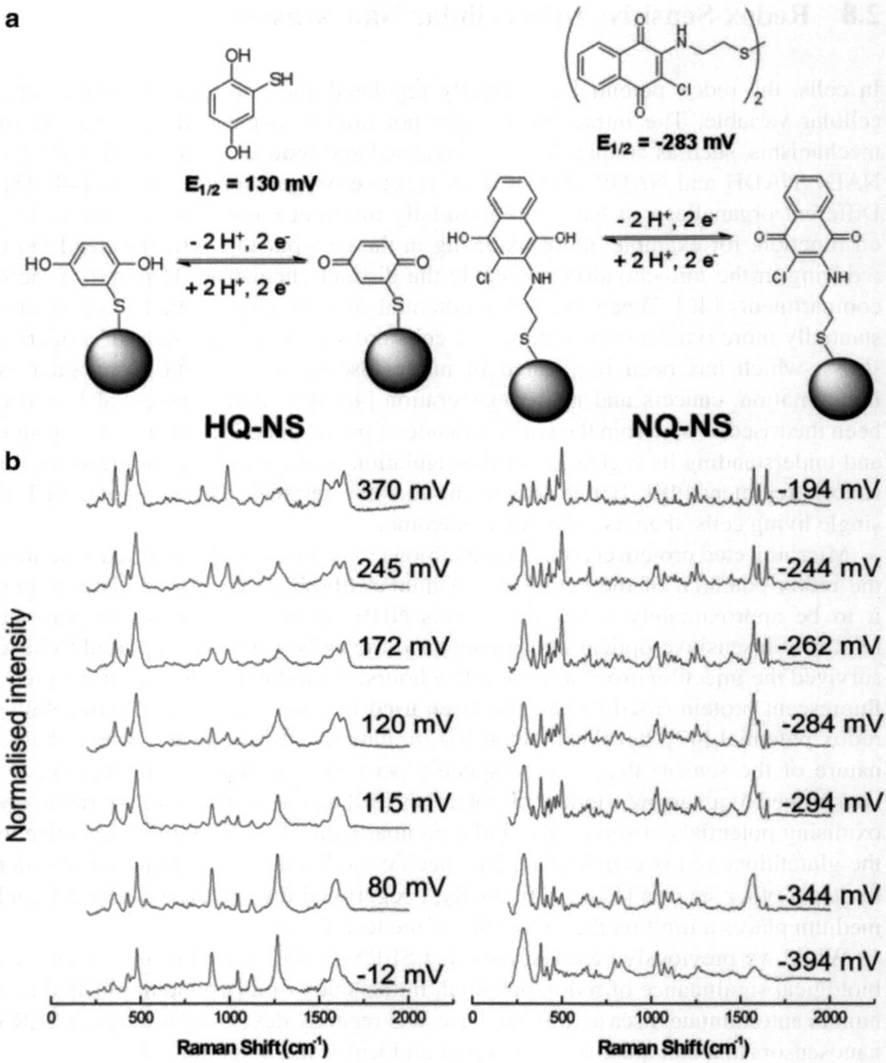

Fig. 2.13 Redox-active SERS nanosensors

potential which result either from exposure to reactive oxygen species or as a result of apoptosis. Figure 2.14 shows the excellent correlation between redox potential measured using SERS nanosensors and the activity of caspase 3/7 (a protease responsible for the execution of apoptosis) in cells undergoing apoptosis.

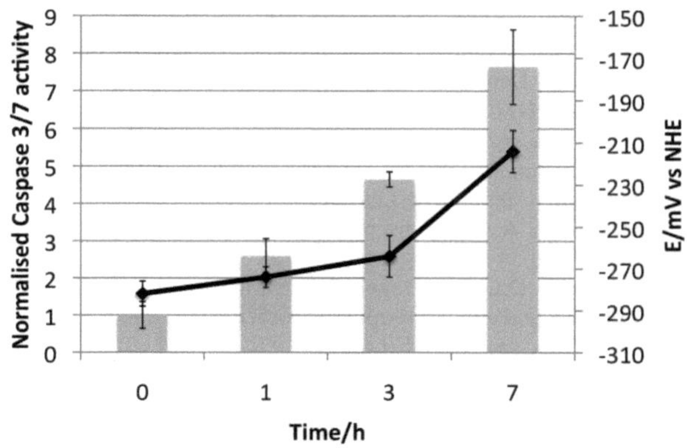

Fig. 2.14 Correlation between redox potential and caspase activity in cells undergoing apoptosis

2.9 Conclusions

Raman spectroscopy of functionalised nanoparticles is an emerging tool in cell biology. There are a variety of nanoparticle architectures and improvements to choose from, with gold nanoshells offering particular promise as single-particle sensing systems with subcellular resolution. The sensors are simple to construct from well-understood gold–thiol ligation chemistry, and reporter molecules and nanoparticles alike are synthetically accessible or commercially available. Nanosensors assembled in this way are taken up readily into cells and show no toxicity or adverse effects over a time span of several days. A wide range of information can be obtained from the use of SERS nanosensors in the study of biological systems, from protein concentration and enzyme activity to intracellular variables such as pH and redox potential. pH sensing is particularly well established, with subcellular resolution data available as accurate as 0.1 pH units. The use of SERS for redox potential monitoring is an emerging technique, which promises to further our understanding of the importance of redox potential in cells and its implication in numerous disease models and cellular processes.

References

1. Watson, W.H., Pohl, J., Montfort, W.R., Stuchlik, O., Reed, M.S., Powis, G., Jones, D.P.: D. P. Redox potential of human thioredoxin 1 and identification of a second dithiol/disulfide motif. J. Biol. Chem. **278**, 33408 (2003)
2. Stolee, J.A., Shrestha, B., Mengistu, G., Vertes, A.: Observation of subcellular metabolite gradients in single cells by laser ablation electrospray ionization mass spectrometry. Angew. Chem. Int. Ed. **51**, 10386 (2012)
3. Lee, Y.E., Kopelman, R.: Optical nanoparticle sensors for quantitative intracellular imaging. Wiley Interdiscip. Rev. Nanomed. Nanobiotechnol. **1**, 98 (2009)
4. Benjaminsen, R.V., Sun, H., Henriksen, J.R., Christensen, N.M., Almdal, K., Andresen, T.L.: Evaluating nanoparticle sensor design for intracellular pH measurements. ACS Nano **5**, 5864 (2011)

5. Rayleigh, L.: On the transmission of light through an atmosphere containing small particles in suspension, and on the origin of the blue of the sky. Philos. Mag. Ser. 5, **47**, 375 (1899)
6. Smith, E., Dent, G., Wiley, J.: Modern Raman Spectroscopy: A Practical Approach. Wiley, Hoboken, NJ (2005)
7. Nobelprize.org, vol. 2013
8. Fleischmann, M., Hendra, P.J., McQuillan, A.J.: Raman spectra of pyridine adsorbed at a silver electrode. Chem. Phys. Lett. **26**, 163 (1974)
9. McQuillan, A.J.: The discovery of surface-enhanced Raman scattering. Notes Rec. R. Soc. **63**, 209 (2009)
10. Jeanmaire, D.L., Van Duyne, R.P.: Surface raman spectroelectrochemistry: part I. Heterocyclic, aromatic, and aliphatic amines adsorbed on the anodized silver electrode. J. Electroanal. Chem. Interf. Electrochem. **84**, 1 (1977)
11. Albrecht, M.G., Creighton, J.A.: Anomalously intense Raman spectra of pyridine at a silver electrode. J. Am. Chem. Soc. **99**, 5215 (1977)
12. Le Ru, E.C., Blackie, E., Meyer, M., Etchegoin, P.G.: Surface enhanced Raman scattering enhancement factors: a comprehensive study. J. Phys. Chem. C **111**, 13794 (2007)
13. Nie, S., Emory, S.R.: Probing single molecules and single nanoparticles by surface-enhanced Raman scattering. Science **275**, 1102 (1997)
14. Qian, X.M., Nie, S.M.: Single-molecule and single-nanoparticle SERS: from fundamental mechanisms to biomedical applications. Chem. Soc. Rev. **37**, 912 (2008)
15. Lombardi, J.R., Birke, R.L.: A unified view of surface-enhanced Raman scattering. Acc. Chem. Res. **42**, 734 (2009)
16. Creighton, J.A., Eadon, D.G.: Ultraviolet-visible absorption spectra of the colloidal metallic elements. J. Chem. Soc. Faraday Trans. **87**, 3881 (1991)
17. Oldenburg, S.J., Averitt, R.D., Westcott, S.L., Halas, N.J.: Nanoengineering of optical resonances. Chem. Phys. Lett. **288**, 243 (1998)
18. Wang, Y., Yan, B., Chen, L.: SERS Tags: Novel Optical Nanoprobes for Bioanalysis. Chem. Rev. **113**, 1391–1428 (2013)
19. Kneipp, K., Wang, Y., Kneipp, H., Perelman, L.T., Itzkan, I., Dasari, R.R., Feld, M.S.: Single molecule detection using surface-enhanced Raman scattering (SERS). Phys. Rev. Lett. **78**, 1667 (1997)
20. Jarvis, R.M., Goodacre, R.: Characterisation and identification of bacteria using SERS. Chem. Soc. Rev. **37**, 931 (2008)
21. Kneipp, K., Haka, A.S., Kneipp, H., Badizadegan, K., Yoshizawa, N., Boone, C., Shafer-Peltier, K.E., Motz, J.T., Dasari, R.R., Feld, M.S.: Surface-enhanced Raman spectroscopy in single living cells using gold nanoparticles. Appl. Spectrosc. **56**, 150 (2002)
22. Ochsenkühn, M.A., Jess, P.R.T., Stoquert, H., Dholakia, K., Campbell, C.J.: Nanoshells for surface-enhanced Raman spectroscopy in eukaryotic cells: cellular response and sensor development. ACS Nano **3**, 3613 (2009)
23. Bain, C., Biebuyck, H., Whitesides, G.: Comparison of self-assembled monolayers on gold: coadsorption of thiols and disulfides. Langmuir **5**, 723 (1989)
24. Bain, C.D., Troughton, E.B., Tao, Y.T., Evall, J., Whitesides, G.M., Nuzzo, R.G.: Formation of monolayer films by the spontaneous assembly of organic thiols from solution onto gold. J. Am. Chem. Soc. **111**, 321 (1989)
25. Rouhana, L.L., Moussallem, M.D., Schlenoff, J.B.: Adsorption of short-chain thiols and disulfides onto gold under defined mass transport conditions: coverage, kinetics, and mechanism. J. Am. Chem. Soc. **133**, 16080 (2011)
26. Coughlin, S.R.: Thrombin signalling and protease-activated receptors. Nature **407**, 258 (2000)
27. Treasure, T., Hill, J.: NICE guidance on reducing the risk of venous thromboembolism in patients admitted to hospital. J. R. Soc. Med. **103**, 210 (2010)
28. Ochsenkuhn, M.A., Campbell, C.J.: Probing biomolecular interactions using surface enhanced Raman spectroscopy: label-free protein detection using a G-quadruplex DNA aptamer. Chem. Commun. **46**, 2799 (2010)

29. Billinghurst, B.E., Oladepo, S.A., Loppnow, G.R.: pH-dependent UV resonance Raman spectra of cytosine and uracil. J. Phys. Chem. B **113**, 7392 (2009)
30. Moger, J., Gribbon, P., Sewing, A., Winlove, C.P.: Feasibility study using surface-enhanced Raman spectroscopy for the quantitative detection of tyrosine and serine phosphorylation. Biochim. Biophys. Acta **1770**, 912 (2007)
31. Benevides, J.M., Tsuboi, M., Bamford, J.K., Thomas, G.J.: Polarized Raman spectroscopy of double-stranded RNA from bacteriophage phi6: local Raman tensors of base and backbone vibrations. Biophys J **72**, 2748 (1997)
32. Yarasi, S., Billinghurst, B.E., Loppnow, G.R.: Vibrational properties of thymine, uracil and their isotopomers. J. Raman Spectrosc. **38**, 1117 (2007)
33. Ochsenkühn, M.A., Borek, J.A., Phelps, R., Campbell, C.J.: Redox potential dependence of peptide structure studied using surface enhanced Raman spectroscopy. Nano Lett. **11**, 2684 (2011)
34. Manoury, B., Mazzeo, D., Fugger, L., Viner, N., Ponsford, M., Streeter, H., Mazza, G., Wraith, D.C., Watts, C.: Destructive processing by asparagine endopeptidase limits presentation of a dominant T cell epitope in MBP. Nat. Immunol. **3**, 169 (2002)
35. Zou, J., Hannier, S., Cairns, L.S., Barker, R.N., Rees, A.J., Turner, A.N., Phelps, R.G.: Healthy individuals have goodpasture autoantigen-reactive T cells. J. Am. Soc. Nephrol. **19**, 396 (2008)
36. Zou, J., Henderson, L., Thomas, V., Swan, P., Turner, A.N., Phelps, R.G.: Presentation of the goodpasture autoantigen requires proteolytic unlocking steps that destroy prominent T cell epitopes. J. Am. Soc. Nephrol. **18**, 771 (2007)
37. Talley, C.E., Jusinski, L., Hollars, C.W., Lane, S.M., Huser, T.: Intracellular pH sensors based on surface-enhanced raman scattering. Anal. Chem. **76**, 7064 (2004)
38. Bishnoi, S.W., Rozell, C.J., Levin, C.S., Gheith, M.K., Johnson, B.R., Johnson, D.H., Halas, N.J.: All-optical nanoscale pH meter. Nano Lett. **6**, 1687 (2006)
39. Wang, Z., Bonoiu, A., Samoc, M., Cui, Y., Prasad, P.N.: Biological pH sensing based on surface enhanced Raman scattering through a 2-aminothiophenol-silver probe. Biosens Bioelectron **23**, 886 (2008)
40. Kneipp, J., Kneipp, H., Wittig, B., Kneipp, K.: Following the dynamics of pH in endosomes of live cells with SERS nanosensors. J. Phys. Chem. C **114**, 7421 (2010)
41. Bálint, Š., Rao, S., Marro, M., Miškovský, P., Petrov, D.: Monitoring of local pH in photodynamic therapy-treated live cancer cells using surface-enhanced Raman scattering probes. J. Raman Spectrosc. **42**, 1215 (2011)
42. Nowak-Lovato, K.L., Rector, K.D.: Targeted surface-enhanced Raman scattering nanosensors for whole-cell pH imagery. Appl. Spectrosc. **63**, 387 (2009)
43. Nowak-Lovato, K.L., Wilson, B., Rector, K.: SERS nanosensors that report pH of endocytic compartments during FcεRI transit. Anal. Bioanal. Chem. **398**, 2019 (2010)
44. Jones, D.P.: Redox sensing: orthogonal control in cell cycle and apoptosis signalling. J. Intern. Med. **268**, 432 (2010)
45. Dooley, C.T., Dore, T.M., Hanson, G.T., Jackson, W.C., Remington, S.J., Tsien, R.Y.: Imaging dynamic redox changes in mammalian cells with green fluorescent protein indicators. J. Biol. Chem. **279**, 22284 (2004)
46. Menon, S.G., Goswami, P.C.: A redox cycle within the cell cycle: ring in the old with the new. Oncogene **26**, 1101 (2006)
47. Gutscher, M., Pauleau, A.-L., Marty, L., Brach, T., Wabnitz, G.H., Samstag, Y., Meyer, A.J., Dick, T.P.: Real-time imaging of the intracellular glutathione redox potential. Nat. Methods **5**, 553 (2008)
48. Mallikarjun, V., Clarke, D.J., Campbell, C.J.: Cellular redox potential and the biomolecular electrochemical series: a systems hypothesis. Free Radic. Biol. Med. **53**, 280 (2012)
49. Keese, M.A., Saffrich, R., Dandekar, T., Becker, K., Schirmer, R.H.: Microinjected glutathione reductase crystals as indicators of the redox status in living cells. FEBS Lett. **447**, 135 (1999)
50. Lohman, J.R., Remington, S.J.: Development of a family of redox-sensitive green fluorescent protein indicators for use in relatively oxidizing subcellular environments. Biochem. **47**, 8678 (2008)

51. Hwang, C., Sinskey, A.J., Lodish, H.F.: Oxidized redox state of glutathione in the endoplasmic reticulum. Science (New York, NY) **257**, 1496 (1992)
52. Meyer, A.J., Brach, T., Marty, L., Kreye, S., Rouhier, N., Jacquot, J.-P., Hell, R.: Redox-sensitive GFP in Arabidopsis thaliana is a quantitative biosensor for the redox potential of the cellular glutathione redox buffer. Plant J. **52**, 973 (2007)
53. Morgan, B., Sobotta, M.C., Dick, T.P.: Measuring EGSH and H_2O_2 with roGFP2-based redox probes. Free Radic. Biol. Med. **51**, 1943 (2011)
54. Auchinvole, C.A.R., Richardson, P., McGuinnes, C., Mallikarjun, V., Donaldson, K., McNab, H., Campbell, C.J.: Monitoring intracellular redox potential changes using SERS nanosensors. ACS Nano **6**, 888 (2011)

Chapter 3
BioFET-SIM: A Tool for the Analysis and Prediction of Signal Changes in Nanowire-Based Field Effect Transistor Biosensors

Martin R. Hediger, Karen L. Martinez, Jesper Nygård, Mads Brandbyge, Jan H. Jensen, and Luca De Vico

Abstract Biosensors based on nanowire field effect transistor (FET) have received much attention in recent years as a way to achieve ultra-sensitive and label-free sensing of molecules of biological interest. The BioFET-SIM computer model permits the analysis and interpretation of experimental sensor signals through its web-based interface *www.biofetsim.org*. The model also allows for predictions of the effects of changes in the experimental setup on the sensor signal. After an introduction to nanowire-based FET biosensors, this chapter reviews the theoretical basis of BioFET-SIM models describing both single and multiple charges on the analyte. Afterwards the usage of the interface and its relative command line version is briefly shown. Finally, possible applications of the BioFET-SIM model are

---

M.R. Hediger • J.H. Jensen • L. De Vico (✉)
Department of Chemistry, University of Copenhagen, Universitetsparken 5, DK-2100 Copenhagen, Denmark
e-mail: martin@chem.ku.dk; jhjensen@chem.ku.dk; luca@chem.ku.dk

K.L. Martinez
Bionanotechnology and Nanomedicine Laboratory, Department of Chemistry, University of Copenhagen, Universitetsparken 5, DK-2100 Copenhagen, Denmark

Nano-Science Center, University of Copenhagen, Universitetsparken 5, DK-2100 Copenhagen, Denmark
e-mail: martinez@nano.ku.dk

J. Nygård
Niels Bohr Institute, Center for Quantum Devices, University of Copenhagen, Universitetsparken 5, DK-2100 Copenhagen, Denmark

Nano-Science Center, University of Copenhagen, Universitetsparken 5, DK-2100 Copenhagen, Denmark
e-mail: nygard@nbi.ku.dk

M. Brandbyge
DTU Nanotech, Department of Micro and Nanotechnology, Technical University of Denmark, DTU-Building 345 East, DK-2800 Kongens Lyngby, Denmark
e-mail: mads.brandbyge@nanotech.dtu.dk

S. Li et al. (eds.), *Nanoscale Sensors*, Lecture Notes in Nanoscale Science and Technology 19, DOI 10.1007/978-3-319-02772-2_3,

presented. Among the possible uses of the interface, the effects on the predicted signal of pH, buffer ionic strength, analyte concentration, and analyte relative orientation on nanowire surface are illustrated. Wherever possible, a comparison to experimental data available in literature is given, displaying the potential of BioFET-SIM for interpreting experimental results.

List of Abbreviations and Symbols

FET	Field Effect Transistor
r_{nw}	Nanowire radius
r_{ox}	Oxide layer thickness
r_{bf}	Biofunctionalization layer thickness
ε	Free space permittivity
ε_1	Nanowire permittivity
ε_2	Oxide permittivity
ε_3	Buffer permittivity
μ	Nanowire charge carrier mobility
n_0/p_0	Nanowire (electron/hole) charge carrier density
λ_{TF}	Nanowire Thomas–Fermi screening length
λ_{D}	Buffer Debye screening length
e	Unit of charge
$\hbar$	Plank's constant divided by 2π
m^*	Charge carrier effective mass
k_{B}	Boltzmann's constant
N_{A}	Avogadro's number

3.1 Introduction

Nanowire-based field effect transistor (FET) biosensors have recently been the subject of much research and development. A *sensor* is a device that permits the recognition of an analyte in a sample through some form of interaction and the generation of a signal that can be recorded. According to [1] a *bio*sensor is "a self-contained integrated device, which is capable of providing specific quantitative or semi-quantitative analytical information using a biological recognition element (biochemical receptor) which is retained in direct spatial contact with an electrochemical transduction element." Nanoscale materials offer a large surface-to-bulk ratio and thus a high sensitivity towards changes in the charge distribution near the surface and are thus highly interesting for sensing. *Nano*biosensors are characterized by a transduction element of nanoscale dimensions, like nanotubes, nanoribbons, or nanowires. The latter is the subject of this chapter.

Nanowire-based biosensors work as FETs: the current flowing through the semiconducting nanowire material is perturbed by the sensing event involving a charge

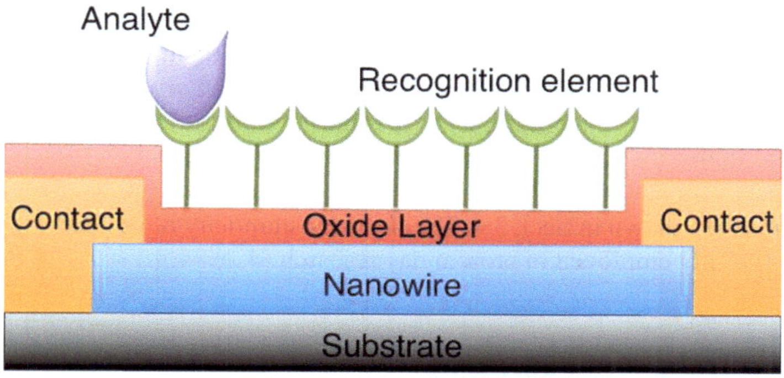

Fig. 3.1 Generic representation of a biofet. An insulating substrate carries a nanowire (transduction element) contacted at its extremes. An oxide layer is deposited on top. Biochemical receptors are chemically bond to form the recognition element capable of specifically recognizing the desired analyte

redistribution at the surface of the nanowire. Generally with an n-type nanowire a positive surface charge produces a positive signal (i.e., higher conductance) and *vice versa*. The opposite is true for a p-type nanowire. In an n-type (p-type) nanowire the charge carriers are electrons (holes). The change in surface potential created by a charge attracts or repels the charge carriers, creating the change in conductance. In a nanowire-based biosensor (for simplicity only called biofet from now on) a nanowire resides on an insulating substrate and is electrically contacted at its extremes, Fig. 3.1. Usually a second insulating layer, commonly constituted by an oxide, is deposited on top. Finally the analyte recognition layer is created through chemical modifications of the surface. The latter layer contains the recognition element and is responsible for specific capturing of the desired analyte and is usually referred to as a biofunctionalization layer. Examples of a biofunctionalization layer are a biotinylated surface for sensing of avidin or streptavidin [2, 3], or an antibody modified surface for the recognition of a given antigen [4].

Biofets have been employed or studied for pH [5], protein [3, 6–8], and DNA sensing [9, 10], for blood analysis [11] and for applications in nanotechnology-based medicine [12]. See [13–17] for a more thorough review of biofets and [18–20] for studies on their performance limits.

In addition to the experimental work, a series of numerical theoretical models have been presented to evaluate the performance of biofets [21–27]. Based on this theoretical work, tools for simulating biofets have been made available [28–30]. Following the work of Sørensen et al. [31] we developed our own model for simulation of the conductance sensitivity of a biofet in the presence of an analyte represented by a single charge [32]. We named our model BioFET-SIM. Section 3.2 gives a description of the theoretical background of BioFET-SIM and the approximations involved. BioFET-SIM has been subsequently extended to simulate the signal given by multiple charges approaching the surface of a biofet [33]. The usage of a multiple charges model proved essential when simulating the capture of analytes with an asymmetric charge distribution. Recently BioFET-SIM has been further extended [34] to include explicitly the biofunctionalization layer and made available through a web interface hosted at *www.biofetsim.org*. The main advantage of the web interface is the possibility to graphically interact with the sensed analyte, in order to define the specific orientation of it with respect to the nanowire surface. Further details on the usage of BioFET-SIM are given in Sect. 3.3 and some possible applications are shown in Sect. 3.4. Thanks to its simplicity of use BioFET-SIM is a tool successfully employed in present day research [4, 35–42].

3.2 Theoretical Background

In this section we will review equations and assumptions behind the BioFET-SIM model.

3.2.1 *Base Conductance*

The BioFET-SIM model describes nanowire-based biosensors with elements as depicted in Fig. 3.2. The BioFET-SIM model predicts the conductance sensitivity $\Delta G/G_0$ where ΔG is the difference between the conductance upon binding, G, and the base conductance, G_0, of the nanowire. The bulk conductance G_0 of a homogeneous cylindrical nanowire can be expressed as:

$$G_0 = \frac{\pi r_{\mathrm{nw}}^2 e (n_0 \mu_n + p_0 \mu_p)}{L} \tag{3.1}$$

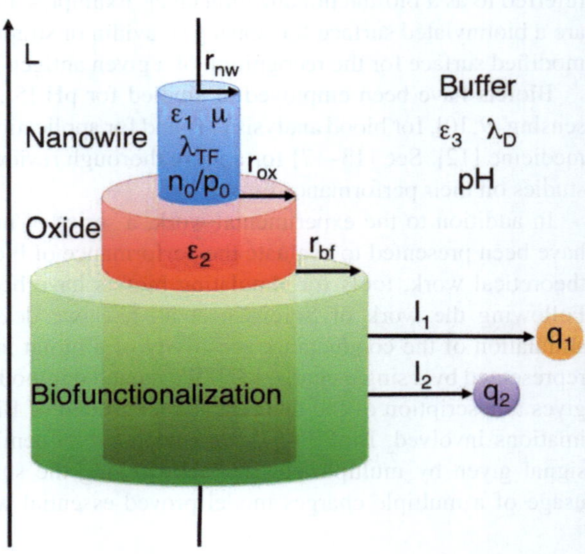

Fig. 3.2 Graphical representation of a generic nano BioFET. A nanowire of length L and radius r_{nw} is coated with an oxide layer of thickness r_{ox}. On the surface a biofunctionalization layer of thickness r_{bf} is responsible for capturing the analyte, represented by the charges q. The entire system is immersed in a buffer solution

where r_{nw} is the radius of the wire, e the elementary charge, n_0 the electron density, μ_n the electron mobility, p_0 the hole density, μ_p the hole mobility, and L the length of the nanowire. For a nanowire material with high enough doping, (3.1) can be approximated by considering only the electron or the hole density (only n_0 or p_0) and mobility (indicated simply as μ). For example, for a doped p-type nanowire, it is possible to rewrite (3.1) as:

$$G_0 = \frac{\pi r_{\mathrm{nw}}^2 e p_0 \mu}{L} \tag{3.2}$$

In our model only one type of charge carrier for the entire nanowire is considered, and the nanowire is assumed to resemble a low density material. $\Delta G/G_0$ is evaluated through a Thomas–Fermi screening model of the charge carriers [43–45].

A major assumption governs the description of the charge carrier concentration, that is the charge carrier concentration is distributed uniformly through the nanowire. This implies that effects of, e.g., charges trapped in the oxide layer are not considered as influencing the charge carrier distribution. One has to remember that, formally, n_0 (p_0) represents the initial charge carrier density.

Moreover, we assume here that effects related to the metal contacts are negligible.

3.2.2 Conductance Sensitivity

Following [31] one can define the generic sensitivity of a p-type doped nanowire as:

$$\frac{\Delta G}{G_0} = -\frac{2}{r_{\mathrm{nw}} e p_0} \Gamma \sigma_{\mathrm{s}} \tag{3.3}$$

where σ_{s} is the "sensed" charge density at the surface of the nanowire and Γ is a dimensionless function quantifying the actual sensitivity of the nanowire. In case of an n-type nanowire (3.3) becomes:

$$\frac{\Delta G}{G_0} = \frac{2}{r_{\mathrm{nw}} e n_0} \Gamma \sigma_{\mathrm{s}} \tag{3.4}$$

In the special case of a cylindrical nanowire it is possible to express Γ analytically as [46]:

$$\Gamma = \frac{\varepsilon_1 K_0\left(\frac{r_{\mathrm{nw}}+r_{\mathrm{ox}}}{\lambda_{\mathrm{D}}}\right) \frac{\lambda_{\mathrm{D}}}{\lambda_{\mathrm{TF}}} I_1\left(\frac{r_{\mathrm{nw}}}{\lambda_{\mathrm{TF}}}\right)}{\left[K_0\left(\frac{r_{\mathrm{nw}}+r_{\mathrm{ox}}}{\lambda_{\mathrm{D}}}\right)\left(\frac{\lambda_{\mathrm{D}}}{r_{\mathrm{nw}}+r_{\mathrm{ox}}}\right) + \ln\left(\frac{r_{\mathrm{nw}}+r_{\mathrm{ox}}}{r_{\mathrm{nw}}}\right) K_1\left(\frac{r_{\mathrm{nw}}+r_{\mathrm{ox}}}{\lambda_{\mathrm{D}}}\right) \frac{\varepsilon_3}{\varepsilon_2}\right] \varepsilon_1 \left(\frac{r_{\mathrm{nw}}}{\lambda_{\mathrm{TF}}}\right) I_1\left(\frac{r_{\mathrm{nw}}}{\lambda_{\mathrm{TF}}}\right) + \varepsilon_3 K_1\left(\frac{r_{\mathrm{nw}}+r_{\mathrm{ox}}}{\lambda_{\mathrm{D}}}\right) I_0\left(\frac{r_{\mathrm{nw}}}{\lambda_{\mathrm{TF}}}\right)} \tag{3.5}$$

where λ_{D} is the Debye screening length of the buffer, λ_{TF} the Thomas–Fermi screening length typical of the nanowire material, ε_1, ε_2, and ε_3 are the relative permittivities of the nanowire, the oxide layer, and the buffer, respectively, r_{ox} is the oxide layer thickness, I_0, I_1, K_0 and K_1 are the modified Bessel functions of first and second kind, respectively [47]. Γ is a dimensionless function with values ranging from zero to unity.

3.2.3 Single Charge Model

The presence of additional charges at a distance l from the nanowire surface gives rise to a charge density σ_b. Equation (3.3) is rewritten as:

$$\frac{\Delta G}{G_0} = -\frac{2}{r_{nw}\, e\, p_0} \Gamma\left(\Gamma_l\, \sigma_b + \sigma_s\right) \tag{3.6}$$

The BioFET-SIM model concerns the simulation of a signal in the conductance when an analyte is attached to the surface. The analyte is represented by σ_b and it is supposed not to modify the charge density σ_s already present at the nanowire surface. The conductance given by σ_s is considered as part of the background signal. For this reason the σ_s term can be dropped from (3.6) that can be written as:

$$\frac{\Delta G}{G_0} = -\frac{2}{r_{nw}\, e\, p_0} \Gamma\left(\Gamma_l\, \sigma_b\right) \tag{3.7}$$

and Γ_l is defined as:

$$\Gamma_l = 2\frac{r_{nw}}{r_{nw}+l}\left(1+\sqrt{\frac{r_{nw}}{r_{nw}+l}}\exp(l/\lambda_D)\right)^{-1} \tag{3.8}$$

where l is the distance of the sensed charge from the nanowire surface, inclusive of the biofunctionalization layer (r_{bf}). If the analyte is approximated as one charge as in Fig. 3.3a, (3.5), (3.7), and (3.8) describe the single charge BioFET-SIM model [32].

3.2.4 Multiple Charges Model

A complex analyte system like a protein can give rise to unexpected signals because of its nonuniform charge distribution [48]. In these cases a single charge description of the sensing event is not sufficient. Considering multiple charges, the sensitivity is evaluated as:

$$\frac{\Delta G}{G_0} = -\frac{2}{r_{nw} e p_0} \Gamma\left[\sum_i^m \left(\Gamma_{l_{i,tot}} \sigma_{b_i}\right)\right] \tag{3.9}$$

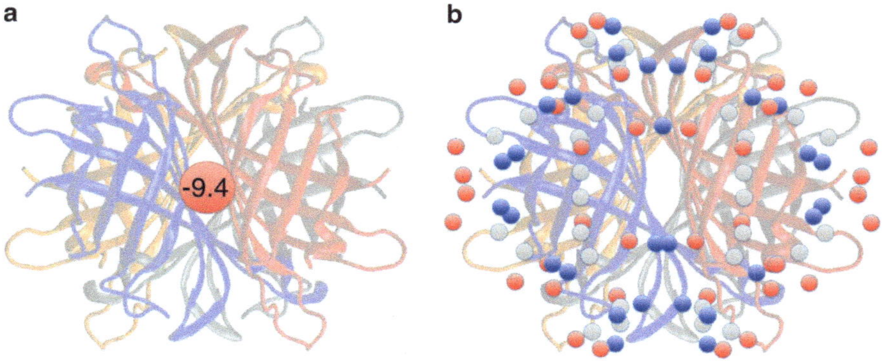

Fig. 3.3 PROPKA computed charges at pH = 7.4 for streptavidin (see Sect. 3.2.8). (**a**) The overall charge is assumed to reside at the center of mass of the protein (single charge model). (**b**) Each ionizable amino acid carries its charge (multiple charges model). *Blue*, *white*, and *red circles* correspond to a positive, neutral, and negative charge, respectively

and $\Gamma_{l_{i,\mathrm{tot}}}$ is defined as:

$$\Gamma_{l_{i,\mathrm{tot}}}(\lambda_\mathrm{D}) = 2\frac{r_\mathrm{nw}}{r_\mathrm{nw} + l_{i,\mathrm{tot}}}\left[1 + \sqrt{\frac{r_\mathrm{nw}}{r_\mathrm{nw} + l_{i,\mathrm{tot}}}}\, e^{l_{i,\mathrm{tot}}/\lambda_\mathrm{D}}\right]^{-1} \tag{3.10}$$

where $l_{i,\mathrm{tot}}$ is the overall distance of the charge q_i from the surface of the nanowire. Considering Fig. 3.2, the distance $l_{1,\mathrm{tot}}$ relative to q_1 is computed as:

$$l_{1,\mathrm{tot}} = r_\mathrm{ox} + r_\mathrm{bf} + l_1 \tag{3.11}$$

Equations (3.5), (3.9), and (3.10) constitute the multiple charges BioFET-SIM model [33]. Figure 3.3b reports the multiple charges description of streptavidin. Each blue, white, and red circles represents a positive, neutral, and negative q_i charge, respectively. A more detailed description of how BioFET-SIM computes $l_{i,\mathrm{tot}}$ is given in [34].

Note that we approximate the charge distributions of the charges as homogeneous over the nanowire surface, and only varying outwards perpendicular to the nanowire surface. This approximation enables analytical solutions, very fast calculations, and high-throughput simulations.

3.2.5 Thomas–Fermi Screening Length

In the BioFET-SIM model we settle on a rough description of the electronic structure in the nanowire reflecting our lack of detailed knowledge about the complex situation at its surface. The possible deactivation of dopants at the surface [49] or the increased dopant concentration near the surface compared to the semiconductor bulk [50] is approximated by the sole Thomas–Fermi screening length λ_{TF} [32]. For the *p*-type (*n*-type) nanowire, the screening length λ_{TF} is related to the charge carrier density p_0 (n_0) through:

$$\lambda_{TF} = \sqrt{\frac{\hbar^2 \varepsilon_r \pi^{4/3}}{m^* e^2 p_0^{1/3}}} \tag{3.12}$$

where ε_r is the relative permittivity of the nanowire material ($\varepsilon_0 \varepsilon_1$) and m^* is the effective mass of the charge carrier in units of an electron mass at rest. p_0 would be replaced by n_0 for an *n*-type nanowire. Equation (3.12) states that λ_{TF} can be interpreted as a measure for the distance between charge carriers (related to the charge carrier density) in the nanowire under no applied bias. Therefore this parameter can be used to simulate the effect of the back gate in an experimental setup. The presented linearized model is not capable of describing nonlinear effects such as inversion mode of operation. However, the model distinguishes between accumulation/depletion mode of operation by allowing to choose between an *n*- or *p*-type material and different values of λ_{TF}.

3.2.6 Debye Screening Length

Buffer characteristics influence device performance, as described for experimental [39, 51, 52] and predicted [32] conductance signals. The screening of the analyte signal by the electrolyte is implemented through (3.8) and (3.10), which depend on the Debye length:

$$\lambda_D = \sqrt{\frac{\varepsilon_0 \varepsilon_3 k_B T}{2 N_A e^2 I}} \tag{3.13}$$

where ε_0 is the free space dielectric constant, *K* is the Boltzmann constant, *T* is the absolute temperature of the experiment, N_A is Avogadro's number, and *I* is the ionic strength of the solution. The expression for the ionic strength is given by:

$$I = 1/2 \sum_i c_i z_i^2 \tag{3.14}$$

where c_i indicates the concentration of ion species *i* and z_i is its formal charge.

This description of the electrolyte assumes that (a) the electrolyte is in equilibrium, i.e. the chemical potential is at a minimum and (b) the value for λ_D used in (3.8) and (3.10) is equal to the Debye length of the bulk buffer. We note that in principle these values can differ due to the biofunctionalization layer [31].

3.2.7 Biofunctionalization Layer

The biofunctionalization layer represents the chemical modifications to the nanowire surface, enabling it to capture the desired analyte. It can be visualized as a forest of linker molecules, Fig. 3.1. These molecules are not densely packed and are supposed to be immersed in the buffer solution. For this reason the biofunctionalization layer is implemented solely as a distance parameter, providing a measure of the spatial extension of the linker molecules. If the linker molecules would be so densely packed to exclude the buffer, an extra dielectric constant would be necessary to describe the effect of the biofunctionalization layer on the signal. Charges on the linker molecules are not considered influencing the signal, but part of the background conductance, see Sect. 3.2.3. In principle the surface functionality of the nanowire is not uniform [6] and requires a combined description of the pH-dependent charge on the linker molecules as well as the oxide where a common description of the charge of the oxide layer is through the site-binding model [53].

3.2.8 Protein Charges

The coverage of the biofet with analyte is described in Sect. 3.3.2. A common usage for biofets is the sensing of proteins [3, 6–8]. BioFET-SIM is specifically tailored for this. The PROPKA method [54–56] is employed to compute the $\mathrm{pK_a}$ values of each ionizable amino acid of a given protein structure. The computed $\mathrm{pK_a}$ values are used to evaluate the protonation state (charge) of each ionizable residue i as a function of pH as:

$$q_i(\mathrm{pH}) = \frac{10^{\mathrm{pK}_\mathrm{a}^i - \mathrm{pH}}}{1 + 10^{\mathrm{pK}_\mathrm{a}^i - \mathrm{pH}}} - p(i) \tag{3.15}$$

where $p(i) = 1$ for $i \in \{\text{Asp, Glu, C-, Tyr, Cys}\}$ and $p(i) = 0$ otherwise. In (3.15), $q_i(\mathrm{pH})$ is the probability for the amino acid to be protonated [57]. In the single charge model (Sect. 3.2.3) the computed charges are summed together and a single charge is placed at the center of mass of the protein (Fig. 3.3a). In the multiple charges model (Sect. 3.2.4) a three-dimensional charge distribution is obtained by placing the charge computed with (3.15) at the average of the coordinates of the terminal atoms of the side chain of residue i (Fig. 3.3b). It is assumed that protein binding to the nanowire does not affect the $\mathrm{pK_a}$ values computed by PROPKA, nor disrupt the overall protein conformation.

3.2.9 Summary

To summarize, the BioFET-SIM model evaluates the conductance sensitivity signal $\Delta G/G_0$ (Sect. 3.2.2) of a biofet as depicted in Fig. 3.2. The model considers a nanowire of length L and radius r_{nw} and described through its permittivity (ε_1), mobility (μ), charge carrier density (n_0/p_0), and Thomas–Fermi screening length (λ_{TF}, Sect. 3.2.5). The nanowire is coated with an oxide layer of thickness r_{ox} and characterized by permittivity ε_2. On top of the oxide layer a biofunctionalization layer (Sect. 3.2.7) of thickness r_{bf} is responsible for capturing the desired analyte. The analyte charges are evaluated according to the buffer pH (Sect. 3.2.8) and placed at a certain distance l from the nanowire. The conductance sensitivity is evaluated with the analyte charges or summed in a single charge (Sect. 3.2.3) or separate (Sect. 3.2.4). The entire system is immersed in a buffer solution described by the Debye screening length (λ_D, Sect. 3.2.6) and its permittivity (ε_3).

3.3 Usage

The BioFET-SIM method is available as a graphical web interface at the address *www.biofetsim.org* where a link to an instruction video to its usage is present. BioFET-SIM is distributed as open source through GitHub, and links to the source code as well as to the command line version of the program are also present on the web interface page. The web interface appears as shown in Fig. 3.4. A detailed description of the web interface and how to use it, along with a description of the uses of the command line version, are given in [34]. The following sections give a brief description of the interface and how the number of sensed analyte is computed. In the last section a simple example of the interface usage is given.

3.3.1 Interface Usage

The upper part of the web interface permits the selection of the protein structure that will act as analyte. The structure can be simply defined through its PDB identifier, to be downloaded from the PDB database (www.pdb.org [58]), or uploaded in PDB format. A pH value for the charges evaluation (Sect. 3.2.8) is also inserted, with default value 7.4.

By initializing the interface, the chosen structure is displayed over a representation of the nanowire surface. In the interface it is possible to regulate the relative orientation of the protein structure to the surface.

In the lower part of the interface it is possible to define all the various parameters needed to characterize the biofet, as previously described in Sects. 3.2.1–3.2.7. The interface permits also to define the number of sensed analyte molecules, as described in Sect. 3.3.2.

Upon performing a BioFET-SIM calculation, the computed value for the sensitivity is displayed, along with a graph showing the change in sensitivity against

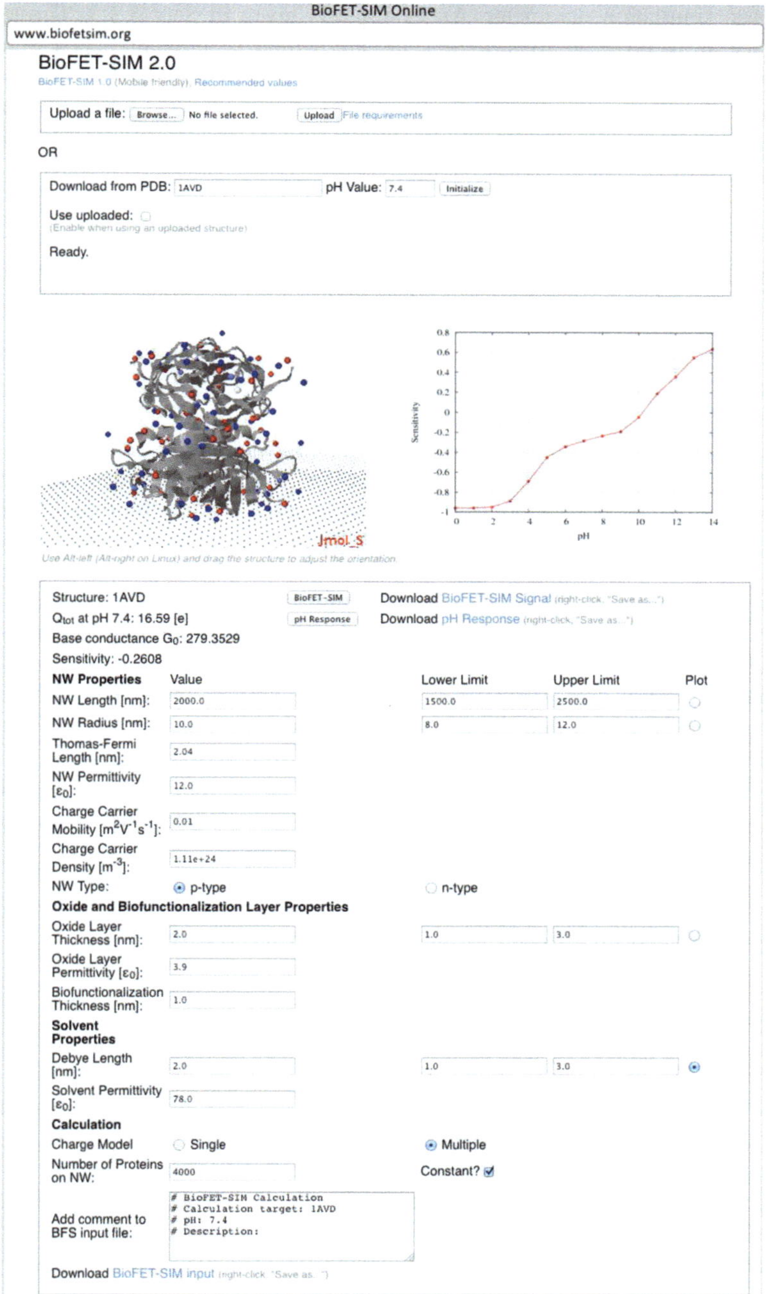

Fig. 3.4 BioFET-SIM web interface at *www.biofetsim.org*. The screenshot shows the interface being used to simulate the pH response to the sensing of 4,000 avidin proteins using default nanowire values

a chosen simulation parameter. An example is given in Sect. 3.3.3. Through the interface it is possible to save a state file which records the chosen relative orientation of the protein and the nanowire surface and an input file for the command line version of the program. The command line version is a powerful tool that allows the user to reproduce a calculation as well as performing many similar ones changing only one parameter at the time.

3.3.2 Analyte Surface Coverage

In order to apply (3.7) or (3.9) the sensed charge density σ_b has to be defined. Once defined the geometry of the nanowire, the charge density can be retrieved from the number of sensed analyte molecules. The web interface has two possible options for inserting the number of sensed molecules: full coverage or user specified.

The full coverage option assumes that the analyte molecules are perfectly packed in order to completely fill the available surface of the biofunctionalization layer. Full coverage of the nanowire has been demonstrated experimentally [59]. Once the relative orientation of the protein to the surface is defined, the interface computes the area on the surface occupied by the molecule. The surface of the biofunctionalization layer is then divided by this area to compute the number of sensed proteins. See [34] for further details.

The user can also specify a fixed number of bound proteins. This can be used, e.g., when it is known that the number of linker molecules on the biofet surface remains constant, even if the proteins bind with different orientations.

It is important to note that this treatment of the protein—biofunctionalization binding assumes that all proteins bind with the same orientation to the surface.

3.3.3 Example: Sensitivity Dependence on Oxide Layer Thickness

In this section we show the basic usage of BioFET-SIM to simulate the effects of different thickness values of the oxide layer. The simulation uses the default values for a *p*-type doped silicon nanowire as found on the web interface and reported in Table 3.1. The conductance sensitivity is evaluated for the sensing of 4,000 avidin proteins (PDB identifier 1AVD) at pH 7.4 and the single charge model.

The computed sensitivity is reported in Fig. 3.5. Avidin carries a positive charge and induces a negative signal on the *p*-type doped nanowire. As expected, to a thicker oxide layer corresponds a diminished (in absolute value) sensitivity. The web interface permits to obtain the plot of Fig. 3.5 with a few simple clicks.

Table 3.1 Parameters defining the simulated biofet

Parameter	Symbol	Value	(Units)
Nanowire length	L	2,000.0	(nm)
Nanowire radius	r_{nw}	10.0	(nm)
Thomas–Fermi length	λ_{TF}	2.04	(nm)
Charge carrier density	p_0	1.11×10^{24}	(m^{-3})
Charge carrier mobility	μ	0.01	($m^2\ V^{-1}\ s^{-1}$)
Nanowire permittivity	ε_1	12.0	(ε_0)
Oxide layer thickness	r_{ox}	2.0	(nm)
Oxide layer permittivity	ε_2	3.9	(ε_0)
Biofunctionalization layer thickness	r_{bf}	1.0	(nm)
Solvent permittivity	ε_3	78.0	(ε_0)
Buffer Debye length	λ_D	2.0	(nm)

We simulate the nanowire as being silicon *p*-type doped

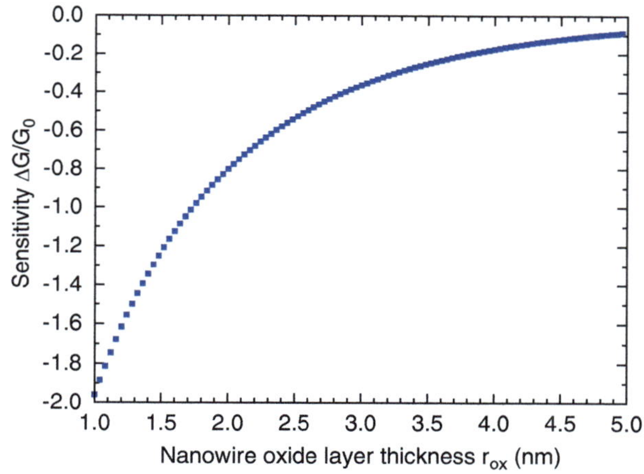

Fig. 3.5 Computed sensitivity response to the sensing of 4,000 molecules of avidin when changing the oxide layer thickness from 1.0 to 5.0 nm

3.4 Applications

The following sections illustrate some of the possible applications of BioFET-SIM. Whenever possible, the computed results are compared to experimental results.

3.4.1 Single Charge Model

For some applications the simpler single charge model (Sect. 3.2.3) is sufficient. In particular the sensing of avidin (PDB identifier 1AVD [60]) or streptavidin (PDB identifier 1STP [61]) is nearly independent from the used model. See Fig. 4

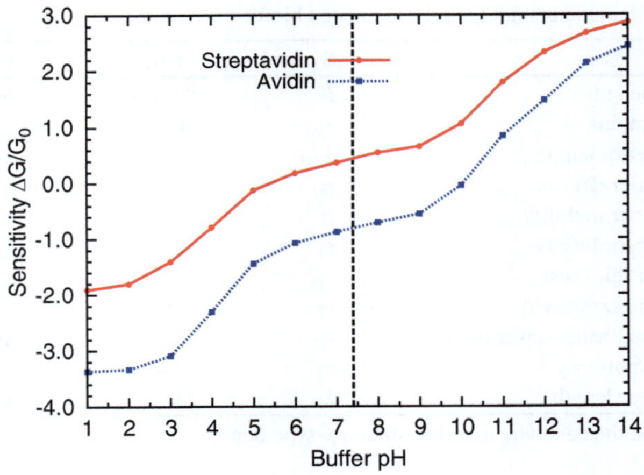

Fig. 3.6 Computed sensitivity response to the sensing of 4,000 molecules of avidin (*dotted line*, square points) or streptavidin (*full line*, round points) on a biotinylated surface with the single charge model. Dependence of the signal on the buffer pH

of [33]. In fact these two proteins are quite symmetrical and do not present any orientation that would favor one type of charges over the other. Another situation when the single charge model is required is represented by the binding of proteins with random orientations (i.e., generic adsorption) over the biofet surface. Random orientations average the protein charge distribution to the protein overall charge.

3.4.1.1 Effect of pH on Signal

At a pH value of 7.4 (physiological pH) avidin and streptavidin have opposite charge. This difference was exploited in [3], where the two proteins could be discerned by their opposite signal. The computed difference in signal for the sensing of avidin and streptavidin is illustrated in Fig. 3.6. At pH 7.4 (vertical dashed line) PROPKA computes a charge of 16.6 and -9.4 e for avidin and streptavidin, respectively (Sect. 3.2.8). This results in a positive signal for streptavidin and a negative signal for avidin.

In the same experiment [3] Stern et al. recorded also the change in signal at different pH values when sensing avidin. Figure 3.6 (dotted line) shows how the expected signal changes with different pH. At low pH values a negative signal is expected. At pH values around 10 the predicted signal is close to zero, while at higher pH values it is possible to obtain a positive signal. Obviously, this simulation does not consider the possible effects of very acid or very basic environments on the structure of the protein [39].

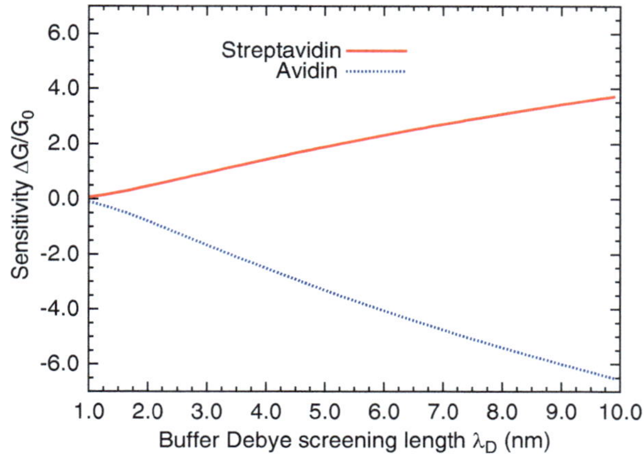

Fig. 3.7 Computed sensitivity response to the sensing of 4,000 molecules of avidin (*dotted line*, square points) or streptavidin (*full line*, round points) on a biotinylated surface with the single charge model. Dependence of the signal on the buffer Debye screening length

3.4.1.2 Signal Dependence on the Debye Screening Length

The intensity of a signal depends also on the ionic strength of the employed buffer solution. At short Debye lengths (high ionic strength, Sect. 3.2.6) a small signal is expected. In [51] Stern et al. analyze the dependence of signal upon dilution of buffer. Figure 3.7 depicts the BioFET-SIM evaluated signal for different Debye lengths for avidin and streptavidin at pH 7.4. It is evident that at longer Debye lengths the difference between the two proteins signals is more and more accentuated. A more complex example on how to use the signal dependence on λ_D to extrapolate information on the system can be found in [4, 34] and in Sect. 3.4.3 (Fig. 3.11).

3.4.1.3 Signal Dependence on Analyte Concentration

Stern et al. in [3] also analyze how the signal changes for different concentrations of analyte streptavidin. The dependence of the number of captured analyte molecules on the bulk analyte concentration is needed to evaluate the corresponding signal. We present here a simple model to evaluate it. A more complex model can be found in [19], but the qualitative results are equivalent. We note that the values used by Nair et al. for the forward and reverse reaction constants in [18, 19] are typical for the binding of single strand DNA but not necessarily for (strept)avidin–biotin complexes.

In the following we consider the sensing reaction as at its equilibrium. The generic sensing of streptavidin on a biotinylated surface can be expressed as:

$$A + R \rightleftharpoons AR \tag{3.16}$$

where A is the analyte (streptavidin), R the receptor and AR the complex. The equilibrium constant K, inverse of the dissociation constant K_D, is given by:

$$K = \frac{[AR]}{[A][R]} \tag{3.17}$$

For the streptavidin–biotin complex a dissociation constant $K_D = 4 \times 10^{-14}$ M has been found [62] and consequently $K = 1/K_D = 2.5 \times 10^{13}\,\mathrm{M}^{-1}$.

The concentration of free analyte $[A]$ in equilibrium with the complex $[AR]$ with respect to the starting concentration of the added sample $[A_{tot}]$ is simply expressed by:

$$[A] = [A_{tot}] - [AR] \tag{3.18}$$

and similarly for the number of free receptors:

$$[R] = [R_{tot}] - [AR] \tag{3.19}$$

Inserting (3.18) and (3.19) in (3.17):

$$K = \frac{[AR]}{([A_{tot}] - [AR])([R_{tot}] - [AR])} \tag{3.20}$$

and using the substitutions $m = 1 + K([A_{tot}] + [R_{tot}])$ and $K' = K[A_{tot}][R_{tot}]$ it is possible to write the equation:

$$K[AR]^2 - m[AR] + K' = 0 \tag{3.21}$$

with solutions

$$\begin{aligned} [AR] &= \frac{1}{2K}\left(m + \sqrt{m^2 - 4KK'}\right) \\ &= \frac{1}{2K}\left(m - \sqrt{m^2 - 4KK'}\right) \end{aligned} \tag{3.22}$$

of which only the solution with the minus sign is significant (the other one giving $[AR] > [R_{tot}]$, which is not possible). Equation (3.22) defines the concentration of analyte-receptor complexes given the starting concentration of analyte in the sample $[A_{tot}]$, the concentration of initial receptor molecules $[R_{tot}]$, and the equilibrium constant K.

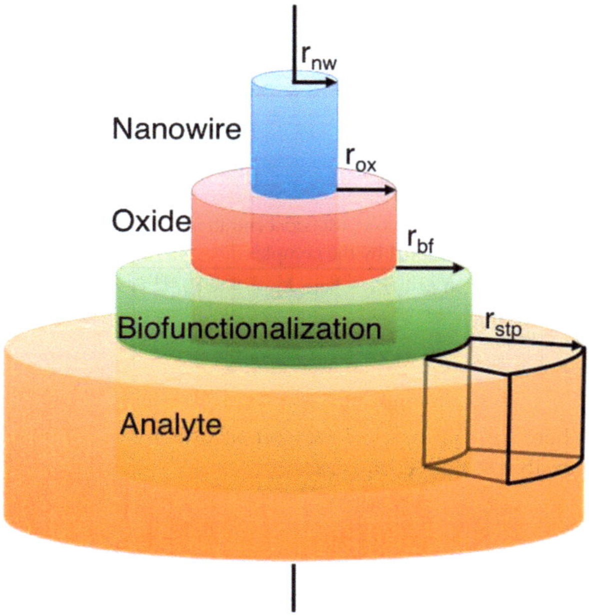

Fig. 3.8 Graphical representation of a generic nano BioFET sensing streptavidin. Supposing full coverage, the analyte creates a layer around the nanowire with thickness r_{stp}. Each streptavidin molecule occupies one cuboid as illustrated

BioFET-SIM requires the number of conjugated molecules, that is the number of formed complexes. This number can be obtained as:

$$n_{AR} = [AR]\, V_{rec}\, N_A \tag{3.23}$$

where V_{rec} is the total volume occupied by the conjugated analyte-receptor molecules. Supposing complete coverage by streptavidin molecules, the total volume V_{rec} occupied by the sensed molecules can be visualized as the outer shell in Fig. 3.8. The analyte occupied layer is supposed to be as thick as the analyte molecule itself (r_{stp}). The occupied volume can be computed as difference between the outer and inner volumes:

$$\begin{aligned} V_{out} &= \pi \left(r_{nw} + r_{ox} + r_{bf} + r_{stp}\right)^2 L \\ V_{in} &= \pi \left(r_{nw} + r_{ox} + r_{bf}\right)^2 L \\ V_{rec} &= V_{out} - V_{in} \\ &= \pi\, r_{stp} \left[2\left(r_{nw} + r_{ox} + r_{bf}\right) + r_{stp}\right] L \end{aligned} \tag{3.24}$$

Alternatively, supposing that each streptavidin molecule occupies a cube of volume r_{stp}^3, as shown in Fig. 3.8, it is possible to express V_{rec} as:

$$V_{rec} = n_{R_{tot}}\, r_{stp}^3 \tag{3.25}$$

Table 3.2 Parameters used to simulate the sensing of streptavidin (1STP) depending on the analyte concentration

Parameter	Symbol	Value	(Units)
Analyte structure		1STP	
Charge at pH 7.4	q	-9.4	(e)
Bounding box side	r_{stp}	5.0	(nm)
Equilibrium constant	K	2.5×10^{13}	(M^{-1})[a]
Total number of receptors concentration	$[R_{\mathrm{tot}}]$	0.013.3	(M)[b]
Volume occupied by the complex	V_{rec}	9.74×10^{-19}	(dm^3)[c]

[a]Computed as $1/K_{\mathrm{D}}$ where $K_{\mathrm{D}} = 4 \times 10^{-14}$ from [62]
[b]From (3.26)
[c]From (3.24)

where $n_{R_{\mathrm{tot}}}$ is the total number of effective receptor molecules. Using (3.25) it is possible to define $[R_{\mathrm{tot}}]$ as:

$$
\begin{aligned}
[R_{\mathrm{tot}}] &= \frac{n_{R_{\mathrm{tot}}}/N_{\mathrm{A}}}{V_{\mathrm{rec}}} \\
&= \frac{n_{R_{\mathrm{tot}}}/N_{\mathrm{A}}}{n_{R_{\mathrm{tot}}}\, r_{\mathrm{stp}}^3} \\
&= \frac{1}{r_{\mathrm{stp}}^3\, N_{\mathrm{A}}}
\end{aligned}
\tag{3.26}
$$

The value obtained with (3.26) and the literature value for K are inserted in (3.22). The concentration of sensed molecules $[AR]$ is so defined as a function of the starting sample analyte concentration $[A_{\mathrm{tot}}]$. $[AR]$ is inserted in (3.23) together with the value obtained with (3.24), to finally obtain the number of sensed analyte molecules n_{AR} to be used as input for BioFET-SIM.

The simulation was performed using the default values of BioFET-SIM (Table 3.1) together with those reported in Table 3.2. The obtained values for the sensitivity are reported in Table 3.3 and plotted in Fig. 3.9. Given experimental conditions as those reported in Tables 3.1 and 3.3, the simulation states that it is possible to expect a quasi-linear behavior in a concentration range 1–10 mM and a sensitivity limit of 0.1 mM. A different setup would be necessary in order to sense smaller concentrations of analyte, e.g. a larger nanowire radius to increase the number of captured molecules. Since sensitivity is inversely proportional to the nanowire radius [32], possibly a longer Thomas–Fermi screening length would be necessary to balance out the effects on sensitivity. It is important to remember that there are many simplifications behind the data of Table 3.3, but the qualitative result of Fig. 3.9 is anyway reliable. Further details can be found in [63].

Table 3.3 Streptavidin sensing simulation data

$[A_{tot}]$ (M)	n_{AR}	$\Delta G/G_0$
1.0×10^0	7,911	0.88584
1.0×10^{-1}	7,911	0.88584
1.2×10^{-2}	7,038	0.80022
1.0×10^{-2}	5,864	0.66637
8.0×10^{-3}	4,692	0.53333
6.0×10^{-3}	3,519	0.40016
4.0×10^{-3}	2,346	0.26723
2.0×10^{-3}	1,173	0.13305
1.0×10^{-3}	586	0.06664
5.0×10^{-4}	293	0.03332
1.0×10^{-4}	58	0.00666
1.0×10^{-5}	6	0.00067
1.0×10^{-6}	1	0.00007

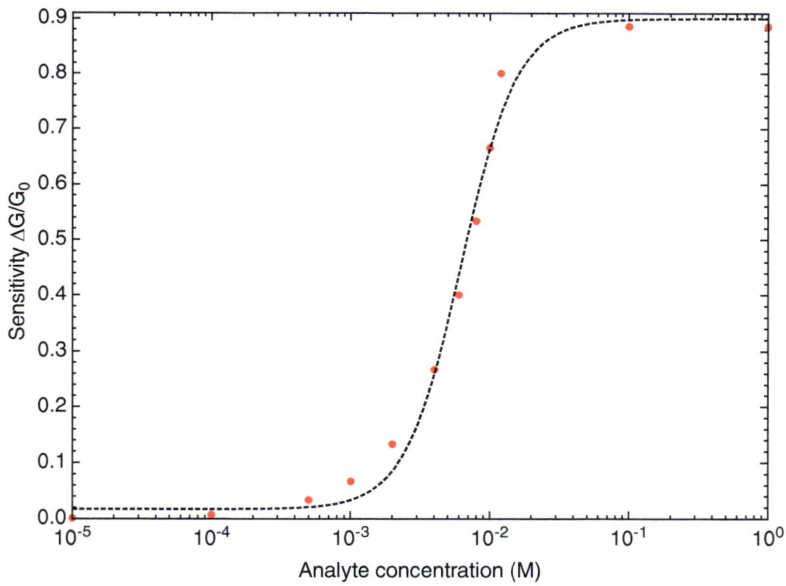

Fig. 3.9 Simulated signal response for the sensing of streptavidin in the concentration range 1×10^{-5} to 1 M with default BioFET-SIM settings. The *dots* correspond to the data of Table 3.3, while the *dashed lines* represent a sigmoid fit to guide the eye

3.4.2 Multiple Charges Model

When the sensed analyte has an asymmetric charge distribution, the single charge model is insufficient to describe its relative signal. However, since the underlying model is based on a linear equation it is possible to obtain the signal from distributions involving multiple charges as a sum of the individual responses.

In the following sections we show the importance of the multiple charges model (Sect. 3.2.4) through the sensing of a model peptide and a more complex example employing nucleocapsid protein.

3.4.2.1 Model Peptide Sensing

In this section we simulate the sensing of a model peptide constituted by two protonated lysine residues at the N-terminus, an eight alanine residues bridge and two deprotonated aspartic acid residues at the C-terminus: KK8ADD. The model was built as a linear peptide using PyMOL [64]. At pH 7.4 PROPKA computes a charge of 0.22 *e*. The peptide has a clear asymmetric charges distribution: three positive charges at the N-terminus and three negative charges at the C-terminus and overall charge nearly zero. The single charge model would predict a nearly zero signal for the sensing of this peptide. If there was the possibility of a biofunctionalization layer capable of specifically capturing aspartic or lysine residues, it would be possible to selectively bind the peptide as represented in Fig. 3.10a, c, respectively. The corresponding simulated signal as function of the Debye screening length is shown in Fig. 3.10d. If otherwise the peptide would be simply adsorbed on the nanowire surface in a horizontal position, the expected signal would be nearly zero. The same zero signal would also be expected if the charges distribution would average out because of random capturing. In fact, in case of random capturing of the analyte no specific direction relative to the nanowire surface would be preferred. In this case the single charge model would be sufficient.

3.4.2.2 Nucleocapsid Protein Sensing

The sensing of nucleocapsid protein (N-protein), as performed by Ishikawa et al. [48], is presented in this section, to exemplify the potential of BioFET-SIM and the multiple charges model.

In their work Ishikawa and co-workers employed In_2O_3 *n*-type doped nanowire-based sensors. Table 3.4 reports the different from the default parameter values used to simulate such devices. At pH 7.4 PROPKA computes a charge of 6.4 *e* for N-protein. The sensing of this protein by an *n*-type doped nanowire-based sensor should produce a positive signal, but a negative one was recorded [48]. It is possible to use BioFET-SIM to investigate the possible causes of this unexpected signal.

N-protein has an asymmetric charge distribution, as shown in Fig. 3.11a, e in the case of pH = 7.4. The protein could be captured by the biofunctionalization layer as reported in Fig. 3.11a, where the bulk of the protein rich in negative charges is closer to the nanowire surface. We refer to this situation as *up*. It is also possible that the protein is captured as depicted in Fig. 3.11e, that is with the positive charges rich hairpin closer to the nanowire surface. We refer to this second possibility as *down*. The two possible orientations *up* and *down* would generate quite different signals at pH = 7.4, as depicted in Fig. 3.11c, g, respectively. In the *up* orientation a small

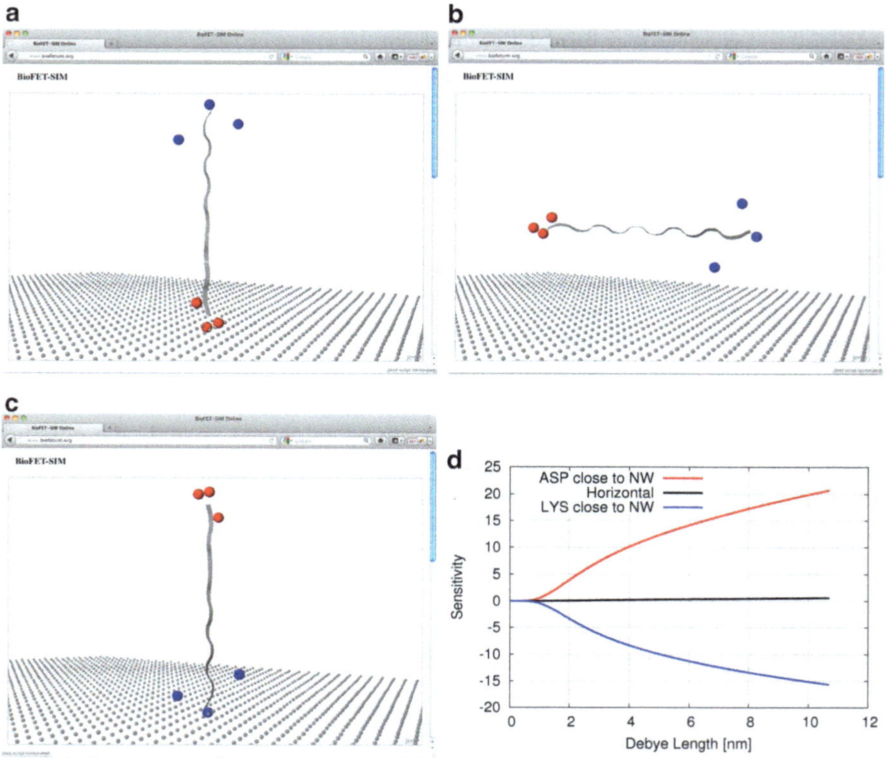

Fig. 3.10 Possible orientations and corresponding signal for the model peptide KK8ADD. The BioFET-SIM simulation was conducted with default values and the multiple charges model. The capturing of the peptide can be specific for (**a**) Asp or (**c**) Lys or (**b**) the peptide could be simply adsorbed horizontally on the surface. (**d**) The simulated signals as function of the Debye screening length for the different peptide—nanowire relative orientations. Figures originally from [34]: (**a**)–(**c**) doi:10.1371/journal.pone.0045379.g004, (**d**) doi:10.1371/journal.pone.0045379.g005

negative signal is expected for Debye screening lengths up to 3 nm (Fig. 3.11c). In the *down* orientation a clear positive signal is expected at any buffer dilution (Fig. 3.11g).

Varying the pH of the buffer solution would change the expected signal. At pH = 3.8 N-protein has a computed charge of 22.3 *e*, ca. three times higher than at pH = 7.4. The protein is now mainly constituted by positive charges and both orientations produce similar positive signals (Fig. 3.11b, f). At pH = 11.0, instead, N-protein has a computed negative charge of −3.2 *e*. The predicted signal for the *up* orientation is clearly negative at any buffer dilution (Fig. 3.11d). For the *down* orientation a nearly zero but positive signal is predicted for Debye screening lengths from 1 to 3 nm (Fig. 3.11h). It is interesting to note that, by inverting the charge of

Table 3.4 Parameters defining the simulated In_2O_3 *n*-type semiconducting nanowire-based device that differs from those of Table 3.1

Parameter	Symbol	Value	(Units)
Nanowire radius	r_{nw}	5.0	(nm)
Oxide layer thickness	r_{ox}	0.0	(nm)
Thomas–Fermi length	λ_{TF}	1.179	(nm)
Charge carrier density	n_0	4.6×10^{25}	(m^{-3})
Charge carrier mobility	μ	0.0078	($m^2\,V^{-1}\,s^{-1}$)
Nanowire permittivity	ε_1	9.0	(ε_0)

N-protein by changing the pH value from 7.4 to 11.0, also the expected signals for the *up* and *down* orientations are inverted.

The different simulated signals for the sensing of N-protein show that the BioFET-SIM model with multiple charges can be employed to study the possible effects of different orientations of the analyte with respect to the nanowire surface, coupled with the effects of different pH values of the buffer solution. In the specific case of N-protein we showed that, given certain conditions (e.g. Fig. 3.11c), an unexpected negative signal is possible.

3.4.3 Antigen Sensing on an Antibody Functionalized Surface

In this section we present a complex application of BioFET-SIM. Following [4], we decided to study the sensing of an antigen captured by an antibody functionalized biofet surface. Differences in the surface functionalization led to different signals, even if stemming from sensing the same antigen in otherwise similar conditions. This is indicative of different average distances between the sensed antigen and the biofet surface. Antibody binding to the surface was made possible through either an N-terminus located on the antigen-binding fragment (Fab) or by one or both C-termini at the antibody base, as shown in Fig. 3.12. N-terminus binding led to a reported antigen—surface distance of 5.9 ± 0.6 nm. C-termini binding corresponded to an antigen—surface distance of 8.4 ± 0.4 nm [4].

We decided to study the effect of possible antibody orientations on the signal. We further developed the formalism presented in Sect. 3.2 in order to obtain a quantity (the relative sensitivity factor) which depends solely on Debye length and distance of the sensed analyte from the surface. The derivation of this quantity is described in Sect. 3.4.3.1. Section 3.4.3.2 describes the necessary steps to prepare the antibody and antigen structures for the BioFET-SIM simulations. Section 3.4.3.3 shows the different orientations of the antibody, and consequently of the sensed antigen, over the biofet surface that we considered in this study. Finally, by performing a series of simulations at different Debye lengths, corresponding to those employed experimentally in [4], it was possible to obtain through a fit the average distance of the sensed antigen from the nanowire surface. These results are presented in Sect. 3.4.3.4.

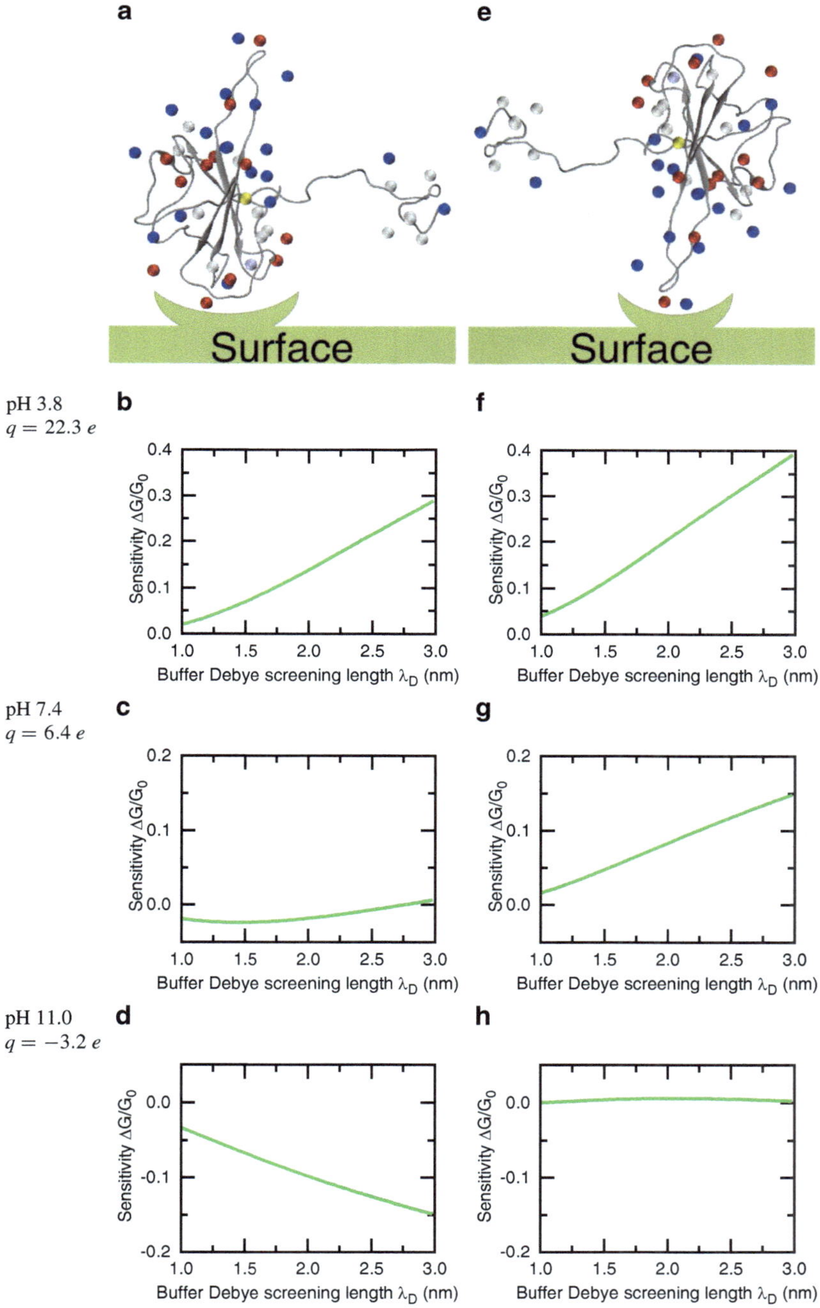

Fig. 3.11 N-protein sensing. The protein can be captured in (**a**) *up* or (**e**) *down* orientations. Different signals when varying the Debye screening length from 1 to 3 nm are expected at the different pH values 3.8, 7.4, and 11.0: (**b**), (**c**), and (**d**), respectively, for the *up* orientation and (**f**), (**g**), and (**h**), respectively, for the *down* orientation

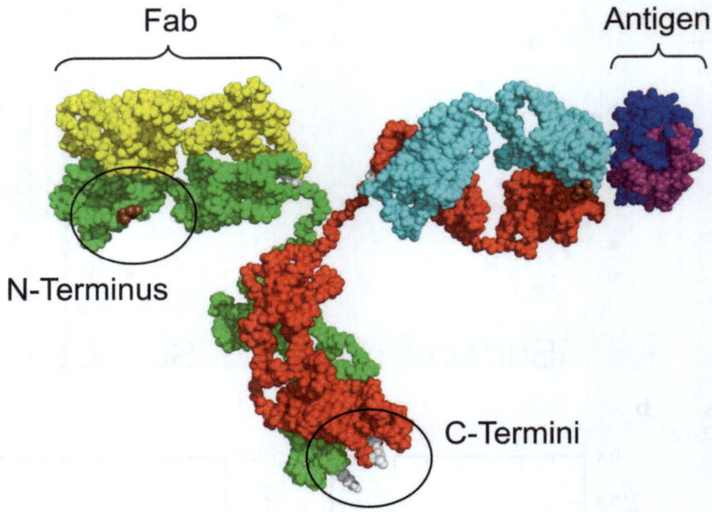

Fig. 3.12 The employed custom-prepared antibody–antigen complex. Different chains are represented with different colors. Moreover, the C-termini at the antibody base are highlighted in *gray* and the N-termini on Fab in *brown*. Figure originally from [34]: doi:10.1371/journal.pone.0045379.g006

3.4.3.1 Theoretical Background

The interface was employed using the multiple charges formalism to take in consideration the different charges of the antigen and the possible effect on the simulated signal of their different orientations. However given the complexity of the studied system the following data treatment was performed through the single charge model formalism. In the following equations l stands for the average distance of the sensed charges from the nanowire surface.

As described in Sects. 3.2.1 and 3.2.3, the base conductance and conductance sensitivity can be described through (3.2) and (3.6). Since the physical and geometrical properties of the nanowire are fixed throughout all the following simulations, G_0 can be considered as constant. It is then possible to express the change in conductivity as:

$$\Delta G = K\Gamma \,\Gamma_l \,\sigma_\mathrm{b} \tag{3.27}$$

where K collects all constant values. We define $\lambda_\mathrm{D}^{\max}$ as the value of the buffer Debye length that we consider as maximum dilution. The maximum change in conductivity at this value is:

$$\Delta G^{\max} = K\Gamma^{\max}\,\Gamma_l^{\max}\,\sigma_\mathrm{b} \tag{3.28}$$

where we assume that changing the buffer dilution does not affect the number of adsorbed analyte molecules even at maximum dilution, hence σ_b is the same as in (3.27). Considering a highly diluted buffer, i.e. $\lambda_D^{max} \gg l$, and substituting in (3.8) we can express Γ_l^{max} as:

$$\Gamma_l^{max} \simeq 2\frac{r_{nw}}{r_{nw}+l}\left(1+\sqrt{\frac{r_{nw}}{r_{nw}+l}}\right)^{-1} \tag{3.29}$$

The ratio between the change in conductivity at a given Debye length and the maximum possible value is:

$$\frac{\Delta G}{\Delta G^{max}} = \frac{K\Gamma\,\Gamma_l\,\sigma_b}{K\Gamma^{max}\,\Gamma_l^{max}\,\sigma_b} = \frac{\Gamma\,\Gamma_l}{\Gamma^{max}\,\Gamma_l^{max}} \tag{3.30}$$

and after reordering it is possible to obtain:

$$\frac{\Delta G\,\Gamma^{max}}{\Delta G^{max}\,\Gamma} = \frac{\Gamma_l}{\Gamma_l^{max}} \tag{3.31}$$

After inserting the explicit expressions, we obtain

$$\frac{\Gamma_l}{\Gamma_l^{max}} = \frac{2\frac{r_{nw}}{r_{nw}+l}\left(1+\sqrt{\frac{r_{nw}}{r_{nw}+l}}\exp(l/\lambda_D)\right)^{-1}}{2\frac{r_{nw}}{r_{nw}+l}\left(1+\sqrt{\frac{r_{nw}}{r_{nw}+l}}\right)^{-1}} \tag{3.32}$$

where we define Γ_l/Γ_l^{max} as the *relative sensitivity factor*. According to (3.32) the relative sensitivity factor depends solely on the Debye length of the simulated buffer and the average distance of the sensed charges, once chosen a value for the nanowire radius.

Using (3.5) it is possible to compute the values of Γ for different Debye lengths, including Γ^{max} for $\lambda_D^{max} = 1{,}000\,\text{nm}$. In practice, the command line version of BioFET-SIM was employed to obtain these values. The BioFET-SIM parameters were set as in Table 3.1. With these parameters we obtained $G_0 = 279.0\,\text{nS}$.

Using BioFET-SIM one obtains the value of ΔG (and ΔG^{max}) simply by multiplying $\frac{\Delta G}{G_0}$ with G_0. It is then possible to plot the l.h.s. of (3.31) for different values of λ_D. This plot is to be fitted to the r.h.s. of (3.32) where l is the fitting parameter. From a series of measures at different Debye lengths, the average distance l of the sensed charge from the nanowire surface is obtained.

3.4.3.2 Antibody and Antigen Preparation

A suitable complex structure of a generic antibody and an antigen as used by Vacic et al. [4] was prepared. Only few full antibody structures have been resolved by X-ray crystallography. We used the structure of an intact IgG2a monoclonal antibody, ascension code 1IGT [65]. For the antigen we used the structure of the SEA domain of human mucin 1, with ascension code MUC1 [66].

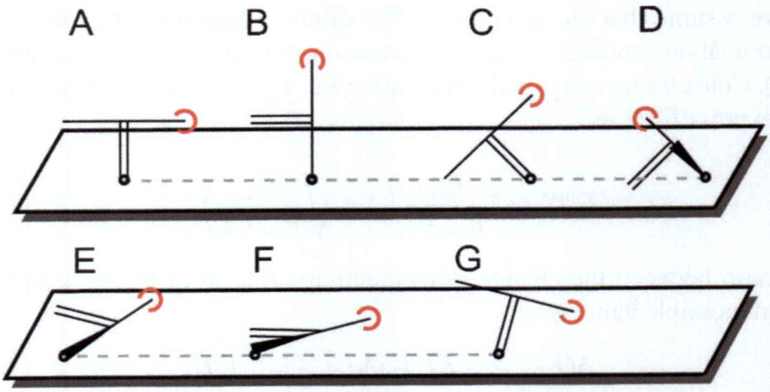

Fig. 3.13 Different orientations of the antibody with respect to the nanowire surface were considered in this study. The antigen is indicated by a *red arc*. The antibody base is indicated by a *double line*. The point of binding between the antibody and biofet surface is indicated by a *small circle*. In orientations A, C, and G the antibody is bound through the C-termini. In orientations B, D, E, and F the antibody is bound through the N-terminus. Figure originally from [34]: doi:10.1371/journal.pone.0045379.g007

The antigen structure was rigidly docked to the antigen-binding site of the antibody through the program AutoDock [67] and visually checked with the program PyMol [64] for a reasonable docking. The scope of this docking was only to obtain a feasible complex structure. The antibody–antigen complex is shown in Fig. 3.12.

Since interested in sensing only the antigen, we made the antibody structure as neutral as possible. All positions in the antibody sequence were changed to glycine using PyMol. Only the side chains were changed, while the backbone spatial disposition was maintained as in the original structure. When BioFET-SIM computes the charges of a protein it assigns a positive charge to the N termini and a negative one to the C termini, according to the PROPKA results. For this reason we changed the four chains N termini to aspartic acids and the C termini to arginines, in order to counterbalance these charges. By using a complex between the neutral antibody and the chosen antigen, we ensured that the simulated signal originated only from the antigen charges.

3.4.3.3 Antibody Orientations

Thanks to the graphical web interface, different orientations of the antibody–antigen complex were tested, simulating binding through the C- or N-termini. The seven orientations that were taken in consideration in this study are reported in Fig. 3.13. Orientations A to C correspond to those considered in [4], while orientations D to G were specifically conceived for this study. The antibody is supposedly bound to the biofet surface through the C-termini in orientations A, C, and G. Conversely, in orientations B, D, E, and F binding is through the N-terminus. C-termini binding

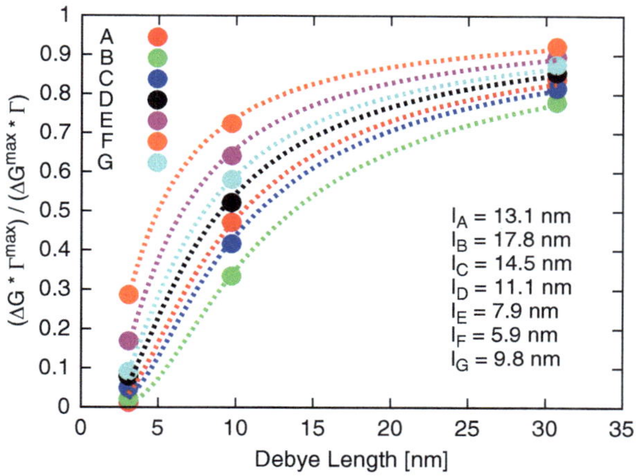

Fig. 3.14 Relative sensitivity factors for the different orientations depending on the Debye length. The computed fitting parameters l are also reported for each orientation

restricts the possible movement of the antibody with respect to the nanowire surface (the antibody remains upright), while binding through the Fab N-terminus allows more freedom of movement (the antibody can nearly lie down on the surface).

3.4.3.4 Antigen Sensing

Different orientations of the neutral antibody–antigen complex were tested, as reported in Fig. 3.13. A biofunctionalization layer of 0.5 and 1.0 nm was added for C- and N-termini binding to the nanowire surface, respectively. We considered a pH of 7.4 and kept fixed the number of proteins to 4,000. The raw data for the BioFET-SIM results for all the orientations for the values of Debye length employed by Vacic et al. (3.07, 9.7 and 30.7 nm) plus $\lambda_D^{max} = 1{,}000$ nm, together with the computed data for the relative sensitivity factor are reported in the supporting material of [34] (doi:10.1371/journal.pone.0045379.s005). Here we present in Fig. 3.14 the plot of the computed relative sensitivity factor for the different orientations, together with the fitting parameters.

Among the orientations relative to C-termini binding, orientation G ($l_G = 9.8$ nm) results in an average antigen—surface distance most similar to the reported experimental value (8.4 nm). For N-terminus binding, orientations E and F ($l_E = 7.9$, $l_F = 5.9$ nm) give an average distance close to the experimental value (5.9 nm). Thus it is possible to postulate that an upright or slightly inclined orientation is preferred for C-termini binding, while when binding through one of the Fab N-termini the antibody is likely to be nearly flat on the biofet surface. Moreover, an orientation like C is quite unlikely ($l_G = 9.8$ nm), as opposed to what suggested in [4].

3.5 Conclusions

BioFET-SIM is a simple but yet powerful tool for the qualitative evaluation of nanowire-based biosensor signals. Its main characteristic is a web interface that permits any user to set up a calculation with a few mouse clicks. Most of all, the graphical interface permits to explore the effect of different orientations of the sensed analyte on the simulated signal. The graphical interface also prepares input files for the command line version of BioFET-SIM, which in turn permits to easily run batch simulations varying any given parameter.

In this chapter we showed the theoretical basis for BioFET-SIM, both in its single charge (Sect. 3.2.3) and multiple charges (Sect. 3.2.4) implementation. This was followed by an explanation of basic usage of the method (Sect. 3.3), where we showed how to study the possible effects of changing a device parameter like the oxide layer thickness on the biofet signal (Sect. 3.3.3).

Among the possible applications of BioFET-SIM we showed in Sects. 3.4.1.1 and 3.4.1.2 the effect on the sensor signal of changes of the buffer solution pH or Debye screening length. As a notable application of BioFET-SIM we mentioned the possibility of simulating a qualitative calibration curve for the signal upon sensing an analyte in different concentrations (Sect. 3.4.1.3). In Sect. 3.4.2.2 we showed the importance of using a multiple charges description of an analyte when different relative orientations of the analyte with respect to the nanowire surface are possible. We demonstrated how, through BioFET-SIM, it is possible to predict the possible signal for different orientations at different buffer pH values. Finally, in Sect. 3.4.3 we showed a complex application of BioFET-SIM aimed at interpreting and rationalizing experimental data. We demonstrated how it is possible to use the graphical web interface to prepare a series of different orientations of an antibody–antigen complex. Consequently the computed data let us postulate which of the considered orientations are those most likely to happen.

Acknowledgments This work has been partially funded by the Danish Research Council for Technology and Production Sciences (FTP), the Danish Natural Science Research Council (FNU), and by UNIK Synthetic Biology program, funded by the Danish Ministry for Science, Technology and Innovation. The authors acknowledge fruitful discussions with Lars Iversen, Noémie Loret, Rune S. Frederiksen, and Shivendra Upadhyay.

References

1. Thévenot, D.R., Toth, K., Durst, R.A., Wilson, G.S.: Electrochemical biosensors: recommended definitions and classification. Biosens. Bioelectron. **16**(1–2), 121 (2001). DOI 10.1016/S0956-5663(01)00115-4
2. Hermanson, G.T.: Bioconjugate Techniques. Academic, San Diego (1996)
3. Stern, E., Klemic, J.F., Routenberg, D.A., Wyrembak, P.N., Turner-Evans, D.B., Hamilton, A.D., LaVan, D.A., Fahmy, T.M., Reed, M.A.: Label-free immunodetection with CMOS-compatible semiconducting nanowires. Nature **445**, 519 (2007). DOI 10.1038/nature05498

4. Vacic, A., Criscione, J.M., Rajan, N.K., Stern, E., Fahmy, T.M., Reed, M.A.: Determination of molecular configuration by Debye length modulation. J. Am. Chem. Soc. **133**(35), 13886 (2011). DOI 10.1021/ja205684a
5. Chen, Y., Wang, X., Erramilli, S., Mohanty, P., Kalinowski, A.: Silicon-based nanoelectronic field-effect pH sensor with local gate control. Appl. Phys. Lett. **89**(22), 223512 (2006). DOI 10.1063/1.2392828
6. Cui, Y., Wei, Q., Park, H., Lieber, C.M.: Nanowire nanosensors for highly sensitive and selective detection of biological and chemical species. Science **293**, 1289 (2001). DOI 10.1126/science.1062711
7. Gao, X.P., Zheng, G., Lieber, C.M.: Subthreshold regime has the optimal sensitivity for nanowire FET biosensors. Nano Lett. **10**(2), 547 (2009). DOI 10.1021/nl9034219
8. Tian, R., Regonda, S., Gao, J., Liu, Y., Hu, W.: Ultrasensitive protein detection using lithographically defined Si multi-nanowire field effect transistors. Lab Chip **11**, 1952 (2011). DOI 10.1039/C0LC00605J
9. Wong, I.Y., Melosh, N.A.: Directed hybridization and melting of DNA linkers using counterion-screened electric fields. Nano Lett. **9**(10), 3521 (2009). DOI 10.1021/nl901710n
10. Dorvel, B.R., Reddy, B., Go, J., Duarte Guevara, C., Salm, E., Alam, M.A., Bashir, R.: Silicon nanowires with high-k Hafnium oxide dielectrics for sensitive detection of small nucleic acid oligomers. ACS Nano **6**(7), 6150 (2012). DOI 10.1021/nn301495k
11. Chang, H.K., Ishikawa, F.N., Zhang, R., Datar, R., Cote, R.J., Thompson, M.E., Zhou, C.: Rapid, label-free, electrical whole blood bioassay based on nanobiosensor systems. ACS Nano **5**(12), 9883 (2011). DOI 10.1021/nn2035796
12. Berthing, T., Sørensen, C.B., Nygård, J., Martinez, K.L.: Applications of nanowire arrays in nanomedicine. J. Nanoneurosci. **1**(1), 3 (2009). DOI 10.1166/jns.2009.001
13. Patolsky, F., Zheng, G., Lieber, C.M.: Nanowire-based biosensors. Anal. Chem. **78**(13), 4260 (2006). DOI 10.1021/ac069419j
14. Shinwari, M.W., Deen, M.J., Landheer, D.: Study of the electrolyte-insulator-semiconductor field-effect transistor (EISFET) with applications in biosensor design. Microelectron. Reliab. **47**(12), 2025 (2007). DOI 10.1016/j.microrel.2006.10.003
15. Curreli, M., Zhang, R., Ishikawa, F.N., Chang, H.K., Cote, R.J., Zhou, C., Thompson, M.E.: Real-time, label-free detection of biological entities using nanowire-based FETs. IEEE Trans. Nanotechnol. **7**(6), 651 (2008). DOI 10.1109/TNANO.2006.880908
16. Neizvestny, I.G.: Semiconductor nanowire sensors. Russ. Microelectron. **38**(4), 223 (2009). DOI 10.1134/S1063739709040015
17. Roy, S., Gao, Z.: Nanostructure-based electrical biosensors. Nano Today **4**(4), 318 (2009). DOI 10.1016/j.nantod.2009.06.003
18. Nair, P.R., Alam, M.A.: Performance limits of nanobiosensors. Appl. Phys. Lett. **88**(23), 233120 (2006). DOI 10.1063/1.2211310
19. Nair, P.R., Alam, M.A.: Screening-limited response of nanoBiosensors. Nano Lett. **8**(5), 1281 (2008). DOI 10.1021/nl072593i
20. Nair, P.R., Alam, M.A.: Theory of "Selectivity" of label-free nanobiosensors: A geometro-physical perspective. J. Appl. Phys. **107**(6), 064701 (2010). DOI 10.1063/1.3310531
21. Heitzinger, C., Klimeck, G.: Computational aspects of the three-dimensional feature-scale simulation of silicon-nanowire field-effect sensors for DNA detection. J. Comput. Electron. **6**, 387 (2007). DOI 10.1007/s10825-006-0139-x
22. Ringhofer, C., Heitzinger, C.: Multi - scale modeling and simulation of field-effect biosensors. ECS Trans. **14**(1), 11 (2008). DOI 10.1149/1.2956012
23. Heitzinger, C., Kennell, R., Klimeck, G., Mauser, N., McLennan, M., Ringhofer, C.: Modeling and simulation of field-effect biosensors (BioFETs) and their deployment on the nanoHUB. J. Phys. Conf. Ser. **107**, 012004 (2008). DOI 10.1088/1742-6596/107/1/012004
24. Chen, H., Mukherjee, S., Aluru, N.: Charge distribution on thin semiconducting silicon nanowires. Comput. Methods Appl. Mech. Eng. **197**(41–42), 3366 (2008). DOI 10.1016/j.cma.2008.02.007
25. Heitzinger, C., Mauser, N.J., Ringhofer, C., Liu, Y., Dutton, R.W.: Proceedings of the Simulation of Semiconductor Processes and Devices (SISPAD 2009) (San Diego, CA, USA, 2009), pp. 86–90 (2009). DOI 10.1109/SISPAD.2009.5290244

26. Heitzinger, C., Mauser, N.J., Ringhofer, C.: Multiscale modeling of planar and nanowire field-effect biosensors. SIAM J. Appl. Math. **70**(5), 1634 (2010). DOI 10.1137/080725027
27. Windbacher, T., Sverdlov, V., Selberherr, S.: Biomedical engineering systems and technologies. In: Fred, A., Filipe, J., Gamboa, H. (eds.) Communications in Computer and Information Science, vol. 52, pp. 85–95. Springer, Berlin (2010). DOI 10.1007/978-3-642-11721-3{_}6
28. Park, H.H., Zeng, L., Buresh, M., Wang, S., Klimeck, G., Mehrotra, S.R., Heitzinger, C., Haley, B.P.: Nanowire, NanoHUB website (2006). DOI 10254/nanohub-r1307.8. URL http://nanohub.org/resources/1307. Accessed Mar 2013
29. Nair, P.R., Go, J., Landells, G.J., Pandit, T.R., Alam, M.A.: BioSensorLab, NanoHUB website (2008). DOI 10254/nanohub-r2929.5. URL http://nanohub.org/resources/2929. Accessed Mar 2013
30. Medici: Two Dimensional Device Simulation Program (2003). Synopsis
31. Sørensen, M.H., Mortensen, N.A., Brandbyge, M.: Screening model for nanowire surface-charge sensors in liquid. Appl. Phys. Lett. **91**(10), 102105 (2007). DOI 10.1063/1.2779930
32. De Vico, L., Sørensen, M.H., Iversen, L., Rogers, D.M., Sørensen, B.S., Brandbyge, M., Nygård, J., Martinez, K.L., Jensen, J.H.: Quantifying signal changes in nano-wire based biosensors. Nanoscale **3**, 706 (2011). DOI 10.1039/C0NR00442A
33. De Vico, L., Iversen, L., Sørensen, M.H., Brandbyge, M., Nygård, J., Martinez, K.L., Jensen, J.H.: Predicting and rationalizing the effect of surface charge distribution and orientation on nano-wire based FET bio-sensors. Nanoscale **3**, 3635 (2011). DOI 10.1039/C1NR10316D
34. Hediger, M.R., Jensen, J.H., De Vico, L.: BioFET-SIM Web Interface: Implementation and Two Applications. PLoS One **7**(10), e45379 (2012). DOI 10.1371/journal.pone.0045379
35. Punzet, M., Baurecht, D., Varga, F., Karlic, H., Heitzinger, C.: Determination of surface concentrations of individual molecule-layers used in nanoscale biosensors by in situ ATR-FTIR spectroscopy. Nanoscale **4**, 2431 (2012). DOI 10.1039/C2NR12038K
36. Hakim, M.M.A., Lombardini, M., Sun, K., Giustiniano, F., Roach, P.L., Davies, D.E., Howarth, P.H., de Planque, M.R.R., Morgan, H., Ashburn, P.: Thin film polycrystalline silicon nanowire biosensors. Nano Lett. **12**(4), 1868 (2012). DOI 10.1021/nl2042276
37. Duan, X., Li, Y., Rajan, N., Routenberg, D., Modis, Y., Reed, M.: Quantification of the affinities and kinetics of protein interactions using silicon nanowire biosensors. Nat. Nanotechnol. **7**(6), 401 (2012). DOI 10.1038/nnano.2012.82
38. Elnathan, R., Kwiat, M., Pevzner, A., Engel, Y., Burstein, L., Khatchtourints, A., Lichtenstein, A., Kantaev, R., Patolsky, F.: Biorecognition layer engineering: Overcoming screening limitations of nanowire-based FET devices. Nano Lett. **12**(10), 5245 (2012). DOI 10.1021/nl302434w
39. Lloret, N., Frederiksen, R.S., Møller, T.C., Rieben, N.I., Upadhyay, S., De Vico, L., Jensen, J.H., Nygård, J., Martinez, K.L.: Effects of buffer composition and dilution on nanowire field-effect biosensors. Nanotechnology **24**(3), 035501 (2013). DOI 10.1088/0957-4484/24/3/035501
40. Magliulo, M., Mallardi, A., Mulla, M.Y., Cotrone, S., Pistillo, B.R., Favia, P., Vikholm-Lundin, I., Palazzo, G., Torsi, L.: Electrolyte-gated organic field-effect transistor sensors based on supported biotinylated phospholipid bilayer. Adv. Mater. **25**(14), 2090 (2013). DOI 10.1002/adma.201203587
41. Hammock, M.L., Knopfmacher, O., Naab, B.D., Tok, J.B.H., Bao, Z.: Investigation of protein detection parameters using nanofunctionalized organic field-effect transistors. ACS Nano **7**(5), 3970 (2013). DOI 10.1021/nn305903q
42. Baumgartnerth, S., Heitzinger, C., Vacic, A., Reed, M.A.: Predictive simulations and optimization of nanowire field-effect PSA sensors including screening. Nanotechnology **24**(22), 225503 (2013). DOI 10.1088/0957-4484/24/22/225503
43. Nitzan, A., Galperin, M., Ingold, G.L., Grabert, H.: On the electrostatic potential profile in biased molecular wires. J. Chem. Phys. **117**(23), 10837 (2002). DOI 10.1063/1.1522406
44. Liang, G.C., Ghosh, A.W., Paulsson, M., Datta, S.: Electrostatic potential profiles of molecular conductors. Phys. Rev. B **69**, 115302 (2004). DOI 10.1103/PhysRevB.69.115302

45. Zhang, X.G., Pantelides, S.T.: Screening in nanowires and nanocontacts: Field emission, adhesion force, and contact resistance. Nano Lett. **9**(12), 4306 (2009). DOI 10.1021/nl902533n, pMID: 19845331
46. Sørensen, M.H.: Nanowires for chemical sensing in a liquid environment. Bachelor Thesis (2007)
47. Olver, F.W.J.: In: Abramowitz, M., Stegun, I.A. (eds.) Handbook of Mathematical Functions with Formulas, Graphs, and Mathematical Tables, pp. 355–436. Dover, New York (1973)
48. Ishikawa, F.N., Chang, H.K., Curreli, M., Liao, H.I., Olson, C.A., Chen, P.C., Zhang, R., Roberts, R.W., Sun, R., Cote, R.J., Thompson, M.E., Zhou, C.: Label-free, electrical detection of the SARS virus N-protein with nanowire biosensors utilizing antibody mimics as capture probes. ACS Nano **3**(5), 1219 (2009). DOI 10.1021/nn900086c
49. Björk, M.T., Schmid, H., Knoch, J., Riel, H., Riess, W.: Donor deactivation in silicon nanostructures. Nat. Nanotechnol. **4**(2), 103 (2009). DOI 10.1038/nnano.2008.400
50. Garnett, E.C., Tseng, Y.C., Khanal, D.R., Wu, J., Bokor, J., Yang, P.: Dopant profiling and surface analysis of silicon nanowires using capacitance–voltage measurements. Nat. Nanotechnol. **4**(5), 311 (2009). DOI 10.1038/nnano.2009.43
51. Stern, E., Wagner, R., Sigworth, F.J., Breaker, R., Fahmy, T.M., Reed, M.A.: Importance of the Debye screening length on nanowire field effect transistor sensors. Nano Lett. **7**(11), 3405 (2007). DOI 10.1021/nl071792z
52. Nair, P.R., Alam, M.A.: Design considerations of silicon nanowire biosensors. IEEE Trans. Electron Devices **54**(12), 3400 (2007). DOI 10.1109/TED.2007.909059
53. Yates, D.E., Levine, S., Healy, T.W.: Site-binding model of the electrical double layer at the oxide/water interface. J. Chem. Soc. Faraday Trans. 1 **70**, 1807 (1974). DOI 10.1039/F19747001807
54. Li, H., Robertson, A.D., Jensen, J.H.: Very fast empirical prediction and rationalization of protein pK_a values. Proteins Struct. Funct. Bioinforma. **61**(4), 704 (2005). DOI 110.1002/prot.20660
55. Bas, D.C., Rogers, D.M., Jensen, J.H.: Very fast prediction and rationalization of pKa values for protein-ligand complexes. Proteins Struct. Funct. Bioinforma. **73**(3), 765 (2008). DOI 10.1002/prot.22102
56. Olsson, M.H.M., Søndergaard, C.R., Rostkowski, M., Jensen, J.H.: PROPKA3: Consistent treatment of internal and surface residues in empirical pKa predictions. J. Chem. Theory Comput. **7**(2), 525 (2011). DOI 10.1021/ct100578z
57. Ullmann, G.M., Knapp, E.W.: Electrostatic models for computing protonation and redox equilibria in proteins. Eur. Biophys. J. **28**, 533 (1999). DOI 10.1007/s002490050236
58. Berman, H.M., Westbrook, J., Feng, Z., Gilliland, G., Bhat, T.N., Weissig, H., Shindyalov, I.N., Bourne, P.E.: The protein data bank. Nucleic Acids Res. **28**(1), 235 (2000). DOI 10.1093/nar/28.1.235
59. Liu, Y., Rieben, N., Iversen, L., Sørensen, B., Park, J., Nygård, J., Martinez, K.: Specific and reversible immobilization of histidine-tagged proteins on functionalized silicon nanowires. Nanotechnology **21**, 245105 (2010). DOI 10.1088/0957-4484/21/24/245105
60. Pugliese, L., Coda, A., Malcovati, M., Bolognesi, M.: Three-dimensional structure of the tetragonal crystal form of egg-white avidin in its functional complex with biotin at 2.7 å resolution. J. Mol. Biol. **231**(3), 698 (1993). DOI 10.1006/jmbi.1993.1321
61. Weber, P., Ohlendorf, D., Wendoloski, J., Salemme, F.: Structural origins of high-affinity biotin binding to streptavidin. Science **243**(4887), 85 (1989). DOI 10.1126/science.2911722
62. Green, N.M.: Avidin-biotin technology. In: Wilchek, M., Bayer, E.A. (eds.) Methods in Enzymology, vol. 184, pp. 51–67. Academic, London (1990). DOI 10.1016/0076-6879(90)84259-J
63. Hediger, M.R.: A perspective on bionanosensor simulation & computational enzyme engineering. Ph.D. Thesis, Department of Chemistry, University of Copenhagen (2013)
64. The PyMOL Molecular Graphics System: Version 1.3.0 Schrödinger, LLC
65. Harris, L.J., Larson, S.B., Hasel, K.W., McPherson, A.: Refined structure of an intact IgG2a monoclonal antibody. Biochemistry **36**(7), 1581 (1997). DOI 10.1021/bi962514+

66. Macao, B., Johansson, D., Hansson, G., Härd, T.: Autoproteolysis coupled to protein folding in the SEA domain of the membrane-bound MUC1 mucin. Nat. Struct. Mol. Biol. **13**(1), 71 (2005). DOI 10.1038/nsmb1035
67. Goodsell, D.S., Morris, G.M., Olson, A.J.: Automated docking of flexible ligands: Applications of autodock. J. Mol. Recognit. **9**(1), 1 (1996). DOI 10.1002/(SICI)1099-1352(199601)9:1<1::AID-JMR241>3.0.CO;2-6

Chapter 4
Semiconductor-Based Nanostructures for Photoelectrochemical Sensors and Biosensors

Hongjun Chen and Lianzhou Wang

Abstract A diverse array of semiconductor-based nanostructures have been synthesized and utilized in many applications, in which photoelectrochemical sensor or biosensor is a new kind of developing analytical technology for the detection of the low concentration of analytes. Due to the parameters of the detection processes in relation to the photocurrent or photopotential and the appearance of new semiconductor-based functional nanostructures, it makes photoelectrochemical analysis with high sensitivity, rapid detection, inherent miniaturization, and easy portability. In this chapter, we will introduce the basic principle of the photoelectrochemical sensor or biosensor and focus on the new progress of photoelectrochemical sensors and biosensors with different semiconductor materials. Meanwhile, a variety of detection mechanisms and the characteristics of photoelectrochemical sensor and biosensor are also introduced and their future developments are also prospected and discussed.

4.1 Introduction

Sensor- and biosensor-related research has experienced rapid growth during the last two decades. A sensor is generally defined as a self-contained integrated device which registers a physical, chemical, or biological change and converts that into a measurable signal [1]. Normally, a sensor is capable of providing specific quantitative or semiquantitative analytical information using a recognition element which is in direct spatial contact with a transducer element. From this definition, it is clearly seen that the main components of a sensor include a recognition

H. Chen • L. Wang (✉)
ARC Centre of Excellence for Functional Nanomaterials, School of Chemical Engineering and Australian Institute for Bioengineering and Nanotechnology, The University of Queensland, St. Lucia, Brisbane, QLD 4072, Australia
e-mail: h.chen11@uq.edu.au; l.wang@uq.edu.au

S. Li et al. (eds.), *Nanoscale Sensors*, Lecture Notes in Nanoscale Science and Technology 19, DOI 10.1007/978-3-319-02772-2_4,

element, a transducer, and a signal processor. If a device contains a biological recognition element like enzymes, antibodies, proteins, nucleic acids, and cells, this kind of sensors is usually named as biosensor. Meanwhile, if a transducer is an electrode and obtains amperometric or potentiometric signals, this kind of devices can be called as electrochemical biosensors. As a subclass of chemical sensors, electrochemical biosensors require biospecific reagent functionalized electrodes, depending on the recognition mechanism, that come in contact with the solution under test and generate the signal that is correlated with the concentration of the target analyte. Combining the advantages of biological recognition mechanism and developed electrochemical techniques, electrochemical biosensors usually have high sensitivity and excellent specificity and low detection limits. Up to now, several reviews about chemical sensors and electrochemical biosensors have been reported [2–8].

In contrast to electrochemical sensor and biosensor, the electrode material for the working electrode of photoelectrochemical sensor and biosensor normally is semiconductor and always needs a lamp for the process of the photoexcitation. When illuminated with photons of energy exceeding the bandgap of the semiconductor, excited charge carriers are formed, charge separation and charge transfer will occur, generating photocurrent or photopotential. The relationship between the photocurrent or photopotential produced and the concentration of the target analyte added construct the basis of photoelectrochemical sensor and biosensor. In order to construct a successful photoelectrochemical sensor especially a biosensor, a number of conditions must be met [4]:

1. The semiconductor photoelectrode should have good photo-response and can produce a detectable photocurrent or photopotential.
2. The recognition element should be stable and highly specific for the analyte.
3. The response should be precise, quick, reproducible, and linear over the concentration range of the analyte. It is desirable for a photoelectrochemical sensor or biosensor to provide real-time analysis.

Obviously, the semiconductor photoelectrode is the key to the operation of the photoelectrochemical sensor and biosensor. The semiconductor photoelectrode is responsible for the adsorption of the incident photon and the generation of the electron–hole charge carriers, while the interface between the semiconductor and the liquid is an important factor in the subsequent chemical steps that leads to charge carrier transfer. Therefore, choosing a suitable semiconductor photoelectrode is the prerequisite for the successful operation of photoelectrochemical sensor and biosensor. There are a variety of semiconductor materials that include Si, TiO_2, ZnO, SnO_2, CdS, and CdSe for choosing [9–11]. In order to further improve photo-to-current conversion efficiency, some composite semiconductor materials like CdS–TiO_2, CdS–SnO_2, and tris(bipyridyl) ruthenium (II) complex–TiO_2 also appeared [12–14], among which TiO_2-based composite has been intensively studied [15].

New input into the field of sensor and biosensor came from the rapidly developed field of nanotechnology at the beginning of the twenty-first century.

Up to now, more and more nanotechnology-related research has been published. Meanwhile, a variety of protocols are also available for the fabrication of shape- and size-controlled nanoparticles [16, 17]. It is well known that the properties of nanoparticles are closely related with their sizes, shapes, components, etc. [18]. Therefore, the unique optical, electronic, and catalytic properties of nanoparticles pave the way to the new generations of devices and materials exhibiting novel functions and properties [19]. For semiconductor nanoparticles, some non-sphere shapes of semiconductor nanostructures appeared like nanotubes, nanorods, nanowires, and nano-plates in addition to the assembly of nanoparticles into 2D and 3D nanoparticle architecture [11, 12, 20–23]. Meanwhile, the functionalized semiconductor nanostructures with monolayer, thin film, or polymer also provide useful building blocks for the assembly of functionalized nanostructures on surface. To further improve the electron capture, the efficient separation of electron and hole, and the charge transfer, some metal nanoparticles and carbon nanostructures are also conjugate with semiconductor nanostructure to form new composite materials in the photoelectrochemical sensor and biosensor [12, 24, 25]. Obviously, the newly developed nanotechnology and plenty of nanostructures provide a huge opportunity for the development of new generations of the photoelectrochemical sensor or biosensor. Meanwhile, miniaturization and easy portability are also the basic requirements for the development of the new generations of smart devices.

In this chapter, the aim is to introduce the basic concept of the electrochemical detection technique and the detection mechanisms in the photoelectrochemical sensor or biosensor. Meanwhile, the recent progress about the functionalized semiconductor nanostructures used for the application of photoelectrochemical sensor or biosensor is also summarized. In contrast to many reviews about electrochemical sensor or biosensor, there are very scarce reviews about the photoelectrochemical sensor or biosensor [13]. Therefore, it is very meaningful to give a systematic introduction about the basic principles and characteristics of photoelectrochemical sensor and biosensor. Meanwhile, their future developments are also prospected and discussed.

4.2 Photoelectrochemical Detection Techniques

Typically in photoelectrochemical sensor or biosensor, the reaction under investigation will either generate a measurable photopotential (potentiometric) or a measurable photocurrent (amperometric) between the electrodes [4]. Based on this, photoelectrochemical detection techniques are generally organized into amperometric and potentiometric techniques. The chapter will introduce these techniques but focus primarily on amperometric technique because the measurement of photocurrent is the most commonly used in the photoelectrochemical sensor or biosensor.

4.2.1 Potentiometric and Amperometric Techniques

Potentiometric technique is characterized by measuring the accumulation of a charge potential at the working electrode (indicator electrode) versus the reference electrode in an electrochemical cell when zero or no significant current flows between them [1, 4, 26, 27]. Whereas the working electrode is in direct contact with the analyte solution, the reference electrode is usually separated from the analyte solution by a salt bridge of various forms. The electrode potential of the indicator electrode is normally directly proportional to the logarithm of the activity of the analyte in the solution. For potentiometric measurements, the relationship between the concentration and the potential follows the Nernst equation:

$$\mathrm{EMF} = K + RT/zF \ln a_{\mathrm{i}}$$

where EMF is the electromotive force (the observed potential at zero current), K is a constant potential contribution that often includes the liquid-junction potential at the reference electrode, a_{i} is the sample activity for the ion I with charge z, and R, T, and F are the gas constant, absolute temperature, and Faraday constant, respectively [28].

Potentiometry is a static technique and of great practical importance. From potentiometry, the information about the sample composition is obtained from measurement of the potential established across a membrane. Since different types of membrane materials possess different ion-recognition processes, these different types of membrane materials thus have been developed to have high selectivity. Therefore, potentiometry has been widely used for direct monitoring of ionic species such as protons or calcium, fluoride, and potassium ions in complex samples through using different potentiometric probes [29]. There are some published reviews about the potentiometric sensors for trace-level analysis [28] and systematic introduction about the transduction principles for potentiometric sensor arrays [30]. Potentiometric technique is currently widely used in many fields, including clinical diagnostics, industrial process control, environmental monitoring, and physiology. The speed at which this field has developed is a measure of the degree to which potentiometric measurements meet the needs of the analytical chemist for rapid, low-cost, and accurate analysis.

Amperometric technique is characterized by applying a potential to a working electrode versus a reference electrode and measuring the current between working electrode and counter electrode. The technique is implemented by stepping the potential directly to the desired value and then measuring the current or fixing the potential at the desired value and flowing samples across the electrode as in flow injection analysis [3]. Because the electrolysis current is limited by the mass transport rate of these compounds to the working electrode, the current response thus generated reflects the concentration profiles of these compounds as they pass through the working electrode. Accordingly, the magnitude of the peak current serves as a measure of the concentration of the electroactive species in the sample.

Normally, the current measured is superimposed on a constant background current due to redox reactions of the mobile phase or carrier solutions [29]. Advantageously, the fixed potential during amperometric detection results in a negligible charging current to minimize the background signal that adversely affects the limit of detection. Meanwhile, choosing a suitable applied potential is also very important; this is because the applied potential affects not only the sensitivity and signal/noise ratio but also the selectivity of amperometric measurements. In general, a lower potential is more selective than a higher one. In order to greatly improve the selectivity, compounds undergoing redox potentials prefer to choose lower potentials for the detection [29]. Amperometric biosensors have additional selectivity in that the oxidation or reduction potential used for detection is characteristic of the analyte species.

4.2.2 The Apparatus and the Characterization of Photoelectrochemical Biosensors

Photoelectrochemical biosensors combine the analytical power of electrochemical techniques with the specificity of biological recognition processes. The aim is to use the physical/chemical interactions between analytes and semiconductor materials to produce a photoelectrical signal that relates to the concentration of an analyte [13]. Because of this, a biospecific reagent is either immobilized or retained on a semiconductor photoelectrode surface, which converts the biological recognition event into a quantitative amperometric or potentiometric response through a suitable photo illumination. The photoelectrochemical sensor also has a very similar detection mechanism. The device of photoelectrochemical sensor or biosensors for carrying out measurement normally requires three-electrode system. A typical three-electrode system consists of a working electrode of semiconductor-modified fluorine-doped tin oxide (FTO) or indium-doped tin oxide (ITO), a reference electrode of Ag/AgCl or saturated calomel electrode (SCE), and a platinum wire auxiliary or counter electrode. The reference electrode is kept at a distance from the reaction site in order to maintain a known and stable potential. The working electrode serves as the transduction element in the biochemical reaction, while the counter electrode establishes a connection to the electrolytic solution so that a current can be applied to the working electrode [4]. FTO or ITO is also called optically transparent electrode (OTE), which is transparent to UV–visible light. Such an electrode is very useful as a transparent conductive support for the fabrication of the working electrode for photoelectrochemical sensor or biosensor [31]. In addition, some other inert electrode materials like glassy carbon (GC), gold, and platinum are also used as conductive support for carrying semiconductor nanostructures. However, using this kind of conductive support in photoelectrochemical sensor or biosensor can only adopt the front-side illumination.

To fabricate a semiconductor photoelectrode, for example, TiO_2 photoelectrode, various techniques can be used for the immobilization of TiO_2 nanoparticles on FTO or ITO substrates, such as sol–gel dip-coating [32, 33], spin-coating [34], chemical vapor deposition [35] sputtering [36], electrodeposition [37], layer-by-layer assembling [38], and in situ hydrothermal growth [22]. After calcination, the TiO_2 nanoparticles should be tightly adhered on the FTO surface. Thus, TiO_2 photoelectrode will be fabricated. In order to further immobilize biospecific reagent on a semiconductor photoelectrode surface, TiO_2 photoelectrode is firstly assembled with a layer of functionalized molecules or polyelectrolyte and then reacted with biospecific reagent through covalent bonding or electrostatic interaction. Obviously, a variety of strategies are adopted for the immobilization of biospecific reagent on semiconductor photoelectrode surface based on different detection mechanism. When the functionalized semiconductor photoelectrode immersed in the electrolyte solution, the space-charge layer (depletion layer) typically 5–200 nm is formed within the semiconductor photoelectrode at the interface. The direction of the electric field is such that any excess holes created in the space-charge region would move toward the surface and any excess electrons would move toward the bulk semiconductor [39]. When the interface is irradiated with light of energy greater than the bandgap of the semiconductor, photons are absorbed and electron–hole pairs are created. The electron excited to the conduction band and the hole left behind in the valence band can separate under the influence of the internal electric field; with the electron drawn into the interior of the semiconductor and the hole driven to the surface (for n type of semiconductor), it can be captured by an acceptor state in solution, driving an electrochemical reaction [31]. The electron can pass round an external circuit to the counter electrode, while some of those electron–hole pairs created beyond the space-charge layer normally recombine together with the evolution of heat [39].

In order to construct the relationship between the generated photocurrent and the concentration of the target analyte, the photocurrent can be tuned by the specific interaction between the biological recognition element and their specific targets. For biocatalytic sensors, enzymes, whole cells, or tissue slices recognize the target analyte and produce electroactive species, which will play as the electron donor or acceptor to involve in the reaction with the photoexcited holes or electrons. So, the more target analyte added in this detection process, the more electroactive species will be produced, which will subsequently lead to much higher increment of the photocurrent. Thus, a serial of the chain reactions for the detection of the target analyte will finally be constructed, while the affinity sensors use the selective and strong binding between the biomolecules of antibody–antigen complexation, DNA hybridization, etc. The properties of high affinity and specificity of the biomolecule for its ligand make these sensors very sensitive and selective. In the process of the molecular recognition for affinity biosensors, the complementary size and shape of the binding site to the analyte of interest is determinable. Based on this, the normal strategy for this kind of detection always use a labeled target, in which the probe is a dye or quantum dot or metal nanoparticle and so on [12]. By virtue of the distance between the probe and the substrate or the quantity of probe on the detection surface, the generated photocurrent will correspondingly increase or decrease to feed back the concentration of the target analyte. The strategy for the recognition and detection in photoelectrochemical sensor or biosensor will be described in more detail in the following section.

In the following steps, the experimental parameters are optimized like applied potential, the pH value, and the temperature. In addition, choosing a suitable wavelength for illumination is also very important for a photoelectrochemical sensor or biosensor. Normally, the functionalized photoelectrode is firstly conducted under a full wavelength to locate the position of the wavelength for peak photocurrent. Obviously, the peak photocurrent facilitates to improve the sensitivity of the device; the wavelength of the peak photocurrent is normally preferred to choose for a photoelectrochemical sensor or biosensor. However, the bioactivity of the biomolecules and the stability of the probe also need to be considered before determining the exact wavelength of the illumination for use. Under optimal experimental condition, the performance of a photoelectrochemical sensor or biosensor is usually experimentally evaluated based on its sensitivity, limit of detection, linear detection ranges, reproducibility, selectivity, and its response to interferences. Other parameters that are often compared include the sensor's response time (i.e., the time after adding the analyte for the sensor response to reach 95 % of its final value), operational and storage stability, ease of use, and portability [3].

4.3 Light Addressable Potentiometric Sensor

As a potentiometric sensor, light addressable potentiometric sensor (LAPS) is a silicon-based device that takes advantage of the photovoltaic effect to selectively determine the point of measurement. Since it was first proposed in 1988 by Hafeman et al. on Journal of Science [40], LAPS became popular in many biological and chemical applications, like the measurement of cell metabolism, the detection of bacterial growth, and the study of mechanisms of drug action on cell physiology [40–43]. Furthermore, the first successfully commercialized system using LAPS is the cytometer microphysiometer system that has been realized in the 1990s by the company Molecular Devices Corporation (USA) [42, 43]. LAPS usually consist of the metal–insulator–semiconductor (MIS) or electrolyte–insulator–semiconductor (EIS) structure. The difference between these two structures is their sensitive "gate" regions, in which MIS structure is sensitive to gases while EIS structure is sensitive to ions. It is possible to determine the spatially resolved surface potential distribution along the interface of the sample and substrate surfaces through scanning with a focused light source. As shown in Fig. 4.1, when a modulated light pointer illuminates LAPS chip, light-induced charge carriers are separated by the internal electric field and thus photocurrent can be produced. An external DC bias voltage is applied to scan the interface of the insulator and semiconductor of LAPS; the amplitude of the photocurrent is changed depending on the local surface potential. Characteristic *I–V* curve of LAPS is obtained demonstrating the relationship

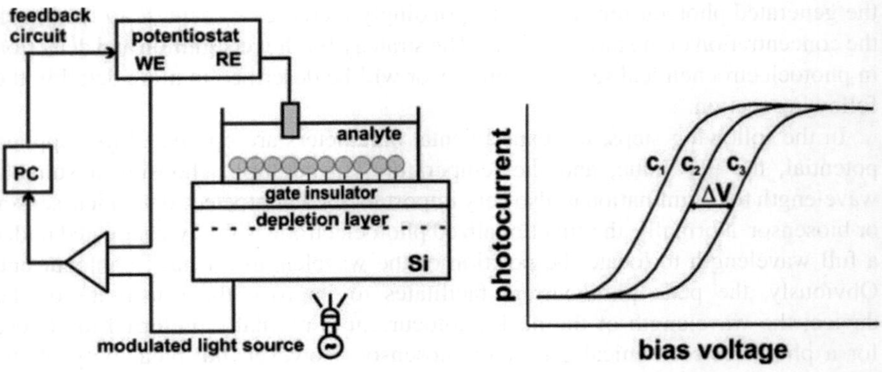

Fig. 4.1 Detection principles of LAPS (*left*) and corresponding change of the sensor signal (*right*) (*WE* working electrode, *RE* reference electrode) [41]

between the generated photocurrent and the applied bias voltage to the sensor chip (Fig. 4.1 right) [41]. In addition, the electric fields within the semiconductor can be modulated by transmembrane potentials when the membranes are appropriately positioned relative to the insulated semiconductor surface, which lead to the change of the voltage and the excursion of the *I–V* curve, as shown in Fig. 4.1 right. Thus, the related analyte concentration can be obtained based on the shift of the *I–V* curve obtained on LAPS.

The LAPS is very sensitive for any kind of potential generation at or near the gate insulator/electrolyte interface. Therefore, combined with the respective chemical or biological recognition element, each biological or chemical reaction can be detected by LAPS only it leads to the chemical or electrical changes at this interface. Generally, the following basic mechanisms of potential generation can be considered [41, 44]:

1. A pH or ion-concentration change
2. Enzymatic reactions
3. Affinity binding of molecules (antigen–antibody affinity reaction or DNA hybridization)
4. Potential changes that are coming from living biological systems as a result of more sophisticated biochemical processes (action potential of nerve cells, dipole potentials, etc.)

Attractive features of LAPS include potentiometric stability, multiparameter, and multipoint determination. Different analytes can be detected simultaneously on different regions of a single semiconductor device [40]. Meanwhile, the ability to address different spatial regions of the semiconductor with light allows multiple potentiometric measurement to be made with a single semiconductor device. For the detailed mechanism and the related device and system, some reviews are published [41, 44–47].

The LAPS has been widely used for ion sensing and pH determination. Chin et al. found that the LAPS with a 2-nm-thick HfO_2 film as an insulating layer after the carbon tetrafluoride (CF_4) plasma treatment has a good response to the concentration of Na^+ ions and the pNa sensitivity achieved 33.9 mV/pNa and the sensitivity was improved by twofold [48]. The LAPS with $Al_2O_3/SiO_2/Si$ structure can be used for the detection of K^+, Ca^{2+}, and Mg^{2+} ions only; ion-recognition element such as valinomycin, 18-crown-6 ether, bis[di(noctylphenyl) phosphato]calcium(II), and ETH1117 were immobilized in the matrix of poly(vinyl chloride) film on the surface of the Al_2O_3 layer [49]. Schöning group reported a novel design of the LAPS which is capable of measuring up to four different ion species (like Cs^+, Li^+, K^+, Mg^{2+}, and Ca^{2+}) by integrating different ion-selective materials on the sensing surface [50–52]. Besides the abovementioned simple metal ions, some heavy metal ions such as Pb^{2+}, Cu^{2+}, Cd^{2+}, and Hg^{2+} can also be detected and measured by the LAPS with the chalcogenide glass membranes [53–58]. Normally the detection limit can reach 10^{-5} even 10^{-7} M. Furthermore, LAPS can also be used for the anion detection. For example, LaF_3 has been studied as the fluoride-sensing layer in the LAPS heterostructure and found that the sensor shows quite linear response in the range of pF1–5 with a sensitivity of 49 mV/pF [59]. In addition, the detection of the anion like NO_3^- [60] and CN^- [61] ions are also reported. For pH determination, Lue et al. reported a differential measurement system using HfO_2-LAPS and PVC/HfO_2-LAPS pair. It is found that the sensitivity of HfO_2-LAPS was as high as 58.6 mV/pH [62]. Ismail et al. propose a simple, low-noise, and low-cost measurement principle for LAPS using a single sensor and two out-of-phase modulated light sources, in which the pH measurement is investigated in difference-signal and bias-difference mode and the pH sensitivity is 57.6 mV/pH [63]. Pulsed laser-deposited Al_2O_3 layer as a pH-sensitive material in a heterostructure of $Al/Si/SiO_2/Al_2O_3$ in a LAPS arrangement has high and nearly stable pH sensitivity (around 56 mV/pH) in the long-time use [64]. Recently, Schöning group found that layer-by-layer assembly of carbon nanotubes and polyamidoamine dendrimers on the LAPS makes the sensors with high pH sensitivity and high sensitivity for penicillin G [65]. Based on the good pH determination, LAPS is used to determine the diffusion coefficient of anions and cations [66]; the pH sensitivity of surface functional group [67]; the H^+ distribution produced by living cells [68]; microscale pH gradient generation by electrolysis [69]; urea, penicillin, and glucose [70]; and surface potential and zeta potential [71]. Furthermore, LAPS can also be developed as a chemical-imaging sensor for the visualization of the two-dimensional distribution of a specific substance like an enzyme of urease immobilized on the sensing surface of the pH-imaging sensor [72] or a redox potential sensor to monitor the enzymatic activity of horseradish peroxidase (HRP) [73]. Now the chemical-imaging system is an important function for LAPS and allows the visualization of the distribution of chemical species on the sensor surface. In order to improve the speed and spatial resolution, many approaches have been reported to improve LAPS like using vertical cavity surface emission laser diodes [74], organic light-emitting diode displays [75], a digital light processing setup [76], and ultrathin Si films deposited on a transparent substrate to reduce the thickness of the semiconductor layer [77, 78] or increasing the doping density [79].

Combining the properties of the high pH sensitivity and the excellent spatial resolution of the chemical imaging, one of the prospective applications of the LAPS is the visualization of biochemical activities in biological systems, like the investigation of the electrophysiological signals of taste receptor cells and the acidic sensation in a biomimetic manner [80], the monitor of the cellular metabolism in a noninvasive way [81, 82], the determination of the extracellular acidification [83–87], and even the cell acidification and extracellular potential simultaneously [88]. In order to improve the adhesion of the biological system to the sensing surface, the porous Si was proposed and tested [89]. Using the enzyme of urease as label on an antifluorescein antibody to form a sandwich immunoassay on the sensing surface of LAPS, human α-acid glycoprotein and asialo-α-acid glycoprotein can be detected to 2 and 250 pg, respectively, according to the rate of pH change [90]. Under the similar mechanism, LAPS is reported to detect $I0^3$–10^4 CFU of cultured *E. coli* O157:H7 in PBS solutions or I CFU/g of *E. coli* 0157:H7 in beef hamburger using urease-labeled antifluorescein antibody and labeled with fluorescein-conjugated anti-*E. coli* OI57:H7 antibodies [91].

4.4 Amperometric Type of Photoelectrochemical Sensor and Biosensor

Being an amperometric sensor, the target analyte normally involved in the electron transfer with the semiconductor photoelectrode as an electron donor or acceptor, which always induces the changes of the generated photocurrent. Based on the changes of the photocurrent, the target analyte can be successfully detected through using the photoelectrochemical method. In recent years, many research works about amperometric type of photoelectrochemical sensor and biosensor are published, especially for the detection of biological materials. In these detection systems, different kinds of semiconductor materials are used to fabricate the semiconductor photoelectrode like TiO_2 nanotubes [92, 93], ZnO nanorods [94], TiO_2-B nanowires [95], bismuth oxyiodide nanoflakes [96], $CdSe_xTe_{1-x}$ $(0<x<1)$ nanocrystal [97] and nanocomposites like CdS–FePt dimers [98], Au-doped TiO_2 nanotubes [99], poly(3-hexylthiophene)/TiO_2 nanoparticle [100], Au–CdS nanoparticle [101], WO_3/TiO_2 nanotube [102], TiO_2/CdS core–shell nanorod [103], and graphene–CdS nanocomposites [104, 105]. Meanwhile, a variety of the target analytes can be detected by monitoring the changes of the photocurrent produced by the photoelectrochemical sensor or biosensor. Up to now, some cationic ions like Cu^{2+} [107, 108] and Hg^{2+} [109–111]; simple molecules like glucose [112–114], hydrogen peroxide [93, 98, 115, 116], ascorbic acid [117],

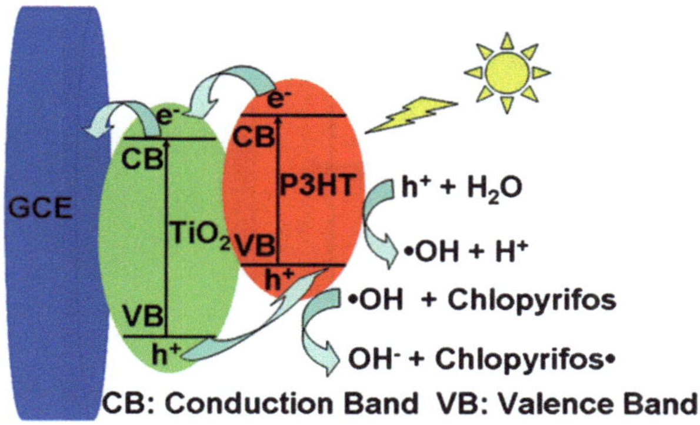

Fig. 4.2 Schematic illustration of proposed photoelectrochemical mechanism for chlorpyrifos oxidation at P3HT/TiO_2 modified GCE [100]

acetylthiocholine [101], guanine [118], glutathione [104], dihydronicotinamide adenine dinucleotide (NADH) [106, 119], binaphthol isomers [120], adenosine triphosphate [121], cocaine [122], and environmental pollutant of pentachlorophenol [97]; some organophosphorus pesticide like chlorpyrifos [100], methyl parathion [96], profenofos [123], and dichlofenthion [124]; and even protein α-synuclein [99], DNA [125–133], and cells [105, 134, 135] are detected through the photoelectrochemical method. In order to clearly introduce the amperometric type of photoelectrochemical sensor or biosensor, we try to sort them into some subclasses based on the different detection mechanisms or the target analytes. Furthermore, all the detection mechanisms will be discussed by one or two typical examples in more details.

4.4.1 The Detection of Simple Electroactive Molecule(s)

The detection mechanism of simple electroactive molecular is based on the principle of photoelectrochemistry. Just as introduced in Sect. 4.2.2, when the semiconductor photoelectrode is illuminated by the photon with the energy larger than its bandgap, the electron–hole pairs are produced and separated; finally the electrons are moved to the counter electrode, while the holes are driven to the surface and reacted with the electron donors in the electrolyte solution (for n type of semiconductor). The vice versa is for the p type of semiconductor. Here, the electroactive molecule(s) plays a role as the electron donor involved in this reaction, which facilitates the hole transfer and inhibits the recombination of electrons and holes. On the macroscopic level, the photocurrent increase after the addition of the electroactive molecule(s). Based on this, photoelectrochemical sensor can be fabricated for the detection of these simple electroactive molecule(s).

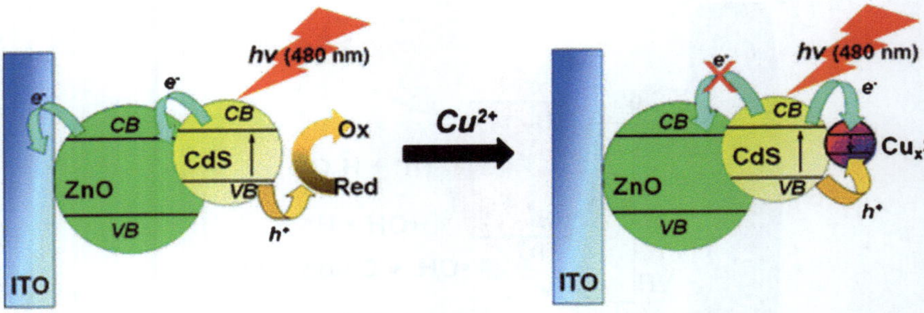

Fig. 4.3 Schematic illustration of proposed photoelectrochemical mechanism for the detection of Cu^{2+} ions at CdS/ZnO modified ITO [107]

As shown in Fig. 4.2, the photoelectrode is composed of poly(3-hexylthiophene) (P3HT)-functionalized TiO_2 nanoparticles on the surface of glassy carbon electrode (GCE), which can efficiently promote the separation of holes and electrons [100]. Under visible light irradiation, P3HT generates the transition from the valence band to the conduction band, delivering the excited electrons into the conduction band of TiO_2 and then to the GCE. Simultaneously, a positive charged hole (h^+) of TiO_2 may form and migrate to the valence band of P3HT, which can react with H_2O to generate $^{\bullet}OH$, and then it converts chlorpyrifos into chlorpyrifos$^{\bullet}$ that promotes the amplifying photocurrent response [100]. From this reaction, it is shown that chlorpyrifos is not directly reacted with the generated hole but through the radical of $^{\bullet}OH$ generated. From this example, it is clearly demonstrated that the indirect reaction between the electroactive molecule(s) and the hole can also be utilized to fabricate the photoelectrochemical sensor. Under optimal conditions, the proposed photoelectrochemical sensor could detect chlorpyrifos ranging from 0.2 to 16 μM with a detection limit of 0.01 μM at a signal-to-noise ratio of 3 [100]. Based on the similar detection mechanism, Wang et al. reported the dopamine-modified TiO_2 photoelectrode for the sensitive detection of NADH with a detection limit of 1.4×10^{-7} M and the detection range up to 1.2×10^{-4} M [119]. Li and coworkers used the similar strategy for the detection of ascorbic acid using an n-silicon electrode modified by platinum and cobalt hexacyanoferrate films [78]. The linear range for the detection of ascorbic acid was 1.0×10^{-6} to 1.0×10^{-3} M, with a detection limit (S/N = 3) of 1.0×10^{-6} M [117]. Also is the similar for the CdS–FePt dimers based photoelectrochemical sensor for the detection of H_2O_2 [98] and the nanometer-sized titania coupled with a screen-printed electrode for the detection of organophosphorus pesticide dichlofenthion (the linear over the 0.02–0.1 and 0.2–1.0 μM ranges with a detection limit of 2.0 nM) [124].

The advantages for this kind of photoelectrochemical sensor are simple, direct, and easily conducted, while the disadvantage is the selectivity sometimes is not satisfactory even need to tune the applied potential [110].

4.4.2 The Detection Based on the Formation of the Insoluble Materials

In some cases, the produced products are insoluble or insulating for certain electron-involved redox reaction. If the products are deposited on the photoelectrode surface, the photocurrent will decrease because the buildup of an insoluble or insulating layer could greatly alter the interfacial electron-transfer feature and thereby significantly influence the photocurrent.

From Fig. 4.3, it is clearly seen that the photoelectrode is composed of CdS-functionalized ZnO on ITO surface. For this photoelectrochemical sensor, the detected analyte is Cu^{2+} ions. The detection mechanism is based on the insoluble product of Cu_xS ($x = 1, 2$) play as electron–hole recombination centers formed on the CdS surface [107]. Therefore, the photocurrent intensity of ZnO/CdS decreases after the addition of the Cu^{2+} ion. Based on this, the linear detection range for Cu^{2+} ions was 0.02–40.0 μM, with a detection limit (S/N = 3) of 0.01 μM. Using the same mechanism, Xu's group also reported a photoelectrochemical sensor for the detection of trace amount of Cu^{2+} using CdS-modified ITO electrode as photoelectrode [108]. This detection mechanism is also effective for the biocatalytic precipitation reaction. Xu and her coworkers reported a photoelectrochemical biosensor with ultrahigh sensitivity and excellent anti-interference ability for the detection of H_2O_2 [116]. In this biosensor, HRP was used to accelerate oxidation of 4-chloro-1-naphthol by H_2O_2 and yield the insoluble product on the photoelectrode surface, which correspondingly reduces the photocurrent. The linear range from 1.0×10^{-9} to 2.0×10^{-5} M and the detection limit of 5.0×10^{-10} M (S/N = 3) was achieved for the determination of H_2O_2 [116].

The detection mechanism of the photoelectrochemical sensor or biosensor in this section can be regarded as the derivative of Sect. 4.4.1. The only difference between these two sections is one leads to the photocurrent increase and the other for the photocurrent decrease just because of the solubility of the product generated. As a special case, this kind of photoelectrochemical sensor or biosensor is only suitable for some specific analytes. Therefore, the general application is somewhat limited.

4.4.3 The Detection of the Substrate of the Enzymatic Reaction

In this section, the detection of substrate is mainly referred to the specific recognition between the enzymes and their corresponding substrates. As a powerful catalyst, the enzyme can convert substrate into product and some electrons are involved in this process. The most commonly used enzymes and their corresponding substrates in biosensing are glucose oxidase (GOx) with glucose, HRP with hydrogen peroxide, urease with urea, and so on. About the detailed structure and recognition mechanism of enzymes, you can refer to some reviews and books [3, 4,

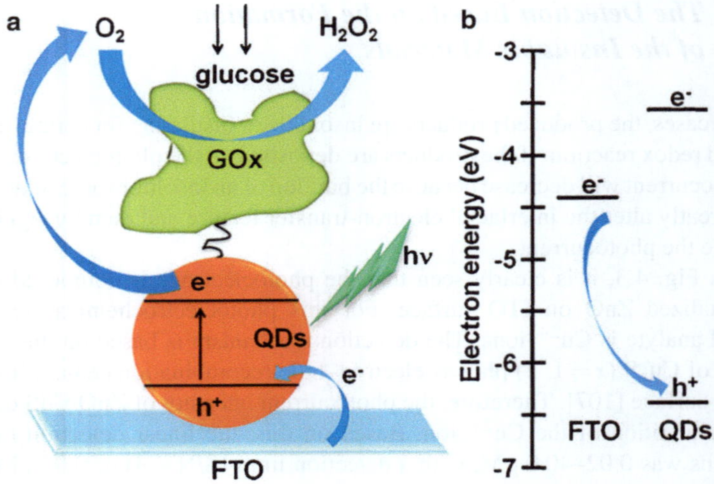

Fig. 4.4 Schematic illustration of the photoelectrochemical strategy for the detection of glucose at GOx/CdTe/FTO photoelectrode [112]

6, 29, 136]. Here, taking glucose oxidase as a typical example, the reaction scheme is given in the following equation:

$$O_2 + \text{glucose} \xrightarrow{\text{Glucose oxidase}} H_2O_2 + \text{gluconic acid}$$

From this equation, it is clearly shown that the detection of the glucose can be realized and quantified through measuring the concentration of H_2O_2 produced in the presence of GOx and O_2 or through controlling the consumption of oxygen during the enzymatic reaction. Recently, Wang et al. reported a visible light-induced photoelectrochemical biosensing platform for the detection of glucose based on oxygen-sensitive near-infrared quantum dots of CdTe [112]. As shown in Fig. 4.4, GOx was covalently attached to the CdTe quantum dots on the surface of FTO. In this photoelectrochemical biosensor, oxygen not only involves in the enzymatic reaction with GOx and glucose to produce hydrogen peroxide but also plays as the electron acceptor to receive the photo-induced electron in the photoelectrochemical reaction. Thus, the photocurrent intensity is suppressed with the consumption of oxygen during the enzymatic reaction, which means that the resulting biosensor has a sensitive response to glucose. Under the visible light irradiation, the photoelectrochemical biosensor could detect glucose ranging from 0.1 to 11 mM with a detection limit of 0.04 mM (S/N = 3) [112].

As one main kind of photoelectrochemical biosensor, many analytes can be sensitively detected through the effective use of the specific recognition between the enzyme and the corresponding substrate. For example, Zhao et al. use xanthine oxidase for the determination of guanine [118]. The obtained photoelectrochemical biosensor has a linear response of guanine ranged from 1.00 to 200 μM and a detection limit of 0.55 μM. Chen et al. use TiO_2 nanotubes immobilized with HRP as the photoelectrode for the detection of H_2O_2 under the visible light illumination [93]. The HRP/TiO_2 nanotubes electrode displayed a linear range of 5.0×10^{-7} to 3.5×10^{-5} M and a low detection limit of 1.8×10^{-7} M for H_2O_2 determination [93]. Gong et al. reported a new, highly sensitive, and selective biosensor for the photoelectrochemical detection of methyl parathion, in which acetylcholinesterase (AChE) functionalized bismuth oxyiodide nanoflake on ITO as the photoelectrode, and use the specific recognition between the AChE and the substrate of acetylthiocholine chloride. In this detection mechanism, the analyte of methyl parathion used as the inhibitor suppresses AChE in recognition of acetylthiocholine chloride, which leads to the decrease of the photocurrent [96]. For this photoelectrochemical biosensor, the detection was proportional to the concentration of methyl parathion in two ranges, from 0.001 to 0.08 mg mL^{-1} and from 0.3 to 1.0 mg mL^{-1}, and the detection limit was about 0.04 ng mL^{-1} [96]. There are still other photoelectrochemical biosensors like formaldehyde dehydrogenase-modified CdS for the detection of formaldehyde [137] and the GOx-functionalized TiO_2/CdSe@CdS nanocomposites as the photoelectrode for the detection of glucose [113].

It should be pointed out that the specific recognition ability is also present in some special molecule besides the enzymes and substrates. Shi et al. use the specific binding ability of *p-tert*-butylcalix [4] arene to profenofos to fabricate a highly selective and sensitive photoelectrochemical sensor for the determination of profenofos with a low detection limit of 1 nM (S/N = 3), in which the sensor is fabricated on highly ordered and vertically aligned TiO_2 nanotubes and *p-tert*-butylcalix [4] arene as recognition element and visible light sensitizer [123].

Due to the specific recognition, the obvious advantage of this kind of photoelectrochemical sensor or biosensor is high selectivity. The requirement of these sensors or biosensors is the recognition element (like enzymes) needs to immobilize on the sensing surface of the photoelectrode. The other feature of these sensors or biosensors is that at least one reactant not only involves in the enzymatic reaction but also attends the photoelectrochemical reaction. Thus, the close combination of these two different reactions makes the sensitive detection of many analytes possible, which also leads the obtained photoelectrochemical sensor and biosensor to have many applications.

4.4.4 Immunosensor and Cytosensor

Compared with the other recognition elements, antibodies are most widely used for the detection of antigen because of the high specificity of the antibody–antigen binding. For the detection, the commonly used method is the enzyme-linked immunosorbent assay (ELISA). A simple scheme of ELISA is depicted like this. After immobilizing antibodies to a surface, an analyte is introduced to which the

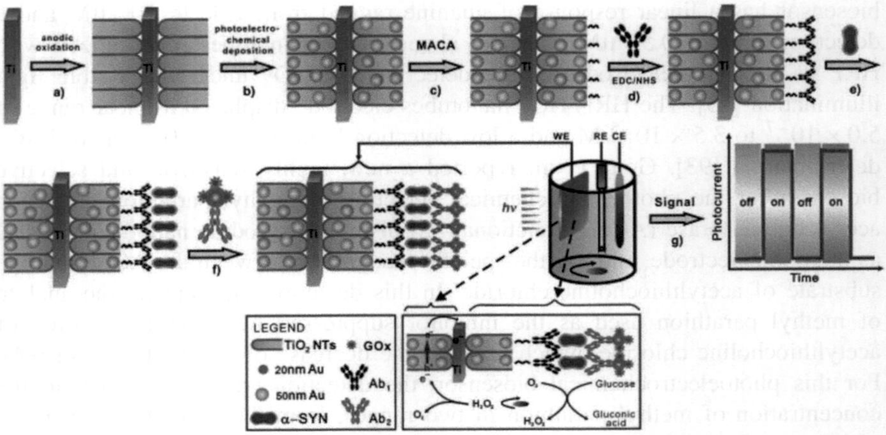

Fig. 4.5 Procedure for the synthesis of the TiO_2-Au-Ab1-α-SYN-{Ab2-Au-GOx} immunosensor and the mechanism for the photoelectrochemical detection of α-SYN [99]

antibodies bind specifically. After that, a secondary labeled antibody then binds to the analyte in order to detect its concentration. The detection antibodies are labeled an enzyme, which allows quantitative measurements of the amount of bound antigens by monitoring the signal (like the color and photoluminance) generated by an enzymatic reaction [4, 138]. Thus, the formed detection structure is also called sandwich structure, one antibody acts as a capture antibody to the analyte, while the second enzyme-labeled antibody is bound to the analyte to form the sandwich structure. In the photoelectrochemical immunosensor, the detection mechanism based on the sandwich structure is also effective.

Zhang and his coworker reported a photoelectrochemical immunosensorusing Au-doped TiO_2 nanotube arrays for the detection of α-synuclein (α-SYN) [99]. As shown in Fig. 4.5, the TiO_2 nanotubes were firstly fabricated by using an electrochemical anodization technique, and then Au nanoparticles were synthesized within the nanotubes to form the Au-doped TiO_2 nanotubes. After that, the Au-doped TiO_2 nanotubes were firstly modified with primary antibodies (Ab_1), then bind to the target α-SYN, and finally bind to secondary antibody (Ab_2)-labeled GOx-functionalized Au nanoparticles to form a sandwich bioconjugate structure. The GOx enzyme immobilized on the prepared immunosensor could catalyze glucose in the detection solution to produce H_2O_2, which acted as a sacrificial electron donor to facilitate the generation of photocurrent. The photocurrents were proportional to the α-SYN concentrations, and the linear range of the developed immunosensor was from 50 pg mL^{-1} to 100 ng mL^{-1} with a detection limit of 34 pg mL^{-1} [99]. Kang et al. reported a label-free photoelectrochemical immunosensor and used a hybrid $CdSe_xTe_{1-x}/TiO_2$ nanotube structure for the detection of pentachlorophenol, in which pentachlorophenol antibodies were covalently conjugated on the TiO_2 nanotube played as the recognition element for the specific detection of pentachlorophenol [97]. For this photoelectrochemical immunosensor, the detection mechanism is based on a decrease in photocurrent due to an increase in the steric hindrances toward the diffusion of quencher molecules and/or photogenerated holes on the photoelectrode interface after the immunoreaction between the analyte of pentachlorophenol and its antibody modified on the $CdSe_xTe_{1-x}/TiO_2$ nanotube [97]. The linear detection range is from 1 nM to 0.3 μM with a limit of detection of 1 pM [97].

For the photoelectrochemical cytosensor, the detection mechanism is based on the blocking effect from the analyte of cells, which prevents the sacrificing electron donor to the photogenerated holes on the photoelectrode and consequently leads to the decrease of photocurrent. This is very similar to the second case of the photoelectrochemical immunosensor introduced above. Zhu group reported a photoelectrochemical cytosensor using folic acid functionalized graphene–CdS nanocomposite as the photoelectrode to detect the HeLa cells. Based on the specific interaction between folic acid and the folate receptor on the HeLa cell surface, the obtained cytosensor could sensitively detect the HeLa cell in the range 1.0×10^2–5.0×10^6 cells mL^{-1}, and the detection limit was about 100 cells mL^{-1} [105]. The utilization of the DNA aptamer for the detection of the Ramos cell was reported by Zhang et al., with the layer-by-layer assembly of CdSe semiconductor nanoparticles used as the photoelectrochemically active species, and the linear range for the detection of Ramos cell was from 160 to 1,600 cells mL^{-1} and the detection limit was 84 cells mL^{-1} [134]. Xu and her coworkers developed a photoelectrochemical cytosensor for the determination of SMMC-7721 human hepatoma carcinoma cells by using a photosensitive CdS–polyamidoamine nanocomposite photoelectrode with a detection limit of 5.0×10^3 cells mL^{-1} [135].

From the above discussion, it is clearly known that the fabrication procedure for the photoelectrochemical immunosensor and cytosensor is a little more complicated. Because the recognition elements in these immunosensors and cytosensors are specific, the detection for the analyte is very selective. For the label-free immunosensors and cytosensors, the detection mechanism is mainly based on the blocking effect; therefore, the relationship between the concentration of the analyte and the decrease of the photocurrent is constructed.

4.4.5 *Photoelectrochemical DNA Sensor and Aptasensor*

Aptamers, as new recognition elements, are folded single-stranded DNA or RNA oligonucleotide sequences with the capacity to recognize various target molecules which are generated in the systematic evolution of ligands by exponential enrichment (SELEX) process [4]. In these detections, the biorecognition relies on either complementary base-pairing between the biosensor's nucleic acid sequence and the analyte of interest or aptamers that recognize and bind to three-dimensional surfaces, such as those of proteins [3]. Nucleic acids are now becoming more popular as the biorecognition agent in the photoelectrochemical biosensors.

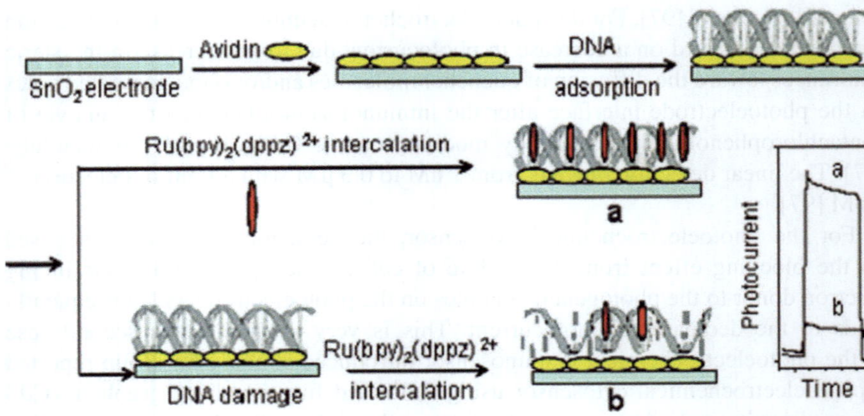

Fig. 4.6 Schematic diagram illustrating the DNA damage detection method based on the photoelectrochemical intercalator [127]

In detection of DNA damage, Guo's group developed one effective method using DNA intercalator as a signal reporter for the detection. As shown in Fig. 4.6, an avidin film and a ds-DNA film are assembled on the SnO_2 photoelectrode. A DNA intercalator, Ru-$(bpy)_2(dppz)^{2+}$ (bpy = 2,2′-bipyridine, dppz = dipyrido[3,2-a:2′,3′-c]phenazine), is employed as the photoelectrochemical signal reporter. After the DNA damage caused by styrene oxide and Fe^{2+}/H_2O_2, the DNA film bound less $Ru(bpy)_2$-$(dppz)^{2+}$, accompanied by a drop in photocurrent [127]. Using the same method, DNA damage induced by enzyme-catalyzed Fenton reaction could also be detected [126]. His group also developed a microplate-based photoelectrochemical DNA biosensor array for high throughput detection of DNA damage [139]. The strategy of using the intercalator as signal reporter was also used for the detection of adenosine triphosphate [121] and Hg^{2+} [109] based on the specific recognition interaction between aptamer and the target of adenosine triphosphate and the formation structure of thymine-Hg^{2+}-thymine, respectively. Besides $Ru(bpy)_2$-$(dppz)^{2+}$, it is also reported that the intercalation of doxorubicin into the DNA–CdS nanostructures also results in the generation of photocurrents upon illumination in the presence of triethanol amine (TEOA) [133].

The other effective method used in the photoelectrochemical DNA sensor or aptasensor is using a label or tag to generate the signal for the target analyte detection. Various materials can be used as the label like ruthenium tris(bipyridine), metal nanoparticles, and quantum dots. As shown in Fig. 4.7, the aptasensors are based on the use of two anticocaine aptamer subunits, where one subunit is assembled on an Au surface and the second aptamer subunit is labeled with CdS-NPs as the readout [122]. After the addition of cocaine, the formation of supramolecular complexes between the CdS-NPs-labeled aptamer subunits and cocaine on an Au surface allows the quantitative analysis of cocaine. The photocurrents generated by the CdS-NPs-labeled aptamer subunits–cocaine complex, in the

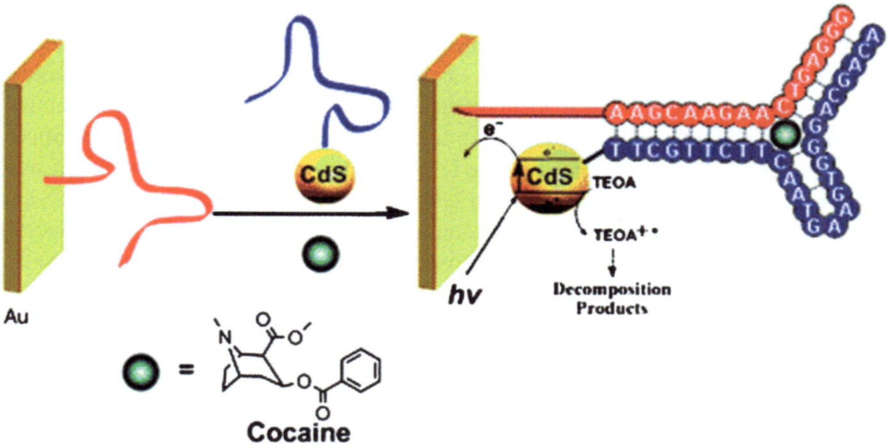

Fig. 4.7 Photoelectrochemical analysis of cocaine through the self-assembly of supramolecular complexes of CdS-NPs-functionalized aptamer subunits and Au-surfaces functionalized with the second aptamer subunit in the presence of cocaine [122]

presence of triethanol amine as a hole scavenger, allow the photoelectrochemical detection of cocaine [122]. The obtained aptasensor enabled the analysis of cocaine with a detection limit of 1×10^{-6} M [122]. Using a similar labeled strategy, Tokudome et al. reported that a nanostructured TiO_2 electrode chemisorbed with probe DNA photoelectrochemically detects a dye-labeled target DNA molecule [131]. Guo's group also reported a ruthenium tris(bipyridine)-labeled avidin film as the indicator for the rapid detection of DNA damage by virtue of the photogenerated Ru(III) oxidized guanidine and adenosine bases more easier in damaged DNA than intact DNA [125–127]. Using the energy transfer between CdS quantum dots and Au nanoparticles, a photoelectrochemical DNA sensor was constructed through modifying the probe DNA functionalized CdS quantum dots on the photoelectrode for the detection of the Au NPs labeled target DNA [129]. The linear range was from 5.0×10^{-15} to 5.0×10^{-12} M and detection limit was 2.0×10^{-15} M [129]. A novel photoelectrochemical DNA biosensor based on target-induced conformational change of DNA was reported by Zhang et al. [128]. In this biosensor, a large conformational change in DNA structure was used to tune the electron-transfer tunneling distance between the photoelectrode and the label of the ruthenium complex. The resulting change in photocurrent was proportional to the concentration of DNA in the range of 1.0×10^{-10} to 8.0×10^{-9} M, and a detection limit of 9.4×10^{-14} M of target DNA was achieved [128]. Furthermore, a photoelectrochemical method was proposed to increase the sensitivity of the detection of the DNA hybridization using Au nanoparticle labeled DNA as one probe on TiO_2 substrate [130]. The labeled strategy was also effective for the quantification of the 8-oxo-7, 8-dihydro-2′-deoxyguanosine lesions in double-stranded DNA using a photoelectrochemical DNA sensor [132].

From this section, it is clearly seen that the design strategy is more flexible for the detection of the target analyte. Whether using DNA intercalator or DNA label, the photocurrent can be tuned to increase or reduce based on the specific detection of the analyte. Like the photoelectrochemical immunosensor, the photoelectrochemical DNA sensor or aptasensor was also characterized by its high affinity and specificity to its targets. In addition, regarding the DNA or aptamer has many advantages over antibodies such as easier deposition on sensing surfaces, higher reproducibility, and easier regeneration and a higher resistance to denaturation [4], the photoelectrochemical DNA sensor or aptasensor shows more applications in clinical diagnosis and pharmaceutical analyzing.

4.4.6 Photoelectrochemical Sensor Based on Molecular Imprinted Polymers

Molecular imprinting is a very attractive and powerful technique to mimic natural molecular recognition through the preparation of the synthetic recognition sites with predetermined selectivity for various target analytes [29, 140]. In this approach, the target analyte is engaged as a template molecule that binds functional monomers by covalent or noncovalent bonding during the following polymerization process. Due to the shape and arrangement of the functional groups, the resulting macroporous polymers containing recognition sites possess the high affinity for the print molecule. The selectivity and affinities acquired from the molecular imprinting process are close to those of biological recognition elements, such as antibodies [29, 140]. The high stability coupled with the long-term stability made the molecular imprinting technique very suitable for the fabrication of sensor.

A novel photoelectrochemical sensor is reported by Zhao group on highly ordered and vertically aligned TiO_2 nanotubes (TiO_2 NTs) with modification of molecularly imprinted polymer for highly sensitive and selective determination of 2,4-dichlorophenoxyacetic acid (2,4-D) in solutions, in which molecularly imprinted polypyrrole (PPy) of 2,4-D is chosen as the recognition element, as shown in Fig. 4.8 [141]. The obtained molecularly imprinted PPy@TiO_2 NTs of the photoelectrochemical sensor demonstrate excellent applicability in 2,4-D determination, with linear range from 0.5 to 13 μM and limit of detection of 10 nM [141]. His group also uses the same technique for the fabrication of molecularly imprinted PPy@TiO_2 NTs for the detection of microcystin-LR and the linear detection range is from 0.5 to 100 $\mu g\,L^{-1}$ and limit of detection is 0.1 $\mu g\,L^{-1}$ [142].

Obviously, molecular imprinting is a relative new technique compared with other techniques in photoelectrochemical sensor. Therefore, there are few publications in photoelectrochemical sensor using this molecular imprinting technique. The advantage is some inert electrochemical analytes, which is difficult to be detected by a simple and direct electrochemical method, can use molecular imprinting for the detection. This also extends the detection scope for the photoelectrochemical sensor.

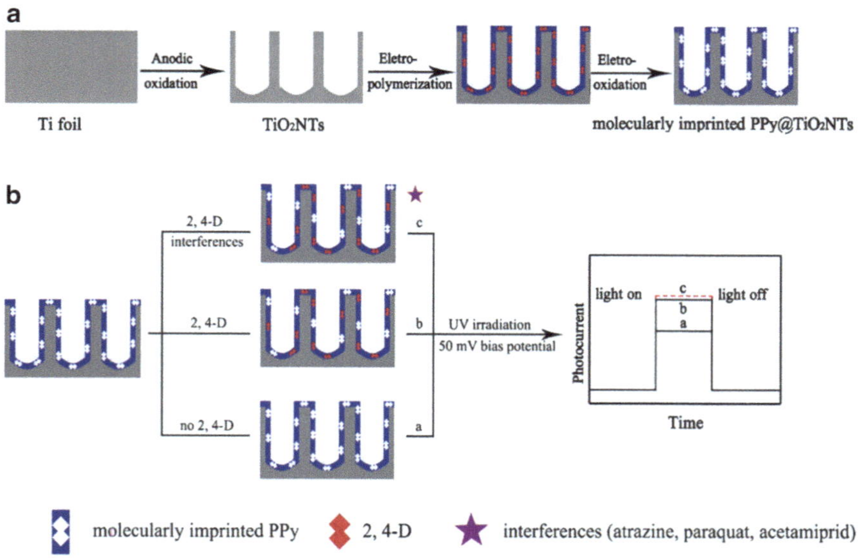

Fig. 4.8 Schematic illustration for (**a**) fabrication and (**b**) detection mechanism of the photoelectrochemical sensor [141]

4.4.7 *Photoelectrochemical Sensor for the Chemical Oxygen Demand*

The chemical oxygen demand (COD) method is preferred for assessing the oxygen demand of organic matters in heavily polluted water bodies. Thus, COD is an important parameter in water-pollution control and is also frequently used as an important index for controlling wastewater treatment plant operation [143]. During the development for the rapid and environmentally friendly new methods for the determination of COD, the photoelectrochemical oxidative degradation principle (PECOD) was proposed by Zhao et al. as a novel rapid methodology for the determination of COD [144]. PECOD possesses the advantages of both the photocatalytic method and the electrocatalytic method while overcoming many of the problems associated with the existing oxygen demand techniques. As shown in Fig. 4.9, in PECOD, the photocatalytic oxidation of organic compounds and water takes place at a TiO_2 nanoparticle-modified working electrode through application of an appropriate positive potential. The photoelectrons are subsequently passed into the external circuit and to the auxiliary electrode, where the reduction of water or oxygen takes place. The photocurrent or charge is monitored and gives a direct measure of the oxidation of organic compounds, which can be used as the analytical signal [144].

$$\mathrm{COD}\left(\mathrm{mg\,L^{-1}\,O_2}\right) = Q/(4FV) \times 32{,}000$$

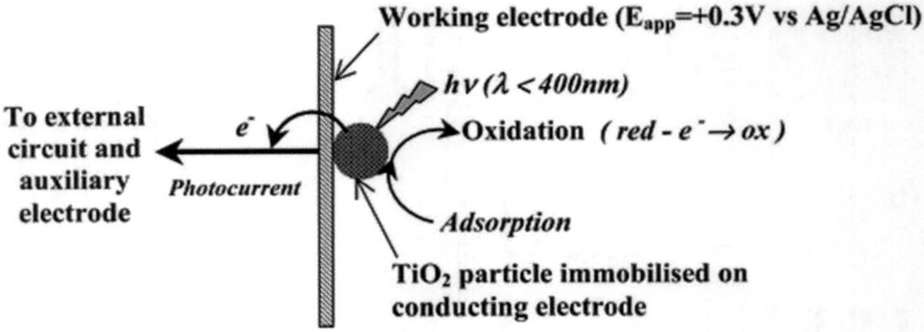

Fig. 4.9 Schematic of the photoelectrocatalytic processes involved in the analytical signal generation [144]

This equation can be used to quantify the COD value of a sample since the charge (*Q*) can be obtained experimentally, while for a given photoelectrochemical cell, the volume (*V*) is a known constant and *F* is Faraday constant [144].

Based on the principle stated above, Zhao's group has done many research works about the improvement of PECOD like using a TiO_2/boron-doped diamond heterojunction photoanode [145] or a portable photoelectrochemical probe [146] for the determination of COD in wastewaters, the development of COD online monitoring system based on a PECOD [147], photocatalytic degradation characteristics of different organic compounds at TiO_2 nanoporous film electrodes with mixed anatase/rutile phases [148], and even using a rapid photoelectrochemical method for in situ determination of effective diffusion coefficient of organic compounds [149]. Zhou and his coworkers used highly ordered and vertically aligned TiO_2 nanotube array as the photoelectrode for the determination of COD [150] and found that the photoelectrocatalytic ability of the TiO_2 nanotube array is superior to that of the standard dichromate method with much lower detection limit [151]. Yuan et al. reported a photoelectrochemical sensor through layer-by-layer assembly of TiO_2 nanoparticles for the determination of COD with a detection limit of 1 mg L^{-1} and a working range of 0–130 mg L^{-1} [38].

In this section, it is clearly known that TiO_2-based semiconductor is the first of choice as the photoelectrode material for the determination of COD. This is because TiO_2-based semiconductor is very stable and the photogenerated holes in the illuminated TiO_2-based semiconductor are a very powerful oxidizing agent for nearly all organic species found in natural waters and wastewaters [144, 151, 152]. At present, most of the modification and improvement accomplished for the determination of COD is in order to circumvent the inherent problems of the standard dichromate method like time-consuming and environmental problem and provide a rapid, in situ, and more sensitive detection for COD with a flow or portable device.

4.5 Conclusion and Outlooks

The photoelectrochemical sensors or biosensors as a main subclass of the chemical sensors have their advantages and manifest their rapid growth and demonstrate a wide and bright potential application prospect in a variety of application areas. Although the photoelectrochemical sensors or biosensors have become a hot topic and attracted increasing interests in the scientific research in the past two decades, the research on the photoelectrochemical sensors especially biosensors is still in its infant stage. Meanwhile, the rapid development of nanotechnology has brought great chance and opportunity in this field, but the further development of the photoelectrochemical sensor or biosensor is still facing great challenges. There are some aspects for the future research trend in the development of photoelectrochemical sensor or biosensor:

1. Developing the new composite semiconductor nanostructure. Although many semiconductor nanoparticles even composite semiconductor nanostructures have been used in the photoelectrochemical sensor or biosensor, there is still a large area unexplored in this field. In order to improve the separation efficiency of the photogenerated electron–hole pairs and improve the absorption property in the visible region, many core–shell nanostructures are reported like Fe_2O_3/ZnO [153, 154], TiO_2/CdS [155], $Fe_2O_3@SnO_2$ [156], and shell-in-shell TiO_2 hollow spheres [157]. Surface-doping is also a powerful tool for manipulating the fundamental optical and photoelectrochemical properties of semiconductor nanostructures. Up to now, many surface-doping semiconductors are published like N-doped TiO_2 [158, 159], N-doped $Cs_{0.68}Ti_{1.83}O_4$ [160], N-doped $CsCa_2Ta_3O_{10}$ [161], N-doped $CsTaWO_6$ [162], N-doped $Sr_2Ta_2O_7$ [163], and hydrogen-treated TiO_2 [164]. In addition, facet-controlled semiconductor as a new kind of shape-controlled nanostructure shows better photoelectrochemical properties [165–167] and will shed light on the new applications in photoelectrochemical sensor or biosensor.
2. Developing the new technique and novel detection method in the photoelectrochemical sensor or biosensor. Although many detecting methods are introduced in this chapter, the developing new technique and novel detection method is still an energetic trend in this field. Especially in the multidiscipline era, many techniques and methods can be combined together to develop new more powerful detection technique. Willner's group recently combines the technique of chemiluminescence into the photoelectrochemical biosensor for the detection of glucose and DNA with no external irradiation [168]. The other successful paradigm is the technique of molecules imprinting introduced in the photoelectrochemical sensor for the detection of some inert molecules [141, 142]. All of these new developed techniques and methods further enlarge the scope of the detection in photoelectrochemical sensor or biosensor and supplement their potential applications.
3. Developing the photoelectrochemical sensor or biosensor for the detection of analyte in the real samples. The aim of the development of the

photoelectrochemical sensor or biosensor is to solve the detection problem which appeared in life. For this, some reported photoelectrochemical sensors for the detection of organophosphorus pesticide [100, 123, 124] construct the good examples. Recent developments in the biotechnology especially in the custom engineer the biorecognition part, more and more aptamer/DNA and antibody can be designed as the biorecognition element for the detection of more specific targets. Therefore, it is more meaningful for the fabrication of the photoelectrochemical biosensor for the detection of some specific target analyte in the real samples like blood, urine, or saliva. Under this condition, it is still a challenge for the technique of the present photoelectrochemical biosensor regarding the complicated interference and detection surrounding in the real samples.

With the rapid development of immunology, biology, and nanotechnology, more and more new techniques and methods will be used in the photoelectrochemical sensor and biosensor. Although many photoelectrochemical sensor and biosensor are still in the development phase, the photoelectrochemical sensor and biosensor will play more and more important roles in many fields like environment monitoring, food safety, drug discovery, and medicine research in the future.

Acknowledgements Australian Research Council is acknowledged for its financial support through DPs to this project, and Dr. H. Chen thanks the University of Queensland for UQ Postdoctoral Fellowship support.

References

1. Eggins, B.R.: Chemical Sensors and Biosensors. John Wiley & Sons, West Sussex (2002)
2. Joo, S., Brown, R.B.: Chemical sensors with integrated electronics. Chem. Rev. **108**, 638–651 (2008)
3. Ronkainen, N.J., Halsall, H.B., Heineman, W.R.: Electrochemical biosensors. Chem. Soc. Rev. **39**, 1747–1763 (2010)
4. Grieshaber, D., MacKenzie, R., Vörös, J., Reimhult, E.: Electrochemical biosensors—sensor principles and architectures. Sensors **8**, 1400–1458 (2008)
5. Skládal, P.: Advances in electrochemical immunosensors. Electroanalysis **9**, 737–745 (1997)
6. Lazcka, O., Campo, F.J.D., Mũnoz, F.X.: Pathogen detection: a perspective of traditional methods and biosensors. Biosens. Bioelectron. **22**, 1205–1217 (2007)
7. Ivnitski, D., Abdel-Hamid, I., Atanasov, P., Wilkins, E.: Biosensors for detection of pathogenic bacteria. Biosens. Bioelectron. **14**, 599–624 (1999)
8. Liu, J., Cao, Z., Lu, Y.: Functional nucleic acid sensors. Chem. Rev. **109**, 1948–1998 (2009)
9. Moshfegh, A.Z.: Nanoparticle catalysts. J. Phys. D Appl. Phys. **42**, 233001 (2009)
10. Rajeshwar, K., de Tacconi, N.R., Chenthamarakshan, C.R.: Semiconductor-based composite materials, preparation, properties, and performance. Chem. Mater. **13**, 2765–2782 (2001)
11. Barth, S., Hernandez-Ramirez, F., Holmes, J.D., Romano-Rodriguez, A.: Synthesis and applications of one-dimensional semiconductors. Progr. Mater. Sci. **55**, 563–627 (2010)
12. Katz, E., Willner, I., Wang, J.: Electroanalytical and bioelectroanalytical systems based on metal and semiconductor nanoparticles. Electroanalysis **16**, 19–44 (2004)

13. Wang, G., Xu, J., Chen, H.: Progress in the studies of photoelectrochemical sensors. Sci. Chin. Ser. B Chem. **52**, 1789–1800 (2009)
14. Kamat, P.V.: Meeting the clean energy demand, nanostructure architectures for solar energy conversion. J. Phys. Chem. C **111**, 2834–2860 (2007)
15. Chen, X., Mao, S.S.: Titanium dioxide nanomaterials, synthesis, properties, modifications, and applications. Chem. Rev. **107**, 2891–2959 (2007)
16. Daniel, M.C., Astruc, D.: Gold nanoparticles, assembly, supramolecular chemistry, quantum-size-related properties, and applications toward biology, catalysis, and nanotechnology. Chem. Rev. **104**, 293–346 (2004)
17. Tiano, A.L., Koenigsmann, C., Santullia, A.C., Wong, S.S.: Solution-based synthetic strategies for one-dimensional metal-containing nanostructures. Chem. Commun. **46**, 8093–8130 (2010)
18. Burda, C., Chen, X., Narayanan, R., El-Sayed, M.A.: Chemistry and properties of nanocrystals of different shapes. Chem. Rev. **105**, 1025–1102 (2005)
19. Hartland, G.V.: Optical studies of dynamics in noble metal nanostructures. Chem. Rev. **111**, 3858–3887 (2011)
20. Shao, M., Ma, D.D.D., Lee, S.: Silicon nanowires—synthesis, properties, and applications. Eur. J. Inorg. Chem. **2010**, 4264–4278 (2010)
21. Roy, P., Berger, S., Schmuki, P.: TiO_2 nanotubes, synthesis and applications. Angew. Chem. Int. Ed. **50**, 2904–2939 (2011)
22. Liu, B., Aydil, E.S.: Growth of oriented single-crystalline rutile TiO_2 nanorods on transparent conducting substrates for dye-sensitized solar cells. J. Am. Chem. Soc. **131**, 3985–3990 (2009)
23. Joo, J., Chow, B.Y., Prakash, M., Boyden, E.S., Jacobson, J.M.: Face-selective electrostatic control of hydrothermal zinc oxide nanowire synthesis. Nat. Mater. **10**, 596–601 (2011)
24. Sheeney-Haj-Khia, L., Basnar, B., Willner, I.: Efficient generation of photocurrents by using CdS/carbon nanotube assemblies on electrodes. Angew. Chem. Int. Ed. **44**, 78–83 (2005)
25. Sheeney-Haj-Ichia, L., Pogorelova, S., Gofer, Y., Willner, I.: Enhanced photoelectrochemistry in CdS/Au nanoparticle bilayers. Adv. Funct. Mater. **14**, 416–424 (2004)
26. Kissinger, P.T., Heineman, W.R.: Laboratory Techniques in Electroanalytical Chemistry. Marcel Dekker Inc, New York, NY (1996)
27. Chaubey, A., Malhotra, B.D.: Mediated biosensors. Biosens. Bioelectron. **17**, 441–456 (2002)
28. Bakker, E., Pretsch, E.: Potentiometric sensors for trace-level analysis. Trends Anal. Chem. **24**, 199–207 (2005)
29. Wang, J.: Analytical Electrochemistry. Wiley, Hoboken, NJ (2006)
30. Bratov, A., Abramova, N., Ipatov, A.: Recent trends in potentiometric sensor arrays—a review. Anal. Chim. Acta **678**, 149–159 (2010)
31. Bard, A.J., Inzelt, G., Scholz, F.: Electrochemical Dictionary. Springer, Berlin (2008)
32. Yu, J., Zhao, X., Zhao, Q.: Photocatalytic activity of nanometer TiO_2 thin films prepared by the sol–gel method. Mater. Chem. Phys. **69**, 25–29 (2001)
33. Qiu, J., Zhang, S., Zhao, H.: Nanostructured TiO_2 photocatalysts for the determination of organic pollutants. J. Hazard. Mater. **211**, 381–388 (2012)
34. Kasanen, J., Suvanto, M., Pakkanen, T.T.: Self-cleaning, titanium dioxide based, multilayer coating fabricated on polymer and glass surfaces. J. Appl. Polym. Sci. **111**, 2597–2606 (2009)
35. Bessergenev, V.G., Pereira, R.J.F., Mateus, M.C., Khmelinskii, I.V., Vasconcelos, D.A., Nicula, R., Burkel, E., Botelho do Rego, A.M., Saprykin, A.I.: Study of physical and photocatalytic properties of titanium dioxide thin films prepared from complex precursors by chemical vapour deposition. Thin Solid Films **503**, 29–39 (2006)
36. Sirghi, L., Aoki, T., Hatanaka, Y.: Hydrophilicity of TiO_2 thin films obtained by radio frequency magnetron sputtering deposition. Thin Solid Films **422**, 55–61 (2002)
37. Karuppuchamy, S., Jeong, J.M., Amalnerkar, D.P., Minoura, H.: Photoinduced hydrophilicity of titanium dioxide thin films prepared by cathodic electrodeposition. Vacuum **80**, 494–498 (2006)

38. Yuan, S., Mao, R., Li, Y., Zhang, Q., Wang, H.: Layer-by-layer assembling TiO_2 film from anatase TiO_2 sols as the photoelectrochemical sensor for the determination of chemical oxygen demand. Electrochim. Acta **60**, 347–353 (2012)
39. Bard, A.J., Faulkner, L.R.: Electrochemical Methods: Fundamentals and Applications. John Wiley & Sons Inc, New York (2001)
40. Hafeman, D.G., Parce, J.W., McConnell, H.M.: Light-addressable potentiometric sensor for biochemical systems. Science **240**, 1182–1185 (1988)
41. Schöning, M.J., Poghossiana, A.: Bio FEDs (field-effect devices), state-of-the-art and new directions. Electroanalysis **18**, 1893–1900 (2006)
42. Owicki, J.W., Bousse, L.J., Hafeman, D.G., Kirk, G.L., Olson, J.D., Wada, H.G., Parce, J.W.: The light-addressable potentiometric sensor, principles and biological applications. Ann. Rev. Biophys. Biomol. Struct. **23**, 87–114 (1994)
43. Hafner, F.: Cytosensor® Microphysiometer, technology and recent applications. Biosens. Bioelectron. **15**, 149–158 (2000)
44. Schöning, M.J.: "Playing around" with field-effect sensors on the basis of EIS structures. LAPS and ISFETs. Sensors **5**, 126–138 (2005)
45. Vlasov, Y.G., Tarantov, Y.A., Bobrov, P.V.: Analytical characteristics and sensitivity mechanisms of electrolyte-insulator-semiconductor system-based chemical sensors—a critical review. Anal. Bioanal. Chem. **376**, 788–796 (2003)
46. Yu, H., Liu, Q., Wang, P.: Light Addressable Potentiometric Sensor as Cell-Based Biosensors for Biomedical Application, Biosensors for Health, Environment and Biosecurity. In: Prof. Pier Andrea Serra (ed.) ISBN: 978-953-307-443-6. InTech (2011)
47. Schöning, M.J., Poghossian, A.: Recent advances in biologically sensitive field-effect transistors (BioFETs). Analyst **127**, 1137–1151 (2002)
48. Chin, C., Lu, T., Wang, J., Yang, J., Lue, C., Yang, C., Li, S., Lai, C.: Effects of CF_4 plasma treatment on pH and pNa sensing properties of light-addressable potentiometric sensor with a 2-nm-thick sensitive HfO_2 layer grown by atomic layer deposition. Jpn. J. Appl. Phys. **50**, 04DL06 (2011)
49. Seki, A., Motoya, K., Watanabe, S., Kubo, I.: Novel sensors for potassium, calcium and magnesium ions based on a silicon transducer as a light-addressable potentiometric sensor. Anal. Chim. Acta **382**, 131–136 (1999)
50. Yoshinobu, T., Schöning, M.J., Otto, R., Furuichi, K., Mourzina, Y., Ermolenko, Y., Iwasaki, H.: Portable light-addressable potentiometric sensor (LAPS) for multisensor applications. Sensor. Actuat. B **95**, 352–356 (2003)
51. Yoshinobu, T., Iwasaki, H., Ui, Y., Furuichi, K., Ermolenko, Y., Mourzina, Y., Wagner, T., Näther, N., Schöning, M.J.: The light-addressable potentiometric sensor for multi-ion sensing and imaging. Methods **37**, 94–102 (2005)
52. Ermolenko, Y., Yoshinobu, T., Mourzina, Y., Levichev, S., Furuichi, K., Vlasov, Y., Schöning, M.J., Iwasaki, H.: Photocurable membranes for ion-selective light-addressable potentiometric sensor. Sensor. Actuat. B **85**, 79–95 (2002)
53. Vlasov, Y.G., Bychkov, E.A.: Ion-selective chalcogenide glass electrodes. Ion-selective Electrode Rev. **9**, 5–93 (1987)
54. Tarantov, Y.A., Vlasov, Y.G., Mesentzev, Y.A., Averyanov, Y.L.: Physical and chemical processes in ISFETS with chalcogenide membranes. Sensor. Actuat. B **1**, 390–394 (1990)
55. Guessous, A., Sarradin, J., Papet, P., et al.: Chemical microsensors based on chalcogenide glasses for the detection of cadmium ions in solution. Sensor. Actuat. B **53**, 13–18 (1998)
56. Schöning, M.J., Schmidt, C., Schubert, J., Zander, W., Mesters, S., Kordos, P., Lüth, H., Legin, A., Seleznev, B., Vlasov, Y.G.: Thin film sensors on the basis of chalcogenide glass materials prepared by pulsed laser deposition technique. Sensor. Actuat. B **68**, 254–259 (2000)
57. Mourzina, Y., Yoshinobu, T., Schubert, J., Lüth, H., Iwasaki, H., Schöning, M.J.: Ion-selective light-addressable potentiometric sensor (LAPS) with chalcogenide thin film prepared by pulsed laser deposition. Sensor. Actuat. B **80**, 136–140 (2001)
58. Schöning, M.J., Kloocka, J.P.: About 20 years of silicon-based thin-film sensors with chalcogenide glass materials for heavy metal analysis, technological aspects of fabrication and miniaturization. Electroanalysis **19**, 2029–2038 (2007)

59. Ismail, A.B., Furuichi, K., Yoshinobu, T., Iwasaki, H.: Light-addressable potentiometric fluoride (F^-) sensor. Sensor. Actuat. B **86**, 94–97 (2002)
60. Mourzina, Y.G., Ermolenko, Y.E., Yoshinobu, T., Vlasov, Y., Iwasaki, H., Schöning, M.J.: Anion-selective light-addressable potentiometric sensors for the determination of nitrate and sulphate ions. Sensor. Actuat. B **91**, 32–38 (2003)
61. Licht, S., Myung, N., Sun, Y.: A light addressable photoelectrochemical cyanide sensor. Anal. Chem. **68**, 954–959 (1996)
62. Lue, C., Lai, C., Wang, J., Wu, C., Yang, C.: Differential light addressable potentiometric sensor with poly(vinyl chloride) and HfO_2 membranes for pH sensors. Jpn. J. Appl. Phys. **49**, 04DL10 (2010)
63. Ismaila, A.B., Sugiharab, H., Yoshinobua, T., Iwasaki, H.: A novel low-noise measurement principle for LAPS and its application to faster measurement of pH. Sensor. Actuat. B **74**, 112–116 (2001)
64. Ismaila, A.B., Harada, T., Yoshinobu, T., Iwasaki, H., Schöning, M.J., Lüth, H.: Investigation of pulsed laser-deposited Al_2O_3 as a high pH-sensitive layer for LAPS-based biosensing applications. Sensor. Actuat. B **71**, 169–172 (2000)
65. Siqueira, J.R., Werner, C.F., Bäcker, M., Poghossian, A., Zucolotto, V., Oliveira, O.N., Schöning, M.J.: Layer-by-layer assembly of carbon nanotubes incorporated in light-addressable potentiometric sensors. J. Phys. Chem. C **113**, 14765–14770 (2009)
66. Yoshinobu, T., Harada, T., Iwasaki, H.: Application of the pH-imaging sensor to determining the diffusion coefficients of ions in electrolytic solutions. Jpn. J. Appl. Phys. **39**, L318–L320 (2000)
67. Yoshimi, Y., Matsuda, T., Itoh, Y., Ogata, F., Katsube, T.: Surface modifications of functional electrodes of a light addressable potentiometric sensor (LAPS), non-dependency of pH sensitivity on the surface functional group. Mater. Sci. Eng. C **5**, 131–139 (1997)
68. Nakao, M., Yoshinobu, T., Iwasaki, H.: Scanning-laser-beam semiconductor pH-imaging sensor. Sensor. Actuat. B **20**, 119–123 (1994)
69. Suzurikawa, J., Nakao, M., Kanzaki, R., Takahashi, H.: Microscale pH gradient generation by electrolysis on a light-addressable planar electrode. Sensor. Actuat. B **149**, 205–211 (2010)
70. Sekia, A., Ikeda, S., Kubo, I., Karube, I.: Biosensors based on light-addressable potentiometric sensors for urea, penicillin and glucose. Anal. Chim. Acta **373**, 9–13 (1998)
71. Bousse, L.J., Mostarshed, S., Hafeman, D.: Combined measurement of surface potential and zeta potential at insulator/electrolyte interfaces. Sensor. Actuat. B **10**, 67–71 (1992)
72. Inoue, S., Nakao, M., Yoshinobu, T., Iwasaki, H.: Chemical-imaging sensor using enzyme. Sensor. Actuat. B **32**, 23–26 (1996)
73. PiTasa, L., Adami, M., Fenu, S., Dovis, M., Nicolini, C.: Immunoenzymatic application of a redox potential biosensor. Anal. Chim. Acta **335**, 127–135 (1996)
74. Wagner, T., Werner, C., Miyamoto, K., Schöning, M., Yoshinobu, T.: A high-density multipoint LAPS set-up using a VCSEL array and FPGA control. Proc. Chem. **1**, 1483–1486 (2009)
75. Miyamoto, K., Kaneko, K., Matsuo, A., Wagner, T., Kanoh, S., Schöning, M.J., Yoshinobu, T.: Miniaturized chemical imaging sensor system using a OLED display panel. Proc. Eng. **5**, 516–519 (2010)
76. Wagnera, T., Werner, C.F., Miyamoto, K., Schöning, M.J., Yoshinobu, T.: Development and characterisation of a compact light-addressable potentiometric sensor (LAPS) based on the digital light processing (DLP) technology for flexible chemical imaging. Sensor. Actuat. B **170**, 34–39 (2012)
77. Yoshinobu, T., Schöning, M.J., Finger, F., Moritz, W., Iwasaki, H.: Fabrication of thin-film LAPS with amorphous silicon. Sensors **4**, 163–169 (2004)
78. Ito, Y.: High-spatial resolution LAPS. Sensor. Actuat. B **52**, 107–111 (1998)
79. George, M., Parak, W.J., Gerhardt, I., Moritz, W., Kaesen, F., Geiger, H., Eisele, I., Gaub, H.E.: Investigation of the spatial resolution of the light-addressable potentiometric sensor. Sensor. Actuat. **86**, 187–196 (2000)

80. Chen, P., Liu, X., Wang, B., Cheng, G., Wang, P.: A biomimetic taste receptor cell-based biosensor for electrophysiology recording and acidic sensation. Sensor. Actuat. B **139**, 576–583 (2009)
81. Hu, N., Ha, D., Wu, C., Zhou, J., Kirsanov, D., Legin, A., Wang, P.: A LAPS array with low cross-talk for non-invasive measurement of cellular metabolism. Sensor. Actuat. A **187**, 50–56 (2012)
82. Stein, B., George, M., Gaub, H.E., Behrends, J.C., Parak, W.J.: Spatially resolved monitoring of cellular metabolic activity with a semiconductor-based biosensor. Biosensor. Bioelectron. **18**, 31–41 (2003)
83. Werner, C.F., Krumbe, C., Schumacher, K., Groebel, S., Spelthahn, H., Stellberg, M., Wagner, T., Yoshinobu, T., Selmer, T., Keusgen, M., Baumann, M.E.M., Schöning, M.J.: Determination of the extracellular acidification of Escherichia coli by a light-addressable potentiometric sensor. Phys. Status Solidi A **208**, 1340–1344 (2011)
84. Stein, B., George, M., Gaub, H.E., Parak, W.J.: Extracellular measurements of averaged ionic currents with the light-addressable potentiometric sensor. Sensor. Actuat. B **98**, 299–304 (2004)
85. Seki, A., Kawakubo, K., Iga, M., Nomura, S.: Microbial assay for tryptophan using silicon-based transducer. Sensor. Actuat. B **94**, 253–256 (2003)
86. Liu, Q., Hu, N., Zhang, F., Wang, H., Ye, W., Wang, P.: Neurosecretory cell-based biosensor, monitoring secretion of adrenal chromaffin cells by local extracellular acidification using light-addressable potentiometric sensor. Biosensor. Bioelectron. **35**, 421–424 (2012)
87. Werner, C.F., Groebel, S., Krumbe, C., Wagner, T., Selmer, T., Yoshinobu, T., Baumann, M.E.M., Keusgen, M., Schöning, M.J.: Nutrient concentration-sensitive microorganism-based biosensor. Phys. Status Solidi A **209**, 900–904 (2012)
88. Yu, H., Cai, H., Zhang, W., Xiao, L., Liu, Q., Wang, P.: A novel design of multifunctional integrated cell-based biosensors for simultaneously detecting cell acidification and extracellular potential. Biosensor. Bioelectron. **24**, 1462–1468 (2009)
89. Yoshinobu, T., Ecken, H., Ismail, A.B., Iwasaki, H., Lüth, H., Schöning, M.J.: Chemical imaging sensor and its application to biological systems. Electrochim. Acta **47**, 259–263 (2001)
90. Dill, K., Bearden, D.W.: Detection of human asialo-alpha(1)-acid glycoprotein using a heterosandwich immunoassay in conjunction with the light addressable potentiometric sensor. Glycoconj. J. **13**, 637–641 (1996)
91. Tu, S.I., Uknalis, J., Irwin, P., Yu, L.S.L.: The use of streptavidin coated magnetic beads for detecting pathogenic bacteria by light addressable potentiometric sensor (LAPS). J. Rapid Meth. Autom. Microbiol. **8**, 95–109 (2000)
92. Shi, H., Zhao, G., Cao, T., Liu, M., Guan, C., Huang, X., Zhu, Z., Yang, N., Williams, O.A.: Selective and visible-light-driven profenofos sensing with calixarene receptors on TiO_2 nanotube film electrodes. Electrochem. Commun. **19**, 111–114 (2012)
93. Chen, D., Zhang, H., Li, X., Li, J.H.: Biofunctional titania nanotubes for visible-light-activated photoelectrochemical biosensing. Anal. Chem. **82**, 2253–2261 (2010)
94. Chiu, Y., Lee, C.: pH Sensor investigation of various-length photoelectrochemical passivated ZnO nanorod arrays. J. Electrochem. Soc. **158**, J282–J285 (2011)
95. Wang, G., Wang, Q., Lu, W., Li, J.: Photoelectrochemical study on charge transfer properties of TiO_2-B nanowires with an application as humidity sensors. J. Phys. Chem. B **110**, 22029–22034 (2006)
96. Gong, J., Wang, X., Li, X., Wang, K.: Highly sensitive visible light activated photoelectrochemical biosensing of organophosphate pesticide using biofunctional crossed bismuth oxyiodide flake arrays. Biosensor. Bioelectron. **38**, 43–49 (2012)
97. Kang, Q., Yang, L., Chen, Y., Luo, S., Wen, L., Cai, Q., Yao, S.: Photoelectrochemical detection of pentachlorophenol with a multiple hybrid $CdSe_xTe_{1-x}/TiO_2$ nanotube structure-based label-free immunosensor. Anal. Chem. **82**, 9749–9754 (2010)
98. Yue, Z., Zhang, W., Wang, C., Liu, G., Niu, W.: CdS-FePt dimers based photoelectrochemical sensor for detection of H_2O_2. Mater. Lett. **74**, 180–182 (2012)

99. An, Y., Tang, L., Jiang, X., Chen, H., Yang, M., Jin, L., Zhang, S., Wang, C., Zhang, W.: A photoelectrochemical immunosensor based on Au-doped TiO_2 nanotube arrays for the detection of α-synuclein. Chem. Eur. J. **16**, 14439–14446 (2010)
100. Li, H., Li, J., Xu, Q., Hu, X.: Poly(3-hexylthiophene)/TiO_2 nanoparticle-functionalized electrodes for visible light and low potential photoelectrochemical sensing of organophosphorus pesticide chlorpyrifos. Anal. Chem. **83**, 9681–9686 (2011)
101. Zayats, M., Kharitonov, A.B., Pogorelova, S.P., Lioubashevski, O., Katz, E., Willner, I.: Probing photoelectrochemical processes in Au-CdS nanoparticle arrays by surface plasmon resonance, application for the detection of acetylcholine esterase inhibitors. J. Am. Chem. Soc. **125**, 16006–16014 (2003)
102. Wang, J., Han, Y., Feng, M., Chen, J., Li, X., Zhang, S.: Preparation and photoelectrochemical characterization of WO_3/TiO_2 nanotube array electrode. J. Mater. Sci. **46**, 416–421 (2011)
103. Cao, C., Hua, C., Shen, W., Wang, S., Tian, Y., Wang, X.: Synthesis and characterization of TiO_2/CdS core-shell nanorod arrays and their photoelectrochemical property. J. Alloys Compd. **523**, 139–145 (2012)
104. Zhao, X., Zhou, S., Shen, Q., Jiang, L., Zhu, J.: Fabrication of glutathione photoelectrochemical biosensor using graphene-CdS nanocomposites. Analyst **137**, 3697–3703 (2012)
105. Zhao, X., Zhou, S., Jiang, L., Hou, W., Shen, Q., Zhu, J.: Graphene–CdS nanocomposites, facile one-Step synthesis and enhanced photoelectrochemical cytosensing. Chem. Eur. J. **18**, 4974–4981 (2012)
106. Wang, K., Wu, J., Liu, Q., Jin, Y., Yan, J., Cai, J.: Ultrasensitive photoelectrochemical sensing of nicotinamide adenine dinucleotide based on graphene-TiO_2 nanohybrids under visible irradiation. Anal. Chim. Acta **745**, 131–136 (2012)
107. Shen, Q., Zhao, X., Zhou, S., Hou, W., Zhu, J.: ZnO/CdS hierarchical nanospheres for photoelectrochemical sensing of Cu^{2+}. J. Phys. Chem. C **115**, 17958–17964 (2011)
108. Wang, G., Xu, J., Chen, H.: Selective detection of trace amount of Cu^{2+} using semiconductor nanoparticles in photoelectrochemical analysis. Nanoscale **2**, 1112–1114 (2010)
109. Zhang, B., Guo, L.: Highly sensitive and selective photoelectrochemical DNA sensor for the detection of Hg^{2+} in aqueous solutions. Biosensor. Bioelectron. **37**, 112–115 (2012)
110. Chamier, J., Crouch, A.M.: Improved photoelectrochemical detection of mercury (II) with a TiO_2-modified composite photoelectrode. Mater. Chem. Phys. **132**, 10–16 (2012)
111. Chamier, J., Leaner, J., Crouch, A.M.: Photoelectrochemical determination of inorganic mercury in aqueous solutions. Anal. Chim. Acta **661**, 91–96 (2010)
112. Wang, W., Bao, L., Lei, J., Tu, W., Ju, H.: Visible light induced photoelectrochemical biosensing based on oxygen-sensitive quantum dots. Anal. Chim. Acta **744**, 33–38 (2012)
113. Zheng, M., Cui, Y., Li, X., Liu, S., Tang, Z.: Photoelectrochemical sensing of glucose based on quantum dot and enzyme nanocomposites. J. Electroanal. Chem. **656**, 167–173 (2011)
114. Wen, D., Guo, S., Wang, Y., Dong, S.: Bifunctional nanocatalyst of bimetallic nanoparticle/TiO_2 with enhanced performance in electrochemical and photoelectrochemical applications. Langmuir **26**, 11401–11406 (2010)
115. Cohen, C.B., Weber, S.G.: Photoelectrochemical sensor for catalase activity based on the in situ generation and detection of substrate. Anal. Chem. **65**, 169–175 (1993)
116. Zhao, W., Yu, P., Xu, J., Chen, H.: Ultrasensitive photoelectrochemical biosensing based on biocatalytic deposition. Electrochem. Commun. **13**, 495–497 (2011)
117. Li, H., Gao, Q., Chen, L., Hao, W.: Photocurrent determination ascorbic acid using an n-silicon electrode modified by platinum and cobalt hexacyanoferrate films. Sensor. Actuat. B **173**, 540–546 (2012)
118. Zhao, C., Zhao, G., Zhang, Z., Liang, J.: Preparation and application of photoelectrochemical guanine sensor. Acta Chim. Sin. **70**, 1401–1406 (2012)
119. Wang, G., Xu, J., Chen, H.: Dopamine sensitized nanoporous TiO_2 film on electrodes, photoelectrochemical sensing of NADH under visible irradiation. Biosensor. Bioelectron. **24**, 2494–2498 (2009)
120. Feng, J., Yang, P., Wang, S., Wang, J.: Photoelectrochemical chiral sensing on the basis of TiO_2–metal complex hybrid film. J. Electroanal. Chem. **674**, 97–102 (2012)

121. Zhang, X., Zhao, Y., Li, S., Zhang, S.: Photoelectrochemical biosensor for detection of adenosine triphosphate in the extracts of cancer cells. Chem. Commun. **46**, 9173–9175 (2010)
122. Golub, E., Pelossof, G., Freeman, R., Zhang, H., Willner, I.: Electrochemical, photoelectrochemical, and surface plasmon resonance detection of cocaine using supramolecular aptamer complexes and metallic or semiconductor nanoparticles. Anal. Chem. **81**, 9291–9298 (2009)
123. Shi, H., Zhao, G., Cao, T., Liu, M., Guan, C., Huang, X., Zhu, Z., Yang, N., Williams, O.A.: Selective and visible-light-driven profenofos sensing with calixarene receptors on TiO_2 nanotube film electrodes. Electrochem. Commun. **19**, 111–114 (2012)
124. Li, H., Li, J., Yang, Z., Xu, Q., Hu, X.: A novel photoelectrochemical sensor for the organophosphorus pesticide dichlofenthion based on nanometer-sized titania coupled with a screen-printed electrode. Anal. Chem. **83**, 5290–5295 (2011)
125. Liang, M., Liu, S., Wei, M., Guo, L.: Photoelectrochemical oxidation of DNA by ruthenium tris(bipyridine) on a tin oxide nanoparticle electrode. Anal. Chem. **78**, 621–623 (2006)
126. Liang, M., Jia, S., Zhu, S., Guo, L.: Photoelectrochemical sensor for the rapid detection of in situ DNA damage induced by enzyme-catalyzed fenton reaction. Environ. Sci. Technol. **42**, 635–39 (2008)
127. Liang, M., Guo, L.: Photoelectrochemical DNA sensor for the rapid detection of DNA damage induced by styrene oxide and the fenton reaction. Environ. Sci. Technol. **41**, 658–664 (2007)
128. Zhang, X., Xu, Y., Zhao, Y., Song, W.: A new photoelectrochemical biosensors based on DNA conformational changes and isothermal circular strand-displacement polymerization reaction. Biosensor. Bioelectron. **39**, 338–341 (2013)
129. Zhao, W., Wang, J., Xu, J., Chen, H.: Energy transfer between CdS quantum dots and Au nanoparticles in photoelectrochemical detection. Chem. Commun. **47**, 10990–10992 (2011)
130. Lu, W., Jin, Y., Wang, G., Chen, D., Li, J.: Enhanced photoelectrochemical method for linear DNA hybridization detection using Au-nanoparticle labeled DNA as probe onto titanium dioxide electrode. Biosensor. Bioelectron. **23**, 1534–1539 (2008)
131. Tokudome, H., Yamada, Y., Sonezaki, S., Ishikawa, H., Bekki, M., Kanehira, K., Miyauchi, M.: Photoelectrochemical deoxyribonucleic acid sensing on a nanostructured TiO_2 electrode. Appl. Phys. Lett. **87**, 213901 (2005)
132. Zhang, B., Guo, L., Greenberg, M.M.: Quantification of 8-OxodGuo lesions in double-stranded DNA using a photoelectrochemical DNA sensor. Anal. Chem. **84**, 6048–6053 (2012)
133. Freeman, R., Gill, R., Beissenhirtz, M., Willner, I.: Self-assembly of semiconductor quantum-dots on electrodes for photoelectrochemical biosensing. Photochem. Photobiol. Sci. **6**, 416–422 (2007)
134. Zhang, X., Li, S., Jin, X., Li, X.: Aptamer based photoelectrochemical cytosensor with layer-by-layer assembly of CdSe semiconductor nanoparticles as photoelectrochemically active species. Biosensor. Bioelectron. **26**, 3674–3678 (2011)
135. Qian, Z., Bai, H., Wang, G., Xu, J., Chen, H.: A photoelectrochemical sensor based on CdS-polyamidoamine nano-composite film for cell capture and detection. Biosensor. Bioelectron. **25**, 2045–2050 (2010)
136. Copeland, R.A.: Enzymes, a Practical Introduction to Structure, Mechanism, and Data Analysis. A John Wiley & Sons, Inc, New York (2000)
137. Curria, M.L., Agostiano, A., Leo, G., Mallardi, A., Cosma, P., Monica, M.D.: Development of a novel enzyme/semiconductor nanoparticles system for biosensor application. Mater. Sci. Eng. C **22**, 449–452 (2002)
138. Haab, B.B.: Applications of antibody array platforms. Curr. Opin. Biotechnol. **17**, 415–421 (2006)
139. Liu, Y., Jia, S., Guo, L.: Development of microplate-based photoelectrochemical DNA biosensor array for high throughput detection of DNA damage. Sensor. Actuat. B **161**, 334–340 (2012)
140. Kriz, D., Ramstrom, O., Mosbach, K.: Molecular imprinting—new possibilities for sensor technology. Anal. Chem. **69**, A345–A349 (1997)

141. Shi, H., Zhao, G., Liu, M., Zhu, Z.: A novel photoelectrochemical sensor based on molecularly imprinted polymer modified TiO_2 nanotubes and its highly selective detection of 2,4-dichlorophenoxyacetic acid. Electrochem. Commun. **13**, 1404–1407 (2011)
142. Chen, K., Liu, M., Zhao, G., Shi, H., Fan, L., Zhao, S.: Fabrication of a novel and simple microcystin-LR photoelectrochemical sensor with high sensitivity and selectivity. Environ. Sci. Technol. **46**, 11955–11961 (2012)
143. Hua, W., Yong, P.: Investigation of Internal Pollutant Loading for a tide-influenced waterbody. Water Environ. Res. **81**, 2437–2446 (2009)
144. Zhao, H., Jiang, D., Zhang, S., Catterall, K., John, R.: Development of a direct photoelectrochemical method for determination of chemical oxygen demand. Anal. Chem. **76**, 155–160 (2004)
145. Han, Y., Qiu, J., Miao, Y., Han, J., Zhang, S., Zhang, H., Zhao, H.: Robust TiO_2/BDD heterojunction photoanodes for determination of chemical oxygen demand in wastewaters. Anal. Methods **3**, 2003–2009 (2011)
146. Zhang, S., Li, L., Zhao, H.: A portable photoelectrochemical probe for rapid determination of chemical oxygen demand in wastewaters. Environ. Sci. Technol. **43**, 7810–7815 (2009)
147. Zhang, S., Jiang, D., Zhao, H.: Development of chemical oxygen demand on-line monitoring system based on a photoelectrochemical degradation principle. Environ. Sci. Technol. **40**, 2363–2368 (2006)
148. Jiang, D., Zhang, S., Zhao, H.: Photocatalytic degradation characteristics of different organic compounds at TiO_2 nanoporous film electrodes with mixed anatase/rutile phases. Environ. Sci. Technol. **41**, 303–308 (2007)
149. Wen, W., Zhao, H., Zhang, S., Pires, V.: Rapid photoelectrochemical method for in situ determination of effective diffusion coefficient of organic compounds. J. Phys. Chem. C **112**, 3875–3880 (2008)
150. Zheng, Q., Zhou, B., Bai, J., Li, L., Jin, Z., Zhang, J., Li, J., Liu, Y., Cai, W., Zhu, X.: Self-organized TiO_2 nanotube array sensor for the determination of chemical oxygen demand. Adv. Mater. **20**, 1044–1049 (2008)
151. Chen, H., Zhang, J., Chen, Q.: Assessment of a COD analytical method based on the photoelectrocatalysis of a TiO_2 nanotube array sensor. Anal. Methods **4**, 1790–1796 (2012)
152. Kim, Y., Sasaki, S., Yano, K., Ikebukuro, K., Hashimoto, K., Karube, I.: A flow method with photocatalytic oxidation of dissolved organic matter using a solid-phase (TiO_2) reactor followed by amperometric detection of consumed oxygen. Anal. Chem. **74**, 3858–3864 (2002)
153. Yan, W., Fan, H., Yang, C.: Ultra-fast synthesis and enhanced photocatalytic properties of alpha-Fe_2O_3/ZnO core-shell structure. Mater. Lett. **65**, 1595–1597 (2011)
154. Si, S., Li, C., Wang, X., Peng, Q., Li, Y.: Fe_2O_3/ZnO core–shell nanorods for gas sensors. Sensor. Actuat. B **119**, 52–56 (2006)
155. Cao, C., Hua, C., Shen, W., Wang, S., Tian, Y., Wang, X.: Synthesis and characterization of TiO_2/CdS core–shell nanorod arrays and their photoelectrochemical property. J. Alloys Compd. **523**, 139–145 (2012)
156. Zhu, L., Bing, N., Yang, D., Yang, Y., Liao, G., Wang, L.: Synthesis and photocatalytic properties of core–shell structured α-Fe_2O_3@SnO_2 shuttle-like nanocomposites. CrystEngComm. **13**, 4486–4490 (2011)
157. Wu, X., Lu, G., Wang, L.: Shell-in-shell TiO_2 hollow spheres synthesized by one-pot hydrothermal method for dye-sensitized solar cell application. Energy Environ. Sci. **4**, 3565–3572 (2011)
158. Asahi, R., Morikawa, T., Ohwaki, T., Aoki, K., Taga, Y.: Visible-light photocatalysis in nitrogen-doped titanium oxides. Science **293**, 269–271 (2001)
159. Beranek, R., Kisch, H.: Tuning the optical and photoelectrochemical properties of surface-modified TiO_2. Photochem. Photobiol. Sci. **7**, 40–48 (2008)
160. Liu, G., Wang, L., Sun, C., Yan, X., Wang, X., Chen, Z., Smith, S.C., Cheng, H., Lu, G.: Band-to-band visible-light photon excitation and photoactivity induced by homogeneous nitrogen doping in layered titanates. Chem. Mater. **21**, 1266–1274 (2009)

161. Zong, X., Sun, C., Chen, Z., Mukherji, A., Wu, H., Zou, J., Smith, S.C., Lu, G., Wang, L.: Nitrogen doping in ion-exchangeable layered tantalate towards visible-light induced water oxidation. Chem. Commun. **47**, 6293–6295 (2011)
162. Mukherji, A., Marschall, R., Tanksale, A., Sun, C., Smith, S.C., Lu, G., Wang, L.: N-doped $CsTaWO_6$ as a new photocatalyst for hydrogen production from water splitting under solar irradiation. Adv. Funct. Mater. **21**, 126–132 (2011)
163. Mukherji, A., Seger, B., Lu, G., Wang, L.: Nitrogen doped $Sr_2Ta_2O_7$ coupled with graphene sheets as photocatalysts for increased photocatalytic hydrogen production. ACS Nano **5**, 3483–3492 (2011)
164. Chen, X., Liu, L., Yu, P.Y., Mao, S.S.: Increasing solar absorption for photocatalysis with black hydrogenated titanium dioxide nanocrystals. Science **331**, 746–750 (2011)
165. Wu, X., Chen, Z., Lu, G., Wang, L.: Nanosized anatase TiO_2 single crystals with tunable exposed (001) facets for enhanced energy conversion efficiency of dye-sensitized solar cells. Adv. Funct. Mater. **21**, 4167–4172 (2011)
166. Pan, J., Liu, G., Lu, G., Cheng, H.: On the true photoreactivity order of 001}, {010}, and {101} facets of anatase TiO2 crystals. Angew. Chem. Int. Ed. **50**, 2133–2137 (2011)
167. Wang, X., Liu, G., Wang, L., Pan, J., Lu, G., Cheng, H.: TiO_2 films with oriented anatase 001 facets and their photoelectrochemical behavior as CdS nanoparticle sensitized photoanodes. J. Mater. Chem. **21**, 869–873 (2011)
168. Golub, E., Niazov, A., Freeman, R., Zatsepin, M., Willner, I.: Photoelectrochemical biosensors without external irradiation, probing enzyme activities and DNA sensing using hemin/G-quadruplex-stimulated chemiluminescence resonance energy transfer (CRET) generation of photocurrents. J. Phys. Chem. C **116**, 13827–13834 (2012)

Chapter 5
ZnO Hydrogen Nanoscale Sensors

Oleg Lupan and Lee Chow

Abstract Hydrogen has the potential to play an important role in energy and automobile industries in our society. Being an explosive gas, hydrogen is also a colorless and odorless gas; therefore, there is a great need to develop hydrogen sensors for public safety reason. Zinc oxide (ZnO) has been used as hydrogen sensors for a number of years. Here we review the synthesis and fabrication of pure and doped ZnO nanoscaled materials for hydrogen sensing applications. First, we will describe both gas-phase growth and solution growth methods. Next, we describe the fabrication of nanoscaled sensors based on ZnO nanostructures and their characteristics. Several methods that were employed to enhance the nanosensor performance, such as increasing surface volume ratio, impurity doping, and surface functionalization, will be discussed as well. The effect of impurity doping on the gas response of ZnO to hydrogen will be discussed. Doped ZnO nanosystem demonstrates an enhanced gas response for the detection of hydrogen at room temperature compared to previously reported nanosensors based on pure single ZnO nanowire (NW) or multiple NWs. Finally, the sensing mechanisms will be discussed, and major conclusions and future directions will be given.

O. Lupan (✉)
Department of Physics, University of Central Florida, Orlando, FL 32816, USA

Department of Microelectronics and Semiconductor devices, Technical University of Moldova, 168 Stefan cel Mare Blvd., Chisinau MD-2004, Republic of Moldova
e-mail: lupan@physics.ucf.edu

L. Chow
Department of Physics, University of Central Florida, Orlando, FL 32816, USA
e-mail: Lee.Chow@ucf.edu

S. Li et al. (eds.), *Nanoscale Sensors*, Lecture Notes in Nanoscale Science and Technology 19, DOI 10.1007/978-3-319-02772-2_5,

5.1 Introduction

Detection of hydrogen (H_2) is an important issue to human health and safety [1–4], since it is an invisible, odorless, insipid, and extremely flammable gas. Thus, it is necessary to detect early and accurate hydrogen leakage in the environment. Hydrogen sensors are used in combustion systems of hydrogen internal combustion engine cars to monitor hydrogen leakage, to detect certain types of bacterial infections, and also in the petroleum, chemical, or semiconductor industries [5–7]. In addition, H_2 is extending quickly in fuel refinement to supplant hydrocarbons and becomes "the common fuel of the future" [8] or a possible major future fuel among other renewable energies. There is a strong need to develop novel hydrogen sensors, especially nanoscale sensors with improved performances for such applications, because it will require numerous hydrogen sensors for each unit. A nanotechnology offers possibilities to fabricate such low-dimensional sensorial structures with improved performances, low weight, and cost-efficiency by using nanoscale building blocks, e.g., metal oxide nanowires/nanorods and tripods. Metal oxide one-dimensional nanostructures show great potential for nanodevice applications [2, 3, 9–14] based on their large surface-to-volume ratio and controlled flow of confined charge carriers.

ZnO is the most promising and cost-effective materials for gas sensors, especially for H_2 sensing [15–19]. Nano-ZnO possesses large surface area, thermal and mechanical stability [20], and radiation hardness [21, 22]. The physical properties of the nano-ZnO depend on the microstructure of the materials, including crystal size, morphology, orientation, aspect ratio, and crystalline density [23]. In addition, new physical and chemical properties can be obtained upon nanocrystal doping [23–28], enabling new applications in the fields of nanoelectronics [28–30]. Sensing properties of ZnO are directly related to its preparation methodology, particle size, morphology, and operating temperature according to numerous reports. The sensor signal consists of relative resistance changes due to gas adsorption on the surface of nano-ZnO and permits real-time detection. Thus, bio-safe and biocompatible ZnO nanoarchitectures are most attractive candidates for this type of sensing and requirements and are intensively investigated nowadays. From another side, the importance of development of a nanosensor with ability to selectively detect hydrogen at room temperature is driven by the fact that, in most of the field applications, hydrogen is to be detected at the presence of other combustible gases, such as hydrocarbons and volatile organic compounds [31], with the presence of air in the ambient. Recently, the main effort of H_2 sensor development has been the improvement of H_2 gas sensitivity as well as selectivity and to decrease the operating temperature. Working at high temperatures should be avoided because the H_2 may burn in air before reaching the surface of the sensing material at high temperatures ($>500\,^{\circ}$C) [32]. Hence, the operating temperature have been decreased below 400 °C [3, 13]. Thus, zinc oxide micro/nanostructures have to prove improved performances, IC compatibility, and ability to work at room temperature in H_2 sensor applications.

New bottom-up approaches to fabricate nanosensors based on conceptually new synthesis techniques of nanoscale building blocks have been the subject of many research groups worldwide [1–4, 15–19]. However, for extensive use in an industrialized process of individual nanorod/nanowire as building nanoblock in compact, reliable, inexpensive, and low power consumption hydrogen nanosensors, a simple, inexpensive, efficient, and bio-safe synthesis process is required [33–36].

Many types of nano-ZnO-based H_2 sensors have been demonstrated [2, 3, 13, 14, 33]. For example, hydrogen-selective sensing in nano-zinc oxide-based sensors was observed on pure and Pd-coated multiple ZnO nanorods [36, 37], Pt-coated ZnO thin films and nanorods [38], ZnO nanorod arrays and networks [15, 39], multiple ZnO nanorods [15], single nanowires of metal oxides [40], metallic catalyst coatings (Pd, Pt, Au, Ag, Ti, and Ni) on multiple ZnO nanorods [41], single nanorod/nanowire ZnO [2, 3, 13], and SnO_2-coated ZnO nanorods [42]. For example, Sonawane et al. [43] reported gas-sensing properties of nanocrystalline ZnO:Cu for different concentrations of Cu and found for 1 wt% Cu in zinc oxide a higher response and selectivity to H_2. More recently, Zhao et al. [44] used electrospinning technique to fabricate Cu-doped ZnO nanofibers for H_2S sensing application. In this review, we have focused on nanofabrication of transferable single ZnO nanorods/nanowire as the building block for a high-selectivity H_2 sensor that can be operated at room temperature. Uses of catalyst or growth seed layer for the nanorod grows were not needed in our novel nanofabrication technique. We review the fabrication techniques of single ZnO nanorods/nanowire-based hydrogen sensor using focused ion beam in situ lift-out technique and other approaches. Many nanosensors are capable of detection of 100–200 ppm hydrogen in air at room temperature. Although, previous reports of single pure ZnO nanowire/nanorod-based H_2 sensors showed the promise of such devices in commercial sensing applications, their relative slow response and recovery times are major drawbacks for their implementation in *an industrial setting*. Device based on doped ZnO NW was found to display an improved performance compared to previously reported H_2 nanosensors based on pure single ZnO NWs. In particular, by doping the ZnO NWs, a reduction of their operation temperature and enhancement of their gas response and selectivity were observed. We found that Cu–ZnO sensors have enhanced RT sensitivity, faster response time, and good selectivity. Miniaturized Cu–ZnO rod-based sensors can serve as a good candidate for effective H_2 detectors with low power consumption.

5.2 Synthesis of ZnO Nanoscaled Materials

Here we will discuss some of the gas-phase growth and solution growth methods for ZnO nanostructures for nanoscale sensor applications. We will focus on techniques which synthesize nanowires/nanorods that can be transferred from initial substrate to a sensor platform and form an individual NW/NR device.

In the synthesis process reported by Lupan et al. [45], the amorphous glass or crystalline quartz were used as substrates. The cleaning procedure reported by [45] produced a uniformly wettable substrate surface. The synthesis of ZnO nanoarchitectures was carried out using zinc sulfate [$Zn(SO_4) \cdot 7H_2O$] 0.1–0.15 M and ammonia (NH_4OH) (29.6 %). The chemicals were mixed with 100 ml DI-water until complete dissolution at room temperature and became colorless. The glass, quartz, or Si substrates were placed inside an aqueous solution in a reactor and mounted on a hot plate. The temperature was quickly increased to 90–95 °C and was kept constant for 15 min without any stirring. Afterward, the whole system was left on the hot plate (turned off) and slowly cooled off from 90 to 40 °C in 30 min. The substrates with ZnO nanorod depositions were rinsed in deionized water for 2 min, and then the samples were dried in air at 150 °C for 5 min. Technological process parameters such as precursors concentrations, temperature and time, type of substrate, and the reactor design are important for the synthesis of ZnO nanorod arrays with thin diameter and transferable 1D-, 2D-, and 3D-nanoarchitectures. Due to the different growth rates for different planes: $V_{(0001)} > V_{(10\bar{1}1)} > V_{(10\bar{1}0)}$, the controlled growth of preferred nanoarchitecture for specific applications can be realized [45, 46]. The crystal synthesis on a specific surface in the aqueous growth solution is based on heterogeneous nucleation and subsequent growth. Individual ZnO rod demonstrates a hexagonal basis and its dimensions were found to depend on growth parameters.

Figure 5.1 presents SEM images of the ZnO nanorod arrays and individual microrods obtained using this method. Figure 5.1a indicates that the obtained products consist of nanorods with an average diameter of 300 nm and an average length of 2 μm. Figure 5.1b shows that ZnO nanorods with diameters of 100 nm have also been synthesized on quartz with this method. Quasi-1D, 2D, and 3D ZnO-branched microrods are presented in Fig. 5.1d–f [45]. It was found that the branched architectures can be easily transferred to other substrates and can be handled by FIB in order to fabricate different nanodevices (Fig. 5.1f). The XRD studies [45] demonstrated a crystalline ZnO with the hexagonal wurtzite structure (space group: $P6_3mc(186)$; $a = 0.3249$ nm, $c = 0.5206$ nm). No characteristic peaks of impurity phases such as Zn, S, or $Zn(OH)_2$ are observed, and no diffraction peaks except ZnO are found, which indicates that only single-phase hexagonal ZnO is present. More detailed studies of such structures can be found in [45]. As well, X-ray photoelectron spectroscopy (XPS), transmission electron microscopy (TEM), and micro-Raman spectroscopy have been used to characterize the samples, and details can be found in work [45].

The main advantages of the developed synthesis process are its simplicity and fast growth rates (15 min versus several hours reported by previous researchers using aqueous synthesis methods or others) [34, 45, 46]. The strength of this nanotechnology is that any substrate can be used to synthesize ZnO nanorods and individual 2D- or 3D-branched rods, which are transferable and can be used in nanosensors fabrication.

In a similar cost-effective technological procedure self-assembled ZnO-branched nanorods are obtained [1] for sensors. Such approach is a fast synthesis route using an aqueous solution method and efficient for near-future sensor applications,

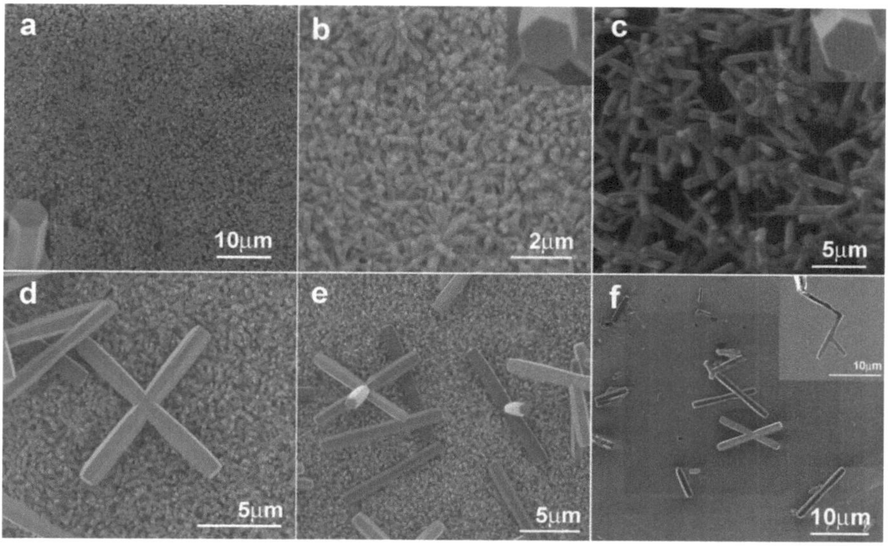

Fig. 5.1 The SEM images of the ZnO nanorods chemically grown from $ZnSO_4$ and ammonia aqueous bath on (**a**) glass substrates and the *inset* showed hexagonal rod images, (**b**) on quartz substrates, and (**c**) on Si substrates. The *insets* in figures (**a**), (**b**), and (**c**) display single ZnO rod with hexagonal basis. Image (**d**) corresponds to 2D ZnO crosses; (**e**) to branched 1D, 2D, and 3D microrods with ZnO nanorod arrays as background; and (**f**) to 1D and 2D branched microrods transferred to another Si substrate. The *inset* in (**f**) the *right-upper corner* shows an individual tripod picked-up by the in situ lift-out needle in the focused ion beam system [Reprinted from Materials Science and Engineering: B, 145, O. Lupan, L. Chow, G. Chai, B. Roldan Cuenya, A. Naitabdi, A. Schulte, H. Heinrich, Nanofabrication and characterization of ZnO nanorod arrays and branched microrods by aqueous solution route and rapid thermal processing, 57–66, Copyright (2007), with permission from Elsevier]

because it permits branched nanorods to be easily transferred to other substrates. Such a flexibility of substrate choice opens the possibility of using FIB system for handling nanostructures.

Another approach to grow transferable ZnO nanowires is the chemical vapor deposition (CVD) procedure at a low pressure (Fig. 5.2) [47]. For this purpose, metallic zinc of high purity (99.999 %) and an oxygen–argon mixture (15 vol.% of oxygen) were the starting reactants [47, 48]. Schematically representation of the CVD setup used for the growth of ZnO nanowires on Si substrates is shown in Fig. 5.2.

A flowing type two-zone quartz reactor was used to growth ZnO nanowires on Si. In the first zone, zinc was evaporated at 670 °C (position 2 in Fig. 5.2), and in the second zone at 650 °C, zinc vapors interacted with oxygen on silicon [(100)-oriented wafer] substrates. An oxygen–argon mixture was fed to the reactor at a rate of 1 liter/h with the consumption of zinc of 20–28 g/h. The growth of ZnO nanowires was for 30 min on substrates spaced at different distances from the zinc source. The

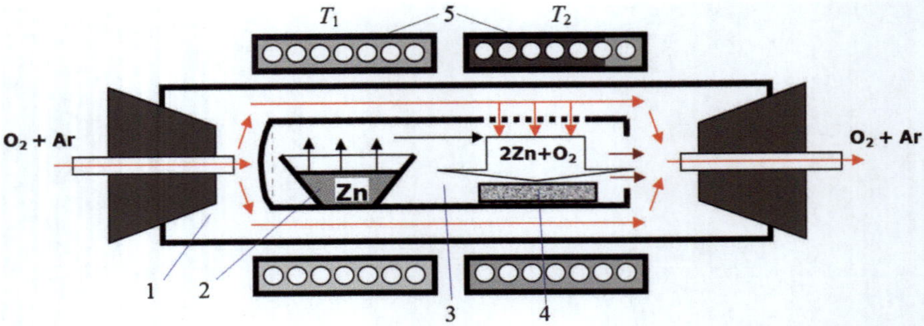

Fig. 5.2 The setup of chemical vapor deposition for the growth of ZnO nanowires on Si substrate: *1*, flowing quartz reactor; *2*, Zn source; *3*, internal quartz retort; *4*, substrates; and *5*, electric heaters [Reprinted from Materials Research Bulletin, 45, O. Lupan, G.A. Emelchenko, V.V. Ursaki, G. Chai, A.N. Redkin, A.N. Gruzintsev, I.M. Tiginyanu, L. Chow, L.K. Ono, B. Roldan Cuenya, H. Heinrich, E.E. Yakimov, 1026–1032, Copyright (2010), with permission from Elsevier]

XRD measurements [47] on the ZnO nanowires indicate the hexagonal wurtzite ZnO structure with lattice constants in accordance with values in the standard card (JCPDS number 36-1451 for ZnO). To observe the entire length of ZnO nanowires and whether it is possible to separate each one from another, cross-sectional images of the depositions were performed. Figure 5.3a shows morphology with high densities of 200 nm ZnO nanowires grown over the entire ZnO seed layer on the Si substrate. The substrate was positioned at the entrance of the growth zone (higher O/Zn ratio, Fig. 5.2). According to a cross-sectional SEM view (Fig. 5.3b), it can be concluded that the samples grown on Si substrate positioned at the exit zone (smaller O/Zn ratio, Fig. 5.2) show similar morphology. Top-view SEM images of the samples grown on the Si substrate positioned at the exit of the growth zone (position 4 in Fig. 5.2) is presented in Fig. 5.3c, d.

In this case, ZnO nanowires possess a diameter of 50–100 nm, and the average length is about 5–10 μm. Such nanostructures are highly interesting for gas nanosensor fabrications. The following peculiarities in the ZnO nanowires/nanorods growth were found [47]: (a) vertical nanowires started to grow at a critical ratio of oxygen/zinc vapors R_c, but for $R < R_c$ only randomly oriented nanowires were synthesized, and (b) transition from a nanorod shape to a nanoribbon shape at larger temperature difference (of about 80 K) between the evaporation zone and deposition zone. The high-resolution transmission electron microscopy (HRTEM) image of a ZnO nanowire grown by CVD is demonstrated in Fig. 5.4a. Selected area electron diffraction (SAED) pattern of a ZnO nanowire grown by CVD is presented in Fig. 5.4b. These studies prove that the ZnO nanowires are uniform and single-crystalline material and it has grown along the direction of the *c*-axis of ZnO. The detailed studies by EDX shows a stoichiometric ZnO composition (within a precision of 1 at.%).

Such ZnO nanowires are transferable to another substrate with pre-patterned external electrical contacts. Details of the systematic characterization of the as-

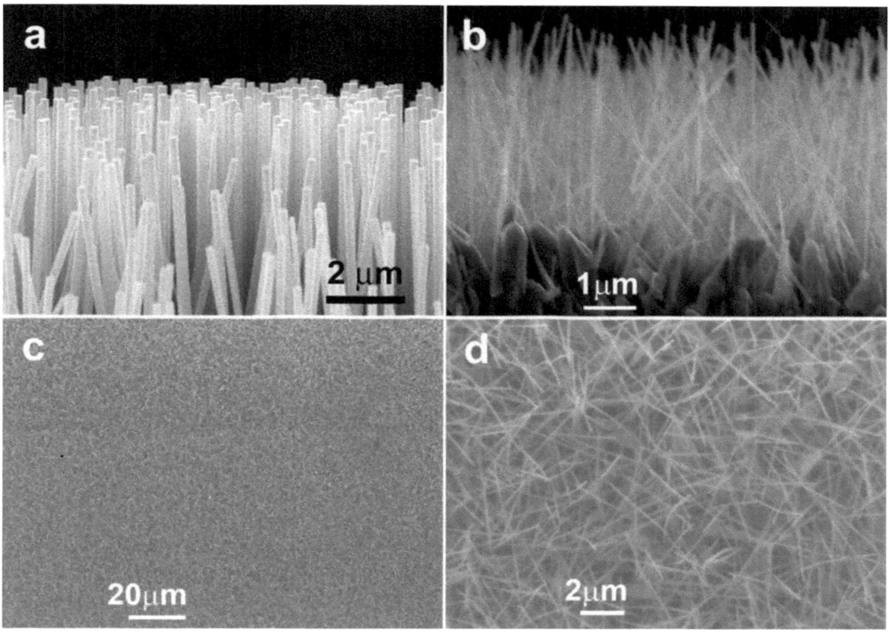

Fig. 5.3 SEM images of the cross section of the ZnO nanowire arrays with respect to the substrate position in the growth zone in the chemical vapor deposition system: (**a**) at the entrance of the growth zone (ratio O/Zn is larger) and (**b**) at the exit of the growth zone (ratio O/Zn is smaller). Next are the SEM *top-view* images: (**c**) of the ZnO nanowires grown on ZnO/Si placed at the exit of zone (ratio O/Zn is smaller) and (**d**) zoom in view image of the ZnO nanowires [Reprinted from Sensors and Actuators B 144, O. Lupan, V.V. Ursaki, G. Chai, L. Chow, G.A. Emelchenko, I.M. Tiginyanu, A.N. Gruzintsev, A.N. Redkin, Selective hydrogen gas nanosensor using individual ZnO nanowire with fast response at room temperature, 56–66 and Materials Research Bulletin, 45, O. Lupan, G.A. Emelchenko, V.V. Ursaki, G. Chai, A.N. Redkin, A.N. Gruzintsev, I.M. Tiginyanu, L. Chow, L.K. Ono, B. Roldan Cuenya, H. Heinrich, E.E. Yakimov,1026–1032, Copyright (2010), with permission from Elsevier]

synthesized zinc oxide nanomaterial for sensor applications can be found in previous work [47]. Such high-quality ZnO nanowires are very important for nanosensors investigations in order to control its properties by doping with different impurities. A third approach reported as successful one to synthesize doped ZnO nanowires for nanosensors is electrochemical deposition (ECD). Lupan et al. [35, 49] demonstrated an efficient ECD technique to growth Cd or Cu-doped ZnO nanowire arrays from which individual NW can be separated for nanosensor applications.

A cross-view SEM image of pure ZnO NWs grown on FTO substrates at 70, 80, and 90 °C for 9,000 s is presented in Fig. 5.5a–c. It was found a marked increase in the aspect ratio of NRs from 3–4 to 16–19 by increasing ECD growth temperature. However, by adding 6 μM $CdCl_2$ in ZnO bath at temperature in the same range (70–90 °C), it was observed that doped nanowires possess a higher aspect ratio

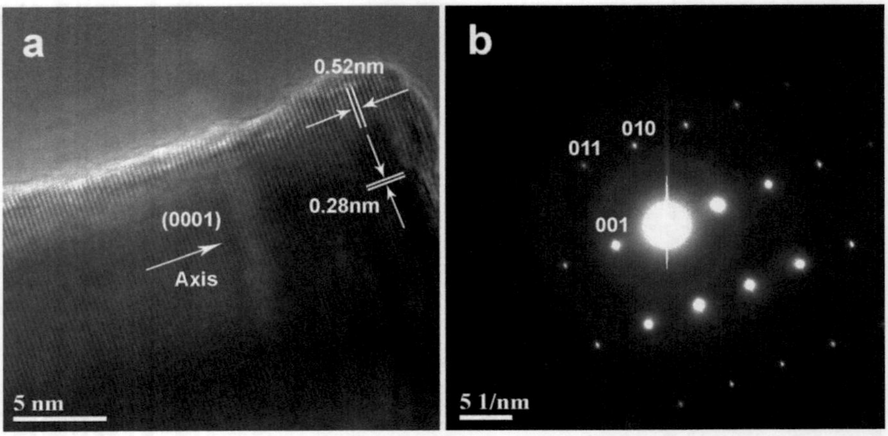

Fig. 5.4 (**a**) The high-resolution transmission electron microscopy (HRTEM) image of ZnO nanowire grown by chemical vapor deposition method and (**b**) its corresponding selected area electron diffraction SAED pattern [Reprinted from Materials Research Bulletin, 45, O. Lupan, G.A. Emelchenko, V.V. Ursaki, G. Chai, A.N. Redkin, A.N. Gruzintsev, I.M. Tiginyanu, L. Chow, L.K. Ono, B. Roldan Cuenya, H. Heinrich, E.E. Yakimov, 1026–1032, Copyright (2010), with permission from Elsevier]

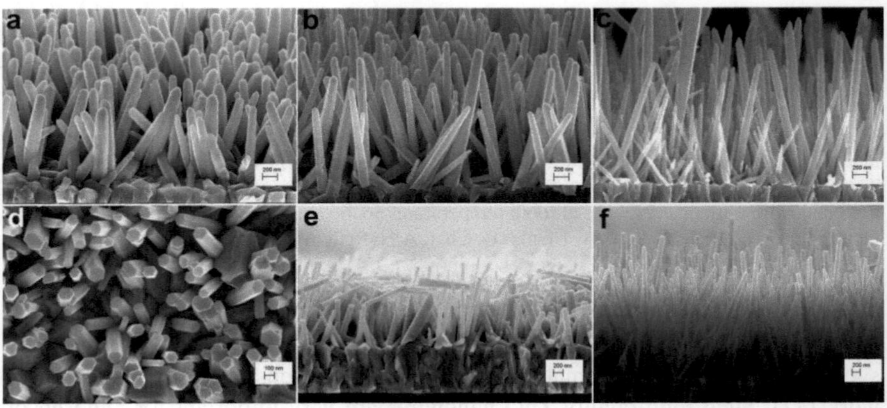

Fig. 5.5 Effect of temperature and of dopant Cd(II) on aspect ratio of ZnO and $Zn_{1-x}Cd_xO$ low-dimensional structures. SEM micrographs (*side view*, except-**d**) of the layers electrodeposited on FTO substrate: (**a**) pure ZnO grown at 70 °C; (**b**) pure ZnO NWs grown at 80 °C; (**c**) pure ZnO NWs grown at 90 °C; (**d**) *top view* of $Zn_{1-x}Cd_xO$ ECD at 70 °C, 6 μM $CdCl_2$; (**e**) *side view* of $Zn_{1-x}Cd_xO$ at 80 °C, 6 μM $CdCl_2$; and (**f**) *side view* of $Zn_{1-x}Cd_xO$ NWs ECD at 90 °C, 6 μM $CdCl_2$ in the electrolyte. ECD durations for (**a**)–(**c**) was 9,000 s and for (**d**)–(**f**) was 7,200 s [Reprinted with permission from J. Phys. Chem. C, 115, 14548–14558. Copyright (2011) American Chemical Society]

(e.g., of about 30 at 90 °C). In addition, doped nanowires (Fig. 5.5d) have a better crystallized morphology with better-defined edges.

We also [35, 49] observed that the presence of Cd(II) in the electrolyte favors the growth along the *c*-axis. Figure 5.5f shows NWs with more uniform diameter distribution and shapes when compared with pure ZnO NWs ECD at 90 °C (Fig. 5.5c). Also, the morphology for ZnO NWs was nearly the same before and after thermal annealing, which is in agreement with previous reports for ZnO [50]. Other approaches to growth transferable pure or doped ZnO NW/NR can be found in works [51, 52].

5.3 Fabrication of a Nanoscale Sensors Based on ZnO Nanostrcutures and Its Characteristics

In this section, we describe the fabrication procedures of nanosensors from individual ZnO nanowire/nanorod grown by gas phase [3, 47] and solution methods [1, 2, 53]. Individual nanowires were released from an agglomeration of nanowires/nanorods grown on the substrate (Fig. 5.3). Initially, the substrate is sonicated in ethanol, and then NWs were transferred to a SiO_2-coated Si substrate as described in [54]. Later, a direct contact technique to transfer ZnO nanowires is developed [3, 47]. Such procedures permit us to reduce the density of ZnO nanowires on the secondary substrate (Fig. 5.6a) for easy nanodevice fabrication (Fig. 5.6b).

Next, the direct contact method was used again to transfer separated ZnO nanowires (Fig. 5.6b) to a substrate with predeposited Cr/Au external electrodes (the separation was about 5 μm). Focused ion beams were used to pattern metal electrodes that contacted both ends of an individual ZnO nanowire (insets of Fig. 5.6b, c), as reported elsewhere [1–3, 55, 56]. Sensors made from branched

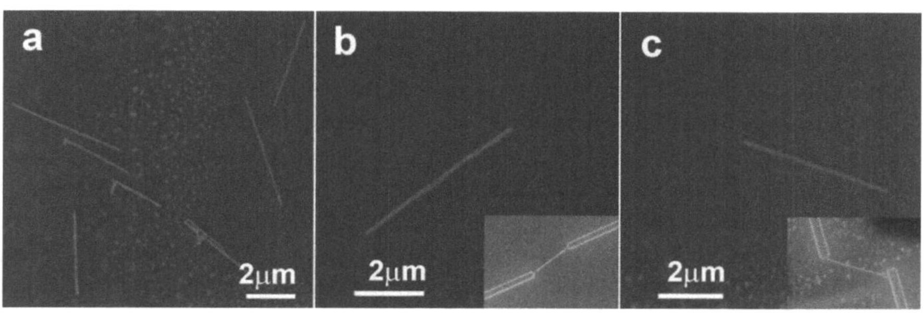

Fig. 5.6 Scanning electron micrographs of (**a**) transferred ZnO nanowires (NWs) to intermediate substrate (Si with ZnO NWs on *top*); (**b**) transferred individual ZnO NW and the *inset* shows a fabricated device based on a single ZnO nanowire of 200 nm in diameter; and (**c**) transferred ZnO NW and the *inset* shows a fabricated device based on a single nanowire of 100 nm in diameter [Reprinted from Sensors and Actuators B 144, O. Lupan, V.V. Ursaki, G. Chai, L. Chow, G.A. Emelchenko, I.M. Tiginyanu, A.N. Gruzintsev, A.N. Redkin, Selective hydrogen gas nanosensor using individual ZnO nanowire with fast response at room temperature, 56–66, Copyright (2010), with permission from Elsevier]

nanorods were developed in 2007 by Lupan et al. [1, 2]. Such a procedure was extended to other nanostructures integration in sensors. Later Lupan et al. [3] made nanodevices based on single nanowires of 200 nm in diameter (inset of Fig. 5.6b) and 100 nm in diameter (inset of Fig. 5.6c).

It is known that in FIB processing of nanostructured zinc oxide material, it is inevitable to expose ZnO nanostructures to ion beam. Thus, it is a very important to point out the characteristics which allows very thin ZnO NW (50–100 nm in diameter) to be connected by using FIB instrument, namely the enhanced radiation hardness of ZnO NWs versus bulk ZnO layers according to previous works [22, 57, 58]. Based on previous research by Look et al. [59], the bombardment by electrons, heavy ions [60], and protons [61] caused much less damages to ZnO material in comparison with other semiconductor materials. The main advantage for using FIB is the localized maskless deposition of metals and insulators, because it works like local chemical vapor deposition (LCVD) with good spatial resolution and lower deposition rate [62]. According to previous works on TEM samples fabrications using FIB setup, the depth of penetration of FIB ions in materials is about 2–20 nm after 2–3 h of continuous exposure [63–66]. In previous works [3], a single ZnO nanowire was exposed to FIB for less than 9 min. According to Weisenberg et al. [67], the FIB Ga ions started to affect the electric properties of ZnO when the dosage of the Ga ion beam exceeds 10^{15} cm^{-2}. During our FIB processing, the Ga ion beam can be very accurately positioned on the individual nanowire and the exposure time to Ga beam can be limited to just a few minutes. We estimated that our ZnO nanostructures were never exposed to Ga ion beam more than 10^{13} cm^{-2}; therefore, the effect of FIB processing on our ZnO nanostructures has been insignificant [66]. This makes the ZnO nanodevices presented below feasible. However, the accidental metal atom deposition on the ZnO nanowire/nanorod during the FIB processing of electrode contact cannot completely rule out.

For such nanodevices a linear behavior of the current–voltage curve is observed [3, 68]. Gas-sensing properties of such sensor are measured with an apparatus consisting of a closed quartz chamber connected to a gas flow system [1–3, 69]. Usually, the gas response is defined as the ratio $|\Delta R/R|$, where $|\Delta R| = |R_{air} - R_{gas}|$ and R_{air} is the resistance of the sensor in dry air and R_{gas} is resistance in the test gas [1–3]. Figure 5.7a shows the transient response of the 100 nm ZnO nanowire gas nanosensor under exposure to H_2 gas. The highest gas response was about 34 % to 100 ppm hydrogen gas at room temperature. As can be seen from Fig. 5.7a, the resistance of the nanowire nanosensor after the exposure to hydrogen does not recover to the initial value due to the presence of adsorbed gas molecules on the surface of ZnO nanowire. That is why, operation at elevated temperature is needed to desorb the gas species from the surface of the sensor material [70–73]. However, according to previous results [1–3], it is desirable to avoid heating ZnO nanowire to high temperature (>300 °C), since its properties will change. It was observed experimentally [3] that the application of heating pulses to the nanowire sensor after H_2 gas exposure leads to permanent changes of its electrical properties. It is evident that this shift is in contradictions with long-term stability requirements for sensors.

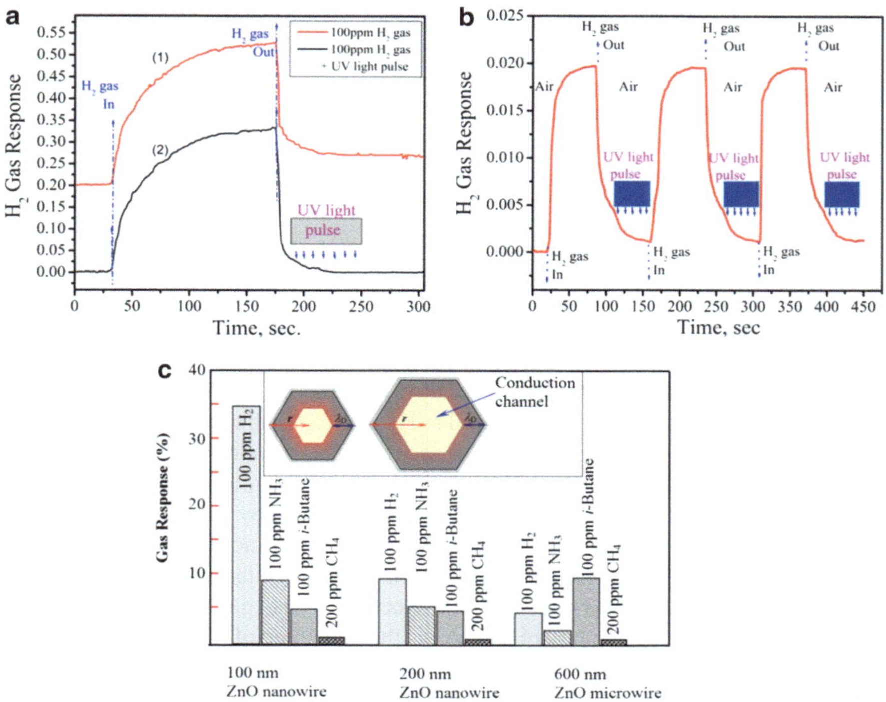

Fig. 5.7 (**a**) Gas response curves of the 100 nm zinc oxide nanowire-based gas sensor under exposure to 100 ppm of H_2 gas at room temperature (22 °C). *Curve* (*1*) is displaced upward by 0.25 to avoid overlapping with *curve* (*2*). (**b**) Dependences of gas responses for 100 nm diameter ZnO NW-based sensor. The response curves of the individual nanowire-based sensor toward 10 ppm H_2 pulses. Before the test, the nanowire was preconditioned in a constant dry air flow. An UV pulse is applied after H_2 gas is turned off. (**c**) Dependences of gas response for different diameter ZnO nanowire-based sensors. The *inset* shows cross-sectional view of ZnO NWs with different radius (r) and same Debye length (λ_D) and conduction channel, respectively [Reprinted from Sensors and Actuators B 144, O. Lupan, V.V. Ursaki, G. Chai, L. Chow, G.A. Emelchenko, I.M. Tiginyanu, A.N. Gruzintsev, A.N. Redkin, Selective hydrogen gas nanosensor using individual ZnO nanowire with fast response at room temperature, 56–66, Copyright (2010), with permission from Elsevier]

For practical purpose, it is recommended to avoid heating sensor in the explosive H_2 environment. As discussed above, the defects in nanomaterials influence the gas sensitivity of ZnO nanowire-based sensors [72, 73].

Lupan et al. [3] implemented a UV irradiation procedure to desorb gas species on the ZnO surface and to enhance the recovery time of the nanowire nanosensor. It was demonstrated (Fig. 5.7a [curve 2]) [3] that the recovery time of the UV radiated nanosensor is much shorter than that of nonradiated sensors; it was also reported for other nanomaterials [74].

Figure 5.7b demonstrates the dynamic performance of the ZnO nanosensor exposed to traces of 10 ppm H_2 gas with response and recovery times

($T = |t_{90\,\%} - t_{10\,\%}|$) much faster compared to Fig. 5.7a. Also, it was reported that the gas response increases with increasing the concentration of H_2 (comparing Fig. 5.7a [curve 2], b) [3].

The mechanism of reducing the sensor recovery time by applying an UV pulse can be explained on facilitated desorption of gas molecules from the surface of ZnO as described in previous work [3]. The gas nanosensors fabricated by FIB technique possess similar or higher gas response compared with sensors based on ZnO nanorod [2], ZnO nanocrystals [75], ZnO-branched rods [1], or single tetrapod [69]. Wang et al. [36] reported on zinc oxide nanorod-based sensor with a sensitivity of 4.2 % at 500 ppm H_2. Kang et al. [19] found that a H_2 gas sensor based on ZnO multiple nanorods has 18 % current change upon exposure to 10 % H_2 in N_2 at 112 °C. Wang et al. [17] reported on enhanced sensitivity to 200 ppm H_2 gas of ZnO nanowires at 250 °C. Lupan et al. [3] developed a single nanowire-based sensor (100 nm in diameter) with similar H_2 gas-sensing characteristics at room temperature. These experimental results suggest that a single ZnO nanowire could be an excellent material for hydrogen gas sensors at room-temperature operation. These works [3, 76] indicate that the gas species adsorbed by ZnO surface influence the charge concentration and the electrical resistance of zinc oxide. As well as the larger surface-to-volume (aspect) ratio of the single-crystalline nanowires offers potential to improve the gas response and selectivity of H_2 gas sensors.

The dependence of the gas response on the diameter of ZnO nanowire-based sensors is presented in Fig. 5.7c. The sensors made from a single ZnO nanowire with 100 nm in diameter possess the highest gas response (about 34 %). Large diameter nanowire (200 nm)-based sensors shows a lower gas response to H_2 (about 10 %) at room temperature. Sensors made from larger diameter ZnO nanorod [2] possess a lower gas response (about 4 %). Lupan et al. [3] demonstrated experimentally the importance of using thinner ZnO nanowires to fabricate H_2 sensors with higher sensitivity that can be operated at room temperature. Figure 5.9 presents the selectivity of the gas sensors to NH_3, *i*-butane, and CH_4.

Next we present the fabrication of a double-rod-based sensor developed by Chow et al. [53] using the in situ lift-out procedure, described in previous works [1–3]. Initially, four external contacts were deposited as shown in Fig. 5.8a. Afterward, the surface of the intermediate Si/SiO_2 substrates was scanned for an intermediate ZnO nanorod and Cu-doped ZnO rod (not shown). A square hole was made on the glass substrate between the external electrodes (Fig. 5.8a) to allow the entire surface of Cu–ZnO nanorod to be involved in the sensing of H_2 gas. Then, pure ZnO nanorod (Fig. 5.8b) or Cu–ZnO nanorod (Fig. 5.8c) was mounted between external electrodes by FIB technique as described by Lupan et al. [2]. The double-rod-based sensor (Fig. 5.8d) was used for comparison reason in sensorial studies [53]. Developed procedure by Chow et al. [53] clearly demonstrated that by using the in situ lift-out procedure it is possible to fabricate in a controlled manner nanosensors with bottom-up technology. The advantage of FIB procedure is a quick verification of the concept and its compatibility with nanoelectronic processes. The single ZnO tetrapod sensor fabrication by FIB technique was developed in 2009 by Lupan

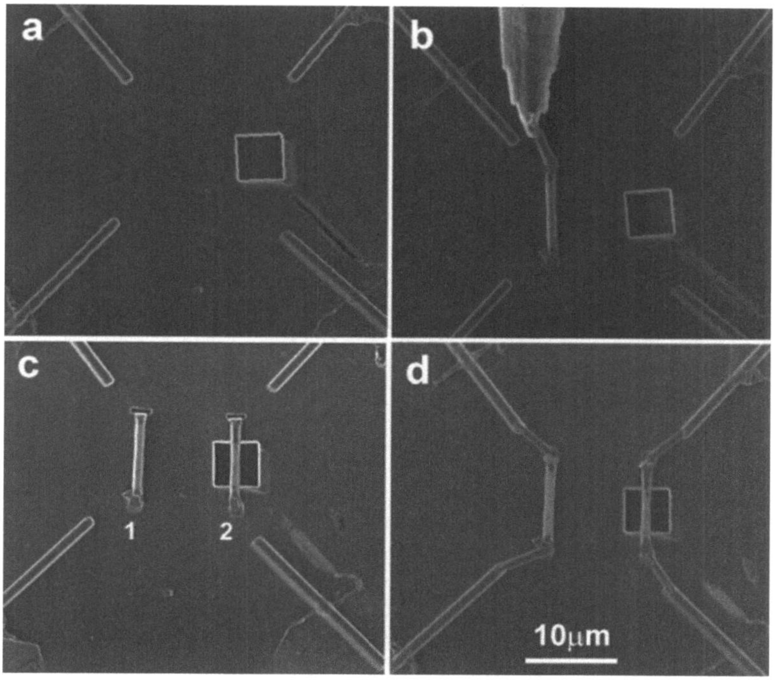

Fig. 5.8 Scanning electron microscope (SEM) images showing the steps of the in situ lift-out fabrication procedure in the FIB/SEM system. (**a**) Four external contacts made on substrate, a *square hole* cut on the glass between two electrodes; (**b**) an intermediate ZnO rod and a single ZnO one-dimensional rod selected for first sensor fabrication picked-up by the FIB needle and positioned between first two electrodes; (**c**) a single ZnO rod *#1* and a single Cu-doped ZnO rod *#2* selected for sensor fabrication placed between contact electrodes) on the same substrate; (**d**) single one-dimensional rod *#1* (ZnO) and *#2* (Cu–ZnO) welded to both electrode and external connections as final double-rod sensor [Reprinted from Sensors and Actuators A 189, L. Chow, O. Lupan, G. Chai, H. Khallaf, L.K. Ono, B. Roldan Cuenya, I.M. Tiginyanu, V.V. Ursaki, V. Sontea, A. Schulte. Synthesis and characterization of Cu-doped ZnO one-dimensional structures for miniaturized sensor applications with faster response, 399–408. Copyright (2013), with permission from Elsevier]

et al. [69]. A SiO_2/Si structure was used as intermediate substrate for ZnO tetrapods transferring and to reduce the density of tetrapods. This step is extremely important for further pickup using in situ lift-out technique. To separate a single ZnO tetrapod for further processing, in situ FIB technique with a micromanipulator needle was used. It was [69] found that the fabrication of individual tetrapod-based device is more difficult in comparison with the fabrication of single nanorod devices. Several requirements for the successful fabrication of tetrapod sensor fabrication have been pointed out: (a) a single ZnO tetrapod should be transferable from one substrate to another (Fig. 5.9a); (b) the slope angles in the FIB chamber must come into the line (Fig. 5.9a, b); (c) FIB needle with single tetrapod (Fig. 5.9b) should be

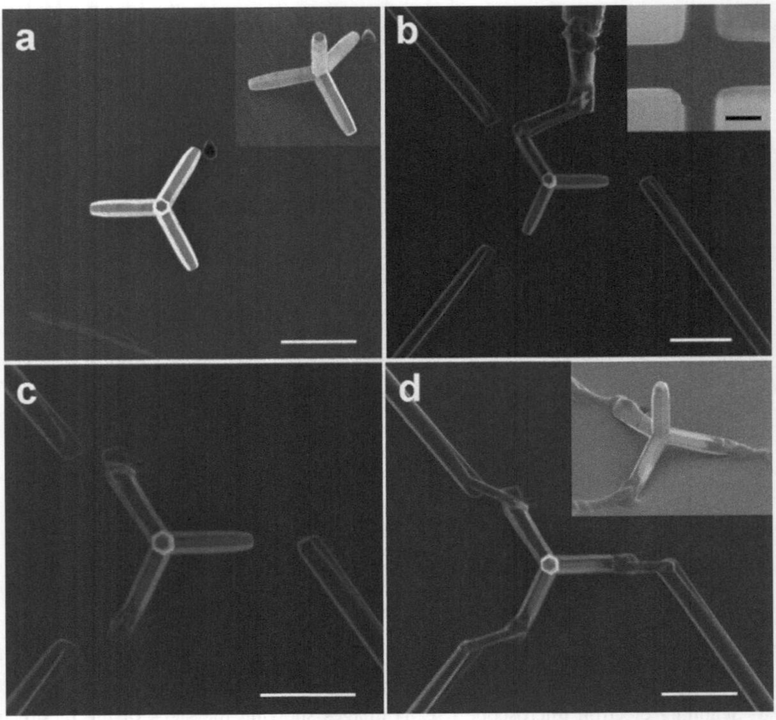

Fig. 5.9 Secondary electron images showing the ZnO tetrapod and the steps of the in situ lift-out fabrication procedure in the FIB/SEM system. (**a**) A single ZnO tetrapod on Si substrate, viewed directly from above along the *z*-axis. (The *inset* shows side view of the ZnO tetrapod.) (**b**) The tungsten needle with an intermediate rod and is attached to one of the legs of a ZnO tetrapod selected for sensor fabrication. It is placed next to external electrical connections. The *inset* is the glass substrate with Cr/Au depositions as contact electrodes (10 μm scale bar). (**c**) The ZnO tetrapod is placed on the substrate and next to external gold electrodes. (**d**) The ZnO tetrapod after connecting its three legs to the three external connections. The *inset* shows side view of a single ZnO tetrapod sensor. The scale bar is 3 μm [Reprinted from Sensors and Actuators B 141, O. Lupan, L. Chow, G. Chai, A single ZnO tetrapod-based sensor, 511–517. Copyright (2009), with permission from Elsevier]

positioned very accurately with nanometer precision to avoid destruction of the tetrapod junction. In Fig. 5.9a, a ZnO tetrapod on a Si substrate is presented. It consists of four hexagonal rods growing from a common core. Then FIB-enhanced CVD was used to deposit three 0.5 μm thick Pt lines to connect with external Au-electrodes and then ZnO tetrapod is positioned near the end of these Pt conducting lines. (Fig. 5.9b and 5.9c). In the final step, the three terminals of the tetrapod are welded with the Pt connection lines using FIB-enhanced CVD. Figure 5.3d presents the fabricated single tetrapod-based sensor with three legs connected to external electrodes. The fourth arm is pointing upward. An advantage of tetrapods is

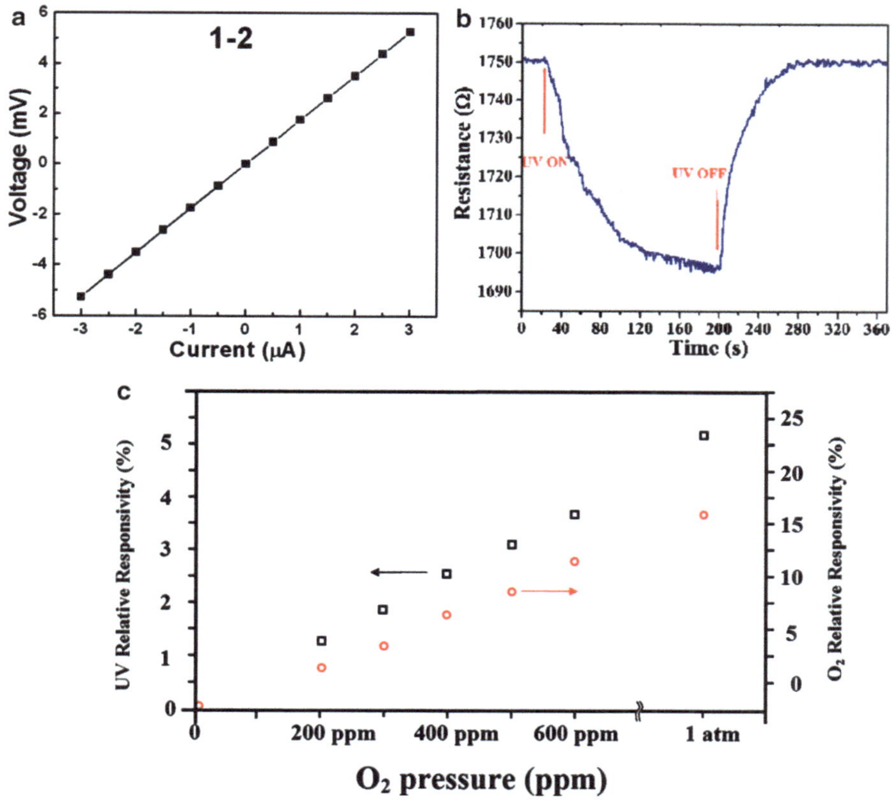

Fig. 5.10 (**a**) I–V characteristics of the single ZnO tetrapod device through leg pair *1–2*. (**b**) The resistance change of the ZnO tetrapod-based sensor under UV irradiation. Characteristics are shown for leg pair *1–2*. (**c**) Sensitivity to UV light and oxygen concentration of a single ZnO tetrapod sensor. Oxygen gas was added with N_2 as buffer gas into the chamber [Reprinted from Sensors and Actuators B 141, O. Lupan, L. Chow, G. Chai, A single ZnO tetrapod-based sensor, 511–517. Copyright (2009), with permission from Elsevier]

an arm directed upward, which may be used for surface-functionalization strategies or localized contacts, etc.

Figure 5.10a shows the current–voltage (I–V) characteristics between the electrodes on ZnO tetrapod legs, which possess a linear behavior. All the connections through different leg pairs displayed similar bias characteristic. Figure 5.10b shows the UV sensitivity of a single zinc oxide tetrapod sensor irradiated with 361 nm UV light perpendicular to the substrate surface. It can be seen in Fig. 5.10b when the UV light was turned on, the electrical resistance shows an exponential decay with a time constant of ~45 s. This phenomenon can be explained by surface process—fast desorption of the chemisorbed oxygen species at the surface of ZnO tetrapod and then by bulk process. It can be seen that the

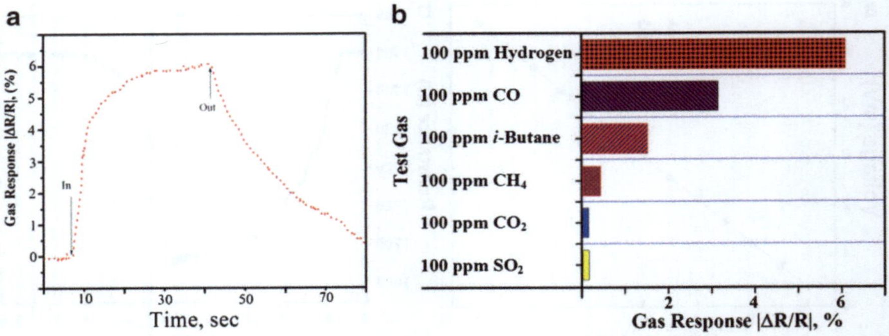

Fig. 5.11 (**a**) Responsivity of the ZnO tetrapod sensor to 100 ppm H_2 gas at the room temperature. (**b**) Gas response of ZnO tetrapod-based sensor to different gases at 100 ppm concentration [Reprinted from Sensors and Actuators B 141, O. Lupan, L. Chow, G. Chai, A single ZnO tetrapod-based sensor, 511–517. Copyright (2009), with permission from Elsevier]

recovery of the resistance seems to be twice as fast (~23 s.) when the UV light was turned off. Such behavior could be explained by two processes—one related to volume process and second one to surface adsorbed species. Such tetrapod device gives rapid responses to the UV light and faster recovery time.

The UV detection occurs after turning on UV light, due to the increase in the charge carrier density as a result of electron–hole (e^-–h^+) pairs generation. However, the increase rate of the resistance is different when turning off the UV radiation in comparison with the decrease rate upon UV illumination. Such phenomenon could be beneficial for further studies of photocatalytic activity. This effect has been observed in several previous reports [77]. This rate is controlled by the oxygen concentration and also affected by the recovery of the modified surface chemistry [77]. ZnO tetrapods have great potential application as chemical gas sensors due to chemisorption. Chemisorbed gas molecules on a surface have chemical bond that will either donate or extract electrons to the ZnO [69]. Relative responsivity can be defined as previously reported [69]. Single tetrapod responsivity under various oxygen pressures are presented in Fig. 5.10c. It is observed that the tetrapod resistance increases monotonically with increasing O_2 at very low concentration. Each point was obtained by waiting 10 min after introducing oxygen. It is found that the tetrapod responsivity at room temperature decreases with decreasing the oxygen pressure. It can be explained as increase in the electrical resistance and of the surface charge depletion region [69]. Such characterizations are important to understand the gas-sensing mechanism of a single tetrapod-based sensor.

The individual ZnO tetrapod-based multiterminal sensor response to 100 ppm H_2 is presented in Fig. 5.11a [69]. It was found that a relative sharp increase is followed by a saturation region and then a slow decrease in the H_2 gas response at the room temperature. The responses of different leg pairs of this tetrapod were found to be similar.

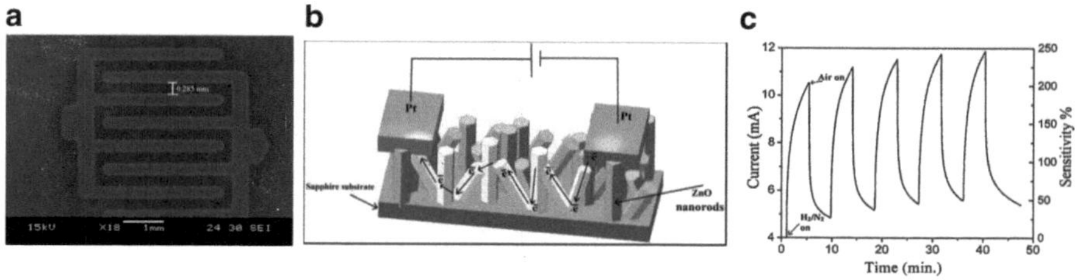

Fig. 5.12 (**a**) SEM image of Pt grid contact deposited on the ZnO nanorods arrays. (**b**) Schematic illustration of sensing mechanism of oblique–vertical nanorods, the sensitivity and the repeatability of ZnO nanorods gas sensor at room temperature to successive pulses of 1,000 ppm H_2 balanced N_2 with different bias voltages: (**c**) 5 V. The H_2 exposure time for (**c**) 5 min [Reprinted from Sensors and Actuators B: Chemical 176, J.J. Hassan, M.A. Mahdi, C.W. Chin, H. Abu-Hassan, Z. Hassan, A high-sensitivity room-temperature hydrogen gas sensor based on oblique and vertical ZnO nanorod arrays, 360–367. Copyright (2013), with permission from Elsevier]

Figure 5.11b represents the room-temperature response of ZnO tetrapod-based sensor in various gas environments, such as 100 ppm H_2, CO, *i*-butane, CH_4, CO_2, and SO_2. The largest response corresponded to the exposure to 100 ppm H_2, which is >6× higher than the response to 100 ppm CH_4, CO_2, and SO_2. Thus, it is suggested that the response to H_2 and CO gas is significantly higher than those of the other gases [69].

Another approach was reported by Hassan et al. [78] by using the networks connections of vertically and oblique nanorod arrays. These authors demonstrated that in such sensor structures with the absence of seed layer the high sensitivity of ZnO nanorods arrays hydrogen sensor at room temperature can be obtained. Figure 5.12a presents the fabricated metal–semiconductor–metal (MSM) gas-sensing device by the radio frequency (RF) sputtering of Pt grid contacts on top of the ZnO nanorod arrays using a shadow mask.

Results presented the above figure reveal that the ZnO nanorods array-based sensor could repeatedly detect hydrogen gas at room temperature over a long period of time. As can be seen from Fig. 5.11c, the sensitivity of the sensor was more stable over time in the instances where biased voltages of 5 V were used as compared for sensors under bias voltage of 0.1 V was used (not shown here) [78].

Currently, the most important use of hydrogen is the ammonia synthesis. That is why doped ZnO is important for sensors since it has sensitivity to H_2 and it can be tuned by doping to sense NH_3 [51].

5.4 Methods Employed to Enhance the Nanosensor Performance

In this part, several approaches to increase the gas-sensing performances such as surface-to-volume ratio, impurity doping, and surface functionalization are discussed. Quasi one-dimensional nanostructures demonstrate great potential for

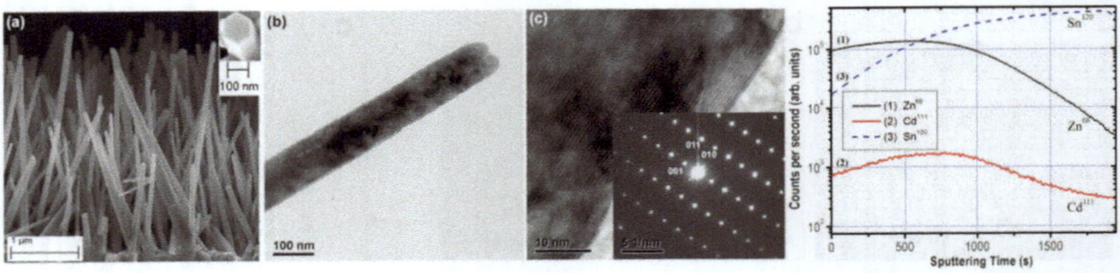

Fig. 5.13 (**a**) SEM micrograph (*side view*) of ZnO–Cd (6 μM $CdCl_2$ in the electrolyte) nanowires electrodeposited on an FTO substrate. (**b**) TEM image of the ZnO–Cd nanowire revealing the surface morphology. (**c**) HRTEM image of the ZnO–Cd nanowire. The *inset* shows a selected area electron diffraction (SAED) pattern. (**d**) SIMS measurements of Cd (6 μM)-doped ZnO. The *dashed curve* (Sn) corresponds to the FTO substrate [Reprinted from Sensors and Actuators B 173, O. Lupan, L. Chow, Th. Pauporté, L.K. Ono, B. Roldan Cuenya, G. Chai, Highly sensitive and selective hydrogen single-nanowire nanosensor, 772–780. Copyright (2012), with permission from Elsevier]

nanodevice applications [2, 3, 13, 14] due to their large surface-to-volume ratio and controlled flow of confined charge carriers. Many ZnO-nanostructured based H_2 sensors have been demonstrated [2, 3, 13, 14], such as ZnO nanorod arrays and networks [15, 39] and single nanorod/nanowire ZnO [2, 3, 13, 14]. However, for scalable sensor technology, a cost-effective process to make highly sensitive H_2 nanosensors is required [34, 35]. This is motivated by the facts that single pure ZnO nanowire/nanorod-based H_2 sensors showed the relative slow response and recovery times. Commercial demands on chemical nanosensors require enhanced selectivity, sensitivity, stability and faster gas responses, which can be achieved via increasing surface-to-volume ratio, impurity doping, and surface functionalization.

Lupan et al. [4] used an electrochemical method for the synthesis of Cd-doped ZnO nanowires with high aspect ratios. In this way, it was possible to show the effects of doping and aspect ratios on H_2 gas sensing with ZnO nanowires. It was found that by doping the ZnO NWs, it is possible to reduce the operation temperature and to enhance the gas response and selectivity [4].

Cadmium-doped ZnO nanowire arrays were electrodeposited on F-doped tin oxide film with a sheet resistance of $10\,\Omega\ \mathrm{sq}^{-1}$ on a glass substrate [34, 35]. It was subjected to thermal annealing in air at 300 °C for 11 h to relax the ZnO lattice. A SEM image of a quasi-aligned high aspect ratio Cd-doped ZnO NW array is presented in Fig. 5.13a. It can be seen that NWs have high aspect ratio of about 30, diameters of 50–200 nm, and are quite uniform in lengths of 2–3 μm [34, 35]. Figure 5.13a inset demonstrates that the NWs are well facetted and have hexagonal cross sections exhibiting $\{100\}$ planes on the sides. Figure 5.13b presents a TEM image of an individual nanowire with uniform diameter. Such nanowires have a single crystal structure and are oriented along the $\langle 0001 \rangle$ *c*- axis according to high-resolution TEM image and the associated electron diffraction (SAED) pattern (Fig. 5.13c). SIMS was used to study the chemical composition of the Cd-doped

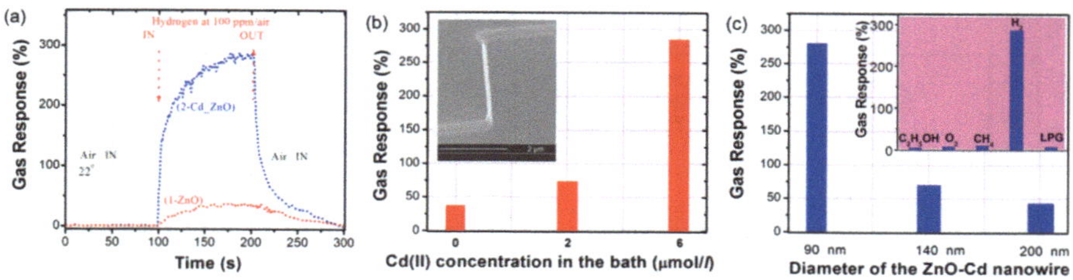

Fig. 5.14 Room-temperature gas responses of nanosensors based on a single pure ZnO NW and a ZnO–Cd (6 μM) NW of ~90 nm diameter versus (**a**) time and (**b**) the concentration of [Cd] (0–6 μM $CdCl_2$) in the electrolyte. (**c**) For the Cd-doped ZnO NW, the gas response is also shown as a function of the NWs' diameter (90, 140, and 200 nm). In all cases, a 100 ppm H_2 gas pulse was used. The *inset* in (**b**) shows an SEM image of a single ZnO–Cd (6 μM) NW connected in a nanosensor configuration. The *inset* in (**c**) shows the gas response of the single Cd–ZnO NW sensor to different gases (100 ppm H_2, 200 ppm others). All measurements were performed at RT [Reprinted from Sensors and Actuators B 173, O. Lupan, L. Chow, Th. Pauporté, L.K. Ono, B. Roldan Cuenya, G. Chai, Highly sensitive and selective hydrogen single-nanowire nanosensor, 772–780. Copyright (2012), with permission from Elsevier]

ZnO NW array (Fig. 5.13d). It was found that the Cd count rate closely follow the Zn count rate over most of the investigated sample depth proving that the Cd dopant has been evenly incorporated into the ZnO nanowire structure.

To fabricate a single nanowire-based nanosensor, the Cd-doped ZnO NWs were first separated from an array of as-grown NWs (see Fig. 5.13a) and transferring to a SiO_2/Si substrate. Lupan et al. [4] fabricated several nanosensors by using Cd–ZnO NWs with lengths between 2 and 3 μm and radii between 40 and 100 nm with reproducible electrical responses. The gas response studied according to ref. [3] presents transient response for ZnO and improved one for Cd-doped ZnO nanowire sensors (curves 1 and 2, respectively in Fig. 5.14a) upon exposure to 100 ppm of H_2 gas at RT. Evidently, a larger and faster gas response was observed in the case of the Cd-doped ZnO NW sensor. Also, the changes in the resistance of the sensor after exposure to hydrogen return to within 10 % of the initial value in about 11 s for the doped sample. Such results suggest that the larger surface-to-volume ratio of the single-crystalline nanowires offers potential to improve the gas response and selectivity for hydrogen. Also, it was observed that different concentrations of Cd in ZnO NWs show different gas response with same diameter of NW (Fig. 5.14b). Thus, a single Cd (6 μM)-doped ZnO NW (90 nm diameter) possesses the highest gas response (about 274 %). For thicker Cd–ZnO NWs (140 and 200 nm), a lower response to H_2 (less than 68 and 40 %) was found.

At the same time, it is important to consider experimental details used for study sensorial properties of nanowires. In such experiments, authors used a picoammeter as power source and a nano-V meter to monitor the voltage drop on the sensor [4]. In such a case, the NW acts as heating and as a sensing element. The self-generated heat locally increases the temperature ($T \sim 120$ °C, evaluated according to [79, 80]) and contributes to the desorption of adsorbed molecules on the NW surface.

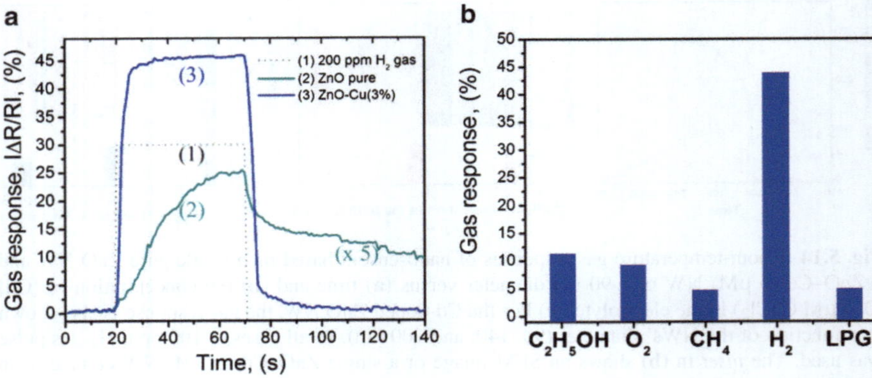

Fig. 5.15 (**a**) The room-temperature relative response (multiplied by 5) of the conductometric single ZnO one-dimensional rod (*curve 2*) and single Cu-doped ZnO (*curve 3*) rod-based sensor structure fabricated by in situ lift-out technique in the FIB system to 200 ppm H_2 gas (*curve 1*). (**b**) Gas response of a Cu–ZnO rod-based sensor to different ambient of 200 ppm concentration [Reprinted from Sensors and Actuators A 189, L. Chow, O. Lupan, G. Chai, H. Khallaf, L.K. Ono, B. Roldan Cuenya, I.M. Tiginyanu, V.V. Ursaki, V. Sontea, A. Schulte. Synthesis and characterization of Cu-doped ZnO one-dimensional structures for miniaturized sensor applications with faster response, 399–408. Copyright (2013), with permission from Elsevier]

Figure 5.14c (inset) presents the selectivity test to H_2 of Cd–ZnO nanosensor, where different responses to C_2H_5OH, O_2, CH_4, and LPG (liquefied petroleum gas) can be seen. The Cd–ZnO nanosensor's response to 200 ppm LPG and 200 ppm CH_4 is much lower in comparison with its gas response to H_2.

Another report on effect of doping on H_2 sensing properties was presented by Chow et al. [53]. Figure 5.15 presents the room-temperature gas response of the single-rod ZnO and single-rod Cu–ZnO sensor to 200 ppm H_2. It was demonstrated experimentally that the response time to H_2 test gas for Cu–ZnO are on the order of 30 ms and after 40 ms the signal reaches the equilibrium value. In this case, the relative resistance changes were about 44 %. Also, the electrical resistance was restored toward the 10 % above the original value within 50–90 ms of the introduction of clean air. For comparison, pure ZnO rod-based sensor gas response shown on the same Fig. 5.15a is about ten times lower and slower. Chow et al. [53] observed an improvement of sensor response of 25 % and 35 % for 0.5 % Cu and 1 % Cu doping, respectively.

It is well known that the gas selectivity is a major issue for the solid-state gas sensors. Although, different approaches have been studied in the past to improve the selectivity of gas sensors, new approaches are still searched by researchers. They are as follows: (a) modulation of the operating temperature, (b) using impedance measurements, and (c) doping with metal impurities, or surface coating, etc. Figure 5.15b presents the results on testing the selectivity to H_2 of Cu–ZnO sensor. One can see that the Cu–ZnO sensor's response to 200 ppm LPG and 200 ppm

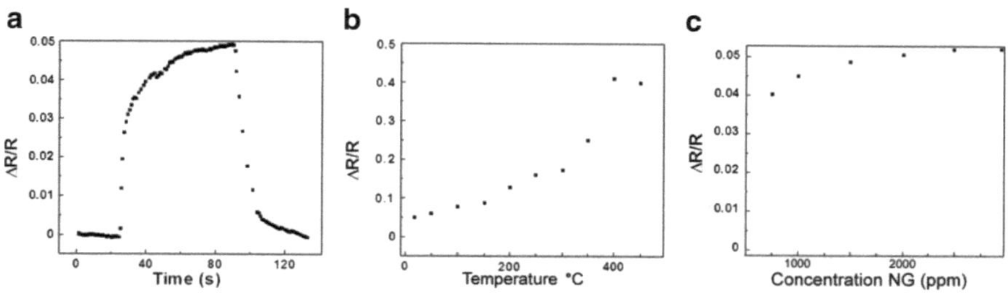

Fig. 5.16 (**a**) Natural gas response measurement of the fabricated single ZnO microwire device at room temperature. (**b**) Temperature dependence of natural gas response measurement of a single ZnO microwire. (**c**) Dependence of gas response versus concentration of the natural gas measured from a single ZnO microwire at room temperature [Reprinted from Sensors and Actuators A: Physical, 176, G.Y. Chai, O. Lupan, E.V. Rusu, G.I. Stratan, V.V. Ursaki, V. Şontea, H. Khallaf, L. Chow, Functionalized individual ZnO microwire for natural gas detection, 64–71. Copyright (2012), with permission from Elsevier]

CH_4 is lower in comparison with its gas response to hydrogen. Such experimental data demonstrates the high selectivity of the fabricated sensor structure and the high prospect of Cu-doped ZnO as material for miniaturized sensors operating at room temperature.

Based on the experimental results presented in Fig. 5.15a, one can conclude that Cu–ZnO sensor has a relative higher gas response and most importantly faster response and recovery times than pure ZnO sensor.

Another approach to improve the gas response is the surface functionalization of nanowires. For example, Chai et al. [81] used individual ZnO microwires dipped in a palladium chloride solution and thermal treated at 400 °C for sensor structure. Such zinc oxide microwire samples were integrated in sensor device between the Au and Ti electrodes. However, the maximum room-temperature gas response (about 5 %) was measured using 2,000 ppm of natural gas (Fig. 5.16a).

Figure 5.16a demonstrates that individual ZnO microwires dipped in a palladium solution and treated at 400 °C can be used as a sensor structure to 2,000 ppm natural gas. Its response increases about 5 % within 30 s (from 10 to 90 % of $\Delta R/R$). The temperature dependence of NG response of the fabricated single ZnO microwire sensor structure (Fig. 5.16b) shows a maximum response at about 400 °C. The phenomena could be explained that at elevated temperature methane is more easily decomposed into species [82] and by catalytic reaction of the gas, which favor increase of response. Chai et al. [81] studied the gas response versus concentration (750–3,000 ppm) to the natural gas (Fig. 5.16c). They observed that the room-temperature value of $\Delta R/R$ increases from 4 to 5.2 % with increasing NG concentration from 750 to 3,000 ppm.

The possible mechanism for natural gas sensing was described by Chai et al. [81]. It was explained by the charge transfer due to adsorption–desorption of an oxygen species on surface. The hydrocarbon was dissociated on ZnO surface or on the surface of noble metal clusters before reacting with surface oxygen. It is believed

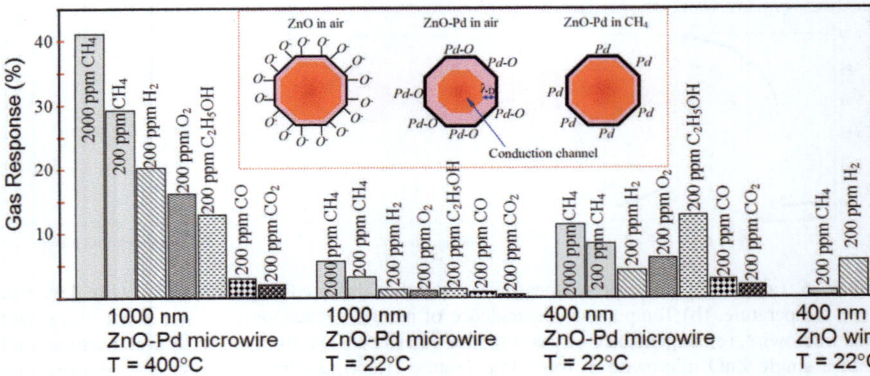

Fig. 5.17 Gas response of the single ZnO microwire gas sensor for different gases, operating temperatures, and diameters. The *inset* shows schematic cross-sectional view of pure ZnO microwire (*left*) and Pd-functionalized ZnO microwire (*middle* and *right*) with different length of the depleted region (λ_D). [Reprinted from Sensors and Actuators A: Physical, 176, G.Y. Chai, O. Lupan, E.V. Rusu, G.I. Stratan, V.V. Ursaki, V. Şontea, H. Khallaf, L. Chow, Functionalized individual ZnO microwire for natural gas detection, 64–71. Copyright (2012), with permission from Elsevier]

that palladium on the surface of zinc oxide microwire can improve the gas response and the rate of response due to catalytic activity for oxidation of natural gas (see inset in Fig. 5.17). Surface functionalization can increase the depletion region of surface, which will decrease rapidly when exposed to natural gas.

Figure 5.17 presents the gas response at 400 °C and 22 °C for a Pd-functionalized individual ZnO microwire and for a pure ZnO microwire (400 nm in diameter). The sensing mechanism of an individual ZnO microwire gas sensor exposed to air and to CH_4 can be found in previous works [3, 81]. Chai et al. [81] demonstrated the highest gas response of about 41 % at 400 °C and about 5.2 % at 22 °C from a sensor made from a single Pd-functionalized ZnO microwire (1,000 nm in diameter). The existence of the depletion layer of space charges at the microwires surface is drawn in Fig. 5.17 (inset).

Another approach was demonstrated by Kim et al. [83] by coating zinc oxide (ZnO)-branched tin oxide (SnO_2) nanowires with a Pd shell layer. They investigated the effects of thermal annealing and found that the Pd shell layers were agglomerated to form the nanoparticles, comprised of the cubic Pd phase. Kim et al. [83] demonstrated that the morphology of the Pd nanoparticles affected the sensing capability, in that the abundant and dense Pd nanoparticles obtained at certain temperatures enhanced the sensitivity to NO_2 gas.

Figure 5.18 presents the schematicoutlines of the gas sensors with Pd-coated (ZnO)-branched tin oxide (SnO_2) prior to thermal annealing and after the thermal annealing at 500, 600, and 700 °C. It was found that at 500 °C the agglomeration of the Pd shell has slightly progressed, a considerable part of the nanowire/branch surface has been covered with Pd nanoparticles and is no longer exposed to the

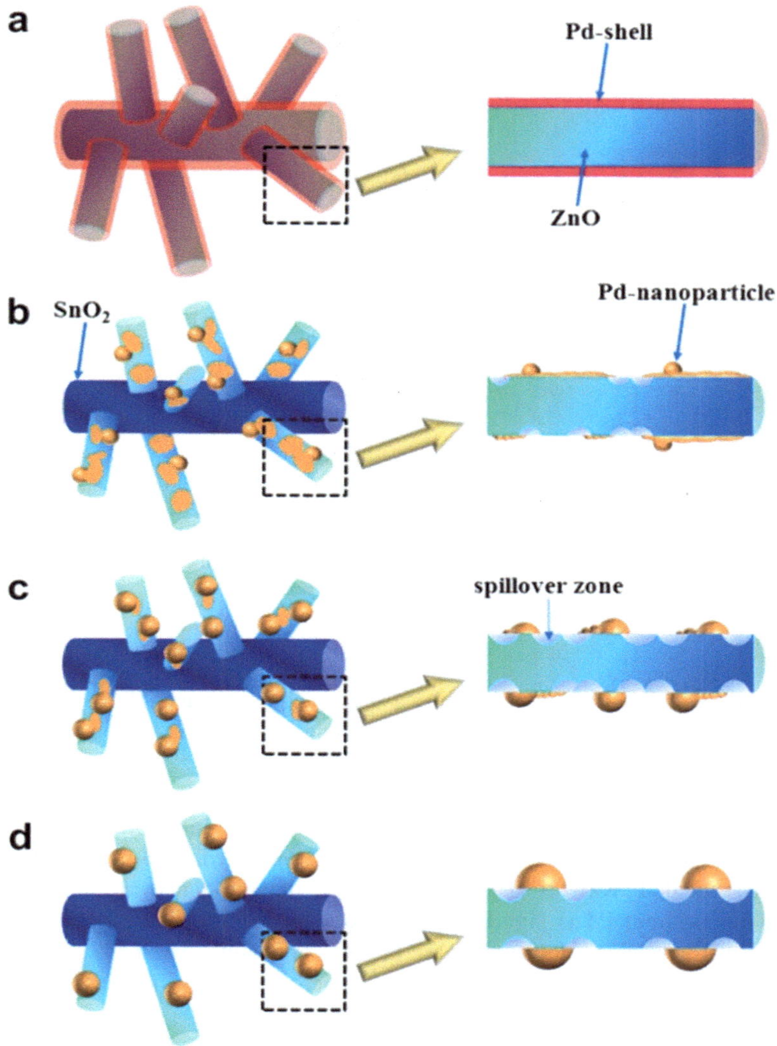

Fig. 5.18 Schematic outlines of Pd-coated-branched nanowires (**a**) prior to thermal annealing and (**b**, **c**) after the thermal annealing at (**b**) 500 °C and (**c**) 700 °C [Reprinted from Microelectronic Engineering 105, S. S. Kim, H. G. Na, S.-W. Choi, D. S. Kwak, H. W. Kim, Decoration of Pd nanoparticles on ZnO-branched nanowires and their application to chemical sensors, 1–7. Copyright (2013), with permission from Elsevier]

atmosphere. However, after annealing at 600 and 700 °C, the agglomeration of the Pd shell has sufficiently progressed, generating numerous Pd nanoparticles with sufficient exposed surfaces of nanowires/branches. Higher annealing temper-

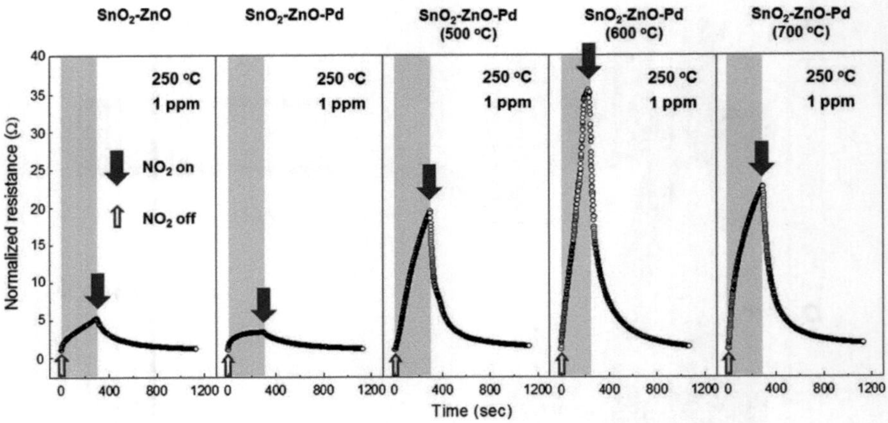

Fig. 5.19 Response curves to NO_2 gas with a concentration of 1 ppm at 250 °C for the sensors fabricated from uncoated-branched nanowires and Pd-coated-branched nanowires, which were as-synthesized, 500 °C-annealed, 600 °C-annealed, and 700 °C-annealed [Reprinted from Microelectronic Engineering 105, S. S. Kim, H. G. Na, S.-W. Choi, D. S. Kwak, H. W. Kim, Decoration of Pd nanoparticles on ZnO-branched nanowires and their application to chemical sensors, 1–7. Copyright (2013), with permission from Elsevier]

ature causes larger particles and thus a larger exposed area. Thus, the NO_2 gas response was controlled by thermal annealing and found dependent on the annealing temperature.

Figure 5.19 presents NO_2 gas response curves with a concentration of 1 ppm, for the sensors fabricated from uncoated-branched nanowires and Pd-coated-branched nanowires, which were as-synthesized, 500 °C-annealed, 600 °C-annealed, and 700 °C-annealed. Kim et al. [83] found that the resistance of the SnO_2/ZnO nanowires without Pd is significantly low and it does not increase by attaching the Pd nanoparticles. However, after thermal annealing of such structures, the normalized resistance has been significantly increased, contributing to the sensing performance.

5.5 The Sensing Mechanisms Proposed for Nanosensors

In order to develop new nanosensors and to improve their detection performances, it is important to understand their gas-sensing mechanism and factors which affect it. Several factors influencing the gas detection mechanism of an individual ZnO nanowire/nanorod and their fundamental mechanisms are still under debate. The main factors governing the gas-sensing performance of single ZnO NW are shown in Fig. 5.20. These factors include (a) the value of the diameter (D_{NW}) and the aspect

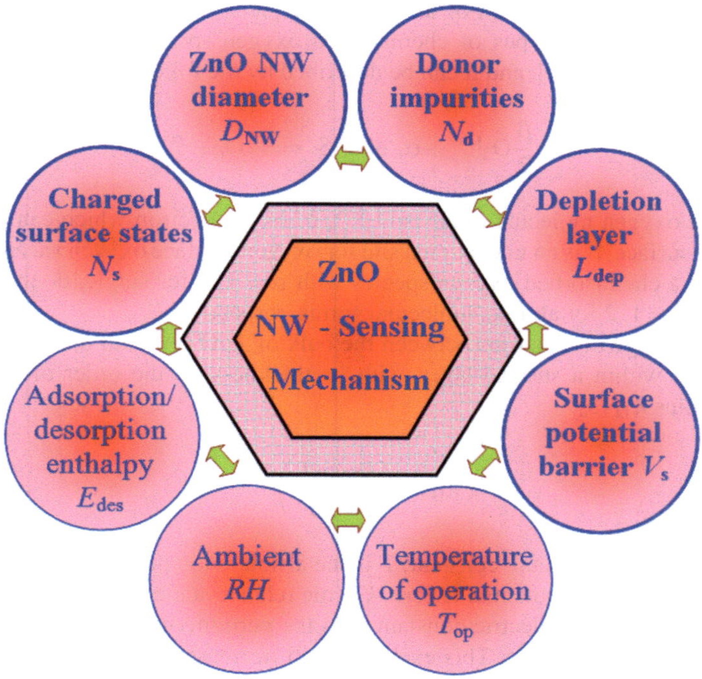

Fig. 5.20 Schematic representation of the major factors influencing the sensing mechanism of a doped single NW ZnO sensor [Reprinted from Sensors and Actuators B 173, O. Lupan, L. Chow, Th. Pauporté, L.K. Ono, B. Roldan Cuenya, G. Chai, Highly sensitive and selective hydrogen single-nanowire nanosensor, 772–780. Copyright (2012), with permission from Elsevier]

ratio (surface/volume) of NW [2, 3, 84], (b) the type of defects and its concentration in ZnO NW [44, 73], (c) the depletion region due to gas adsorption [85, 86] and the surface potential [87], (d) the adsorption–desorption enthalpy E_{des} [88], and (e) the temperature of operation [89]. Two distinguished regions are shown here: one is the conduction channel at the central part, and the other is the depletion region shown as the external hexagonal surrounding area near the surface (Fig. 5.20).

Prades et al. [79] observed experimentally that the main contributions to the H_2 gas-sensing behavior of the NWs are their diameter D_{NW}, aspect ratio, and dopant concentration. Thus, by decreasing the NW diameter, the response of single ZnO NWs to UV light and H_2 molecules have been greatly improved [3, 79]. As reported before, the metal oxide gas sensitivity depends on the interaction between the gas species and the adsorbed oxygen ($O_2{}^-$, O^- and O^{2-}) ions on the material surface [90–92]. When metal oxide nanowire is in air, the adsorbed oxygen on surface extracts electrons from the conduction band:

$$O_2\,(g) \leftrightarrow O_2\,(ad), \quad O_2\,(ad) + e^- \rightarrow O_2{}^-\,(ad), \quad \frac{1}{2}O_2 + e^- \rightarrow O^-\,(ad) \qquad (5.1)$$

and it leads to the formation of an electron depletion region and to the increase of its resistance. The types of chemisorbed oxygen species depend strongly on temperature [76, 90–92], and it can be described by the following [76]:

$$\frac{\beta}{2}{O_2}^{\mathrm{Gas}} + \alpha \cdot \mathrm{e}^- + \mathrm{S} \Longleftrightarrow {O_{\beta S}}^{-\alpha} \tag{5.2}$$

where the oxygen molecule in air is ${O_2}^{\mathrm{Gas}}$, e^- is an electron which can reach the nanowire surface, S is an unoccupied site for oxygen–surface oxygen vacancies, and ${O_{\beta S}}^{-\alpha}$ is a chemisorbed oxygen species with $\alpha = 1, 2$ for single, doubly ionized form and $\beta = 1, 2$ for atomic, molecular forms, respectively.

The exchange of electrons between the bulk of ZnO NW and the surface states takes place within a surface layer, with its thickness on the order of the Debye length/radius λ_{D} [3]:

$$\lambda_{\mathrm{D}} = \left(\frac{k\varepsilon_0 k_{\mathrm{B}} T}{e^2 N_0}\right)^{1/2} \tag{5.3}$$

where k_{B} is the Boltzmann constant, T is the absolute temperature in Kelvin, N_0 is the density where the potential is zero, k is the relative static electric permittivity of the medium, ε_0 is the electric constant or is the permittivity of the free space, and e is the elementary charge. This exchange will decrease of the net carrier density in the nanowire and will lead to band bending near the surface for conduction and valence bands [3]. The change in electrical conductance of the nanowire exposed to gas atmosphere (Figs. 5.14 and 5.15) is determined by the change in electrical charge carriers' concentration Δn [93]:

$$\Delta G = \left(\Delta n_0 \left|e\right| \mu \pi r^2\right)/l \tag{5.4}$$

The gas sensitivity is given by [94]

$$\frac{\Delta G}{G} = \frac{\Delta n_{\mathrm{s}}}{n_0} \tag{5.5}$$

Thus, higher gas sensitivity could be obtained by a larger modulation in the depletion region of ZnO nanowire [3, 94], which can be controlled by the electron density in ZnO nanowire.

The gas response of the resistive nanosensors is given by the following equation [3]:

$$\begin{aligned} S &= \frac{G_{\mathrm{g}} - G_{\mathrm{a}}}{G_{\mathrm{g}}} \cdot 100\,\% = \frac{4}{D}\left(\lambda_{\mathrm{D(a)}} - \lambda_{\mathrm{D(g)}}\right) \cdot 100\,\% \\ &= \frac{4}{D}\left(\left(\frac{k\varepsilon_0 k_{\mathrm{B}} T}{e^2 N_0}\right)^{1/2} - \left(\frac{k\varepsilon_0 k_{\mathrm{B}} T}{e^2 N_0}\right)^{1/2}\right) \cdot 100\,\% \end{aligned} \tag{5.6}$$

or by the following [95]:

$$S = \frac{G_g - G_a}{G_g} = \frac{4}{D}\left(\lambda_{D(a)} - \lambda_{D(g)}\right) = \frac{4}{D}\left(\frac{\varepsilon\varepsilon_0}{en_0}\right)^{1/2}\left(V_{Sa}^{1/2} - V_{Sg}^{1/2}\right) \tag{5.7}$$

where n_0 is carrier concentration in air and G_g and G_a are the conductance of ZnO nanowires in H_2 gas and in air ambient, respectively. V_{Sa} and V_{Sg} are the adsorbance-induced band bending in air and in H_2 gas, respectively. Thus, enhancement of H_2 gas sensitivity can be realized by controlling the geometric factor (4/D), electronic characteristics ($\varepsilon\varepsilon_0/en_0$), and adsorption-induced band bending ($V_{Sa}^{1/2} - V_{Sg}^{1/2}$) due to adsorption on ZnO nanowire surface.

After exposure of a nanosensor to H_2 atmosphere, the adsorption–desorption sensing mechanism can be described by reversible gas chemisorptions on surface of the ZnO nanowire [3]:

$$H_2 + \frac{1}{2}O_2^-\ (ad) \rightarrow H_2O + e^- \tag{5.8}$$

$$H_2 + O^-\ (ad) \rightarrow H_2O + e^- \tag{5.9}$$

The H_2 gas will react with adsorbed oxygen ions on the surface of ZnO nanowire and release electrons, thus increasing the current. Such reaction is exothermic (1.8 kcal mol^{-1}) [96], and the molecular H_2O desorbs quickly from the surface.

Slow recovery times could be explained by fact that not all gas species are desorbed rapidly from the NW surface. Lupan et al. [3] proposed to overcome this drawback through the application of UV radiation pulses to irradiate nanosensor. The mechanisms of improving the dynamic characteristics of the gas sensor by using UV pulses were explained in previous work [3]. The UV light pulse applied to the nanosensor contribute to desorption of oxygen, gas ions, and water molecules from the sensor surface, therefore producing clean ZnO nanowire surface [3, 97, 98]. It stimulates returning of ZnO nanowire sensor resistance to the initial R_a value. Another component of gas sensitivity is related to defects, especially oxygen vacancies can act as adsorption sites for gas species. According to recent reports [73, 99], surface defects such as oxygen vacancies can influence the electronic/chemical properties, adsorption behavior of metal oxides surfaces, and reversibility of sensor characteristics.

Here we have to point out that the adsorption of hydrogen species on the ZnO surface has been in the focus of researchers for more than 40 years [100, 101]. It is of great importance to understand the interaction of hydrogen with ZnO surfaces. This will enable the design of ZnO-based hydrogen sensors, especially by using nanowires as potential candidates for nanoscale devices.

The schematic energy band diagram in the radial direction of a single pure ZnO NW and a Cd-doped ZnO NW indicating the depletion region at the surface is presented in Fig. 5.21 (up) [79]. The structural models of the conduction mechanism

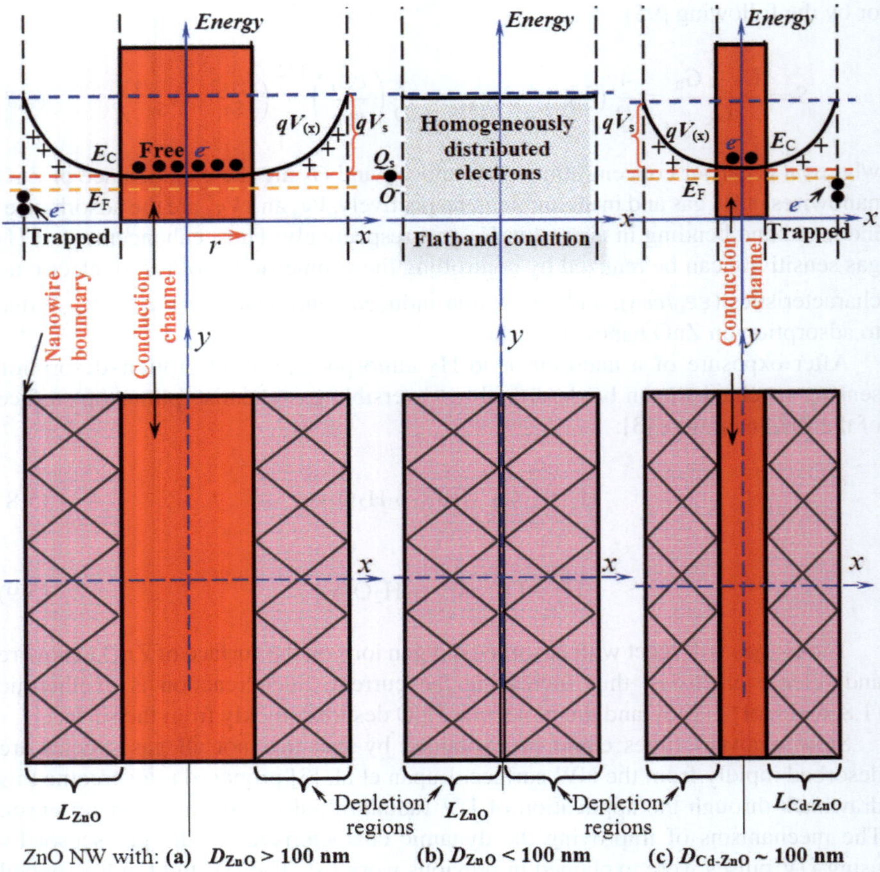

Fig. 5.21 Energy band diagram in the radial direction (*upper*) and structural (*lower*) models of the conduction mechanism of a single ZnO and Cd–ZnO nanowire in air. *x*—denotes depth from the surface. *L*—denotes the depletion layer at the surface. *q* is elemental charge of electron. qV_s is potential barriers at the surface, and $qV_{(x)}$ is the depth profile of the potential energy of electrons. (*upper*) Schematic energy band models in the radial direction of a nanowire indicating the depletion region at the surface and the quasi-neutral core region of radius *r* in the central part of the NW. V_s is the electric potential drop across the depletion layer, E_c is the lower edge of the conduction band, and E_F is the Fermi level. (*lower*) Structural representation of a cross-sectional view of the nanowires along its length: (**a**) pure ZnO NW with diameter $D > 100$ nm, (**b**) pure ZnO NW with diameter $D < 100$ nm, (**c**) and doped ZnO NW with diameter $D \sim 100$ nm [Reprinted from Sensors and Actuators B 173, O. Lupan, L. Chow, Th. Pauporté, L.K. Ono, B. Roldan Cuenya, G. Chai, Highly sensitive and selective hydrogen single-nanowire nanosensor, 772–780. Copyright (2012), with permission from Elsevier]

of single ZnO and Cd–ZnO NWs in air were drawn in Fig. 5.21 (down). Here, the parameter *x* indicates the depth from the nanowire surface, qV_s is the potential barrier at the nanowire surface, and $qV_{(x)}$ is the depth profile of the potential energy of the electrons. As can be seen from the left drawing (thicker NWs) in Fig. 5.21, the depletion layer does not influence the conduction channel significantly, if compared with the case of the thinner ZnO–Cd NWs (right drawing). It is important to note that for pure ZnO NWs with a diameter of about 50–100 nm, the charge depletion region encompasses the entire NW, and flat-band conditions take place (see Fig. 5.6 middle drawing).

However, for cases with controlled doping (the density of the conduction electrons $n \sim N_d$) or functionalization of the surface, it is possible to tune depletion region and obtain an enhanced gas response from ZnO NWs. Lupan et al. [79] reported on proposed mechanism and suggested that $L \sim Q_s$, $L \sim N_s$, $L \sim N_d^{-1}$; thus, it depends on the charge of the surface state (Q_s),the density of charged surface states (N_s), and the concentration of donor impurities (N_d). Such a control can be done by doping or by modulating the operation temperature. By increasing the doping concentration, it is possible to reduce the thickness of the surface depletion layer of the NW and the critical diameter (D_c), as shown in Fig. 5.21. Thus, the current flow through the conduction channel in a ZnO NW can be controlled and, consequently, the gas response. To improve the response and recovery times of NW-based sensors, it is needed to enhance adsorption–desorption kinetics by control of the operation temperature, including the self-heating approach [79, 80, 102, 103], or by doping. The doping of ZnO NWs contributes to the control of the extent of the surface depletion region in the NW, which must be kept small so that the single NW is not completely depleted. Lupan et al. [79] explained why the operation mode of ZnO nanowire devices can be controlled by the modulation of surface states through surface morphology engineering and size control.

5.6 Conclusions

The low-dimensional sensors based on a single ZnO nanowire/nanorod were fabricated by using focused ion beam (FIB/SEM) instrument. The diameter dependence of the gas response and selectivity of a single ZnO nanowire synthesized by chemical vapor-phase growth method was presented. It was reviewed the dependence of the gas response on the diameter of ZnO NW-based sensors and found the highest gas response (about 34 %) from a single ZnO NW with 100 nm in diameter and the highest gas response (about 274 %) from a single Cd (6 μM)-doped ZnO NW (90 nm diameter). For Cd–ZnO NWs with diameters of 140 and 200 nm, a lower response to H_2 less than 68 and 40 % was reported. It can be concluded that the fabricated ZnO NW nanosensor response to 100 ppm NH_3 and *i*-butane or 200 ppm CH_4 is much lower in comparison with the response to H_2 gas. The gas response and selectivity of the ZnO nanowires to H_2, NH_3, *i*-butane, and CH_4 gases at room temperature demonstrate that a single NW holds a high promise for nanoscale sensor applications. This is supported by the capability to operate at room temperature and the possibility to tune the gas response and selectivity by the surface defect concentration and the diameter of ZnO nanowire.

The reduction of the nanosensor's recovery time through the irradiation with an ultraviolet radiation pulse was demonstrated. The sensing mechanisms of ZnO nanowires were elucidated. Furthermore, their low power consumption (<5 nW at 2 mV) makes them suitable for use in portable devices without the need of external microheaters. The fact that our nanosensors require extremely low power to operate represents an important advance in power efficiency and nano-miniaturization.

Acknowledgments The author wishes to thank financial support of the STCU and ASM through Grant 09 STCU.A/5833.

References

1. Lupan, O., Chai, G., Chow, L.: Fabrication of ZnO nanorod-based hydrogen gas nanosensor. Microelectron. J. **38**, 1211–1216 (2007)
2. Lupan, O., Chai, G., Chow, L.: Novel hydrogen gas sensor based on single ZnO nanorods. Microelectron. Eng. **85**, 2220–2226 (2008)
3. Lupan, O., Ursaki, V.V., Chai, G., Chow, L., Emelchenko, G.A., Tiginyanu, I.M., Gruzintsev, A.N., Redkin, A.N.: Selective hydrogen gas nanosensor using individual ZnO nanowire with fast response at room temperature. Sensors Actuators B **144**(1), 56–66 (2010)
4. Lupan, O., Chow, L., Pauporté, T., Ono, L.K., Roldan Cuenya, B., Chai, G.: Highly sensitive and selective hydrogen single-nanowire nanosensor. Sensors Actuators B Chem. **173**, 772–780 (2012)
5. Hübert, T., Boon-Brett, L., Black, G., Banach, U.: Hydrogen sensors—a review. Sensors Actuators B **157**, 329–352 (2011)
6. Rumiche, F., Wang, H.H., Indacochea, J.E.: Development of a fast-response/high-sensitivity double wall carbon nanotube nanostructured hydrogen sensor. Sensors Actuators B **163**, 97–106 (2012)
7. Verhelst, S., Sierens, R.: Hydrogen engine-specific properties. Int. J. Hydrog. Energy **26**, 987–990 (2001)
8. Basic Research Needs to Assure a Secure Energy Future: A Report from the Basic Energy Sciences Advisory Committee. Available at http://www.sc.doe.gov/bes/besac/Basic_Research_Needs_To_Assure_A_Secure_Energy_Future_FEB2003.pdf (2003)
9. Zhou, J., Gu, Y., Fei, P., Mai, W., Gao, Y., Yang, R., Bao, G., Wang, Z.L.: Flexible piezotronic strain sensor. Nano Lett. **8**(9), 3035–3040 (2008)
10. Lupan, O., Pauporté, T., Viana, B.: Low-voltage UV-electroluminescence from ZnO-nanowire array/p-GaN light-emitting diodes. Adv. Mater. **22**, 3298–3302 (2010)
11. Lupan, O., Pauporté, T., Viana, B., Tiginyanu, I.M., Ursaki, V.V., Cortès, R.: Epitaxial electrodeposition of ZnO nanowire arrays on p-GaN for efficient UV-light-emitting diode fabrication. ACS Appl. Mater. Interfaces **2**(7), 2083–2090 (2010)
12. Lupan, O., Pauporté, T., Le Bahers, T., Viana, B., Ciofini, I.: Wavelength-emission tuning of ZnO nanowire-based light-emitting diodes by Cu doping: experimental and computational insights. Adv. Funct. Mater. **21**(18), 3564–3572 (2011)
13. Das, S.N., Kar, J.P., Choi, J.H., Lee, T.I., Moon, K.J., Myoung, J.M.: Fabrication and characterization of ZnO single nanowire-based hydrogen sensor. J. Phys. Chem. C **114**(3), 1689–1693 (2010)
14. Khan, R., Ra, H.-W., Kim, J.T., Jang, W.S., Sharma, D., Im, Y.H.: Nanojunction effects in multiple ZnO nanowire gas sensor. Sensors Actuators B **150**, 389–393 (2010)
15. Rout, C.S., Raju, A.R., Govindaraj, A., Rao, C.N.R.: Hydrogen sensors based on ZnO nanoparticles. Solid State Commun. **138**(3), 136 (2006)

16. LaRoche, J.R., Heo, Y.W., Kang, B.S., Tien, L.C., Kwon, Y., Norton, D.P., Gila, B.P., Ren, F., Pearton, S.J.: Fabrication approaches to ZnO nanowire devices. J. Electron. Mater. **34**(4), 404 (2005)
17. Wang, J.X., Sun, X.W., Yang, Y., Huang, H., Lee, Y.C., Tan, O.K., Vayssieres, L.: Hydrothermally grown oriented ZnO nanorod arrays for gas sensing applications. Nanotechnology **17**, 4995–4998 (2006)
18. Liao, L., Lu, H.B., Li, J.C., He, H., Wang, D.F., Fu, D.J., Liu, C., Zhang, W.F.: Size dependence of gas sensitivity of ZnO nanorods. J. Phys. Chem. C **111**(5), 1900–1903 (2007)
19. Kang, B.S., Heo, Y.W., Tien, L.C., Norton, D.P., Ren, F., Gila, B.P., Pearton, S.J.: Hydrogen and ozone gas sensing using multiple ZnO nanorods. Appl. Phys. A **80**, 1029–1032 (2005)
20. Yonenaga, I.: Thermo-mechanical stability of wide-bandgap semiconductors: high temperature hardness of SiC, AlN, GaN, ZnO and ZnSe. Phys. B Condens. Matter **308–310**, 1150 (2001)
21. Look, D.C.: Recent advances in ZnO materials and devices. Mater. Sci. Eng. B **80**, 383–387 (2001)
22. Burlacu, A., Ursaki, V.V., Lincot, D., Skuratov, V.A., Pauporte, T., Rusu, E., Tiginyanu, I.M.: Enhanced radiation hardness of ZnO nanorods versus bulk layers. Phys Status Solidi Rapid Res Lett **2**, 68–70 (2008)
23. Tian, Z.R., Voigt, J.A., Liu, J., Mckenzie, B., Mcdermott, M.J., Rodriguez, M.A., Konishi, H., Xu, H.: Nat. Mater. **2**(12), 821 (2003)
24. Erwin, S.C., Zu, L., Haftel, M.I., Efros, A.L., Kennedy, T.A., Norris, D.J.: Doping semiconductor nanocrystals. Nature **436**, 91–94 (2005)
25. Chen, W. (ed.): Doped Nanomaterials and Nanodevices, 3-volume set, 800 p. University of Texas at Arlington, USA American Editor, Journal of Nanoscience and Nanotechnology, USA, February, p. 1250 American Scientific Publishers, California, (2010). Hardcover ISBN: 1-58883-110-8 (2010)
26. Gong, H., Hu, J.Q., Wang, J.H., Ong, C.H., Zhu, F.R.: Nano-crystalline Cu-doped ZnO thin film gas sensor for CO. Sensors Actuators B **115**(1), 247–251 (2006)
27. Lupan, O., Shishiyanu, S., Ursaki, V., Khallaf, H., Chow, L., Shishiyanu, T., Sontea, V., Monaico, E., Railean, S.: Synthesis of nanostructured Al-doped zinc oxide films on Si for solar cells applications. Sol. Energy Mater. Sol. Cells **93**, 1417–1422 (2009)
28. Chen, M.T., Lu, M.P., Wu, Y.J., Song, J., Lee, C.Y., Lu, M.Y., Chang, Y.C., Chou, L.J., Wang, Z.L., Chen, L.J.: Near UV LEDs made with in situ doped p-n homojunction ZnO nanowire arrays. Nano Lett. **10**, 4387–4393 (2010)
29. Wang, X., Wang, Y., Aberg, D., Erhart, P., Misra, N., Noy, A., Hamza, A., Yang, J.: Batteryless chemical detection with semiconductor nanowires. Adv. Mater. **23**, 117–121 (2010)
30. Norris, D.J., Efros, A.L., Erwin, S.C.: Doped nanocrystals. Science **319**, 1776–1779 (2008)
31. Wang, Y.D., Ma, C.L., Wu, X.H., Sun, X.A., Li, H.D.: Electrical and gas sensing properties of mesostructured tin oxide-based H sensor. Sensors Actuators B Chem. **85**, 270–276 (2002)
32. Patnaik, P.: A Comprehensive Guide to the Hazardous Properties of Chemical Substances, p. 402. Wiley-Interscience, New Jersey (2007). ISBN 0-471-71458-5
33. Chow, L., Lupan, O., Chai, G.Y.: FIB fabrication of ZnO nanotetrapod and cross-sensor. Phys. Status Solidi B **247**, 1628–1632 (2010)
34. Lupan, O., Chow, L., Ono, L.K., Cuenya, B.R., Chai, G.Y., Khallaf, H., Park, S., Schulte, A.: Synthesis and characterization of Ag- or Sb-doped ZnO nanorods by a facile hydrothermal route. J. Phys. Chem. C **114**, 12401–12408 (2010)
35. Lupan, O., Pauporte, T., Le Bahers, T., Ciofini, I., Viana, B.: High aspect ratio ternary $Zn_{1-x}Cd_xO$ nanowires by electrodeposition for light-emitting diode applications. J. Phys. Chem. C **115**, 14548–14558 (2011)
36. Wang, H.T., Kang, B.S., Ren, F., Tien, L.C., Sadik, P.W., Norton, D.P., Pearton, S.J., Lin, J.: Hydrogen-selective sensing at room temperature with ZnO nanorods. Appl. Phys. Lett. **86**, 243503–243505 (2005)
37. Tien, L.C., Wang, H.T., Kang, B.S., Ren, F., Sadik, P.W., Norton, D.P., Pearton, S.J., Lin, J.: Electrochem. Solid-State Lett. **8**(9), G230 (2005)

38. Tien, L.C., Sadik, P.W., Norton, D.P., Voss, L.F., Pearton, S.J., Wang, H.T., Kang, B.S., Ren, F., Jun, J., Lin, J.: Appl. Phys. Lett. **87**, 222106 (2005)
39. Huh, J., Park, J., Kim, G.T., Park, J.Y.: Highly sensitive hydrogen detection of catalyst-free ZnO nanorod networks suspended by lithography-assisted growth. Nanotechnology **22**, 085502 (2011)
40. Rout, C.S., Kulkarni, G.U., Rao, C.N.R.: J. Phys. D. Appl. Phys. **40**, 2777 (2007)
41. Wang, H.T., Kang, B.S., Ren, F., Tien, L.C., Sadik, P.W., Norton, D.P., Pearton, S.J., Lin, J.: Appl. Phys. A: Mater. Sci. Proc. **81**(6), 1117 (2005)
42. Tien, L.C., Norton, D.P., Gila, B.P., Pearton, S.J., Wang, H.T., Kang, B.S., Ren, F.: Detection of hydrogen with SnO_2-coated ZnO nanorods. Appl. Surf. Sci. **253**(10), 4748–4752 (2007)
43. Sonawane, Y.S., Kanade, K.G., Kale, B.B., Aiyer, R.C.: Electrical and gas sensing properties of self-aligned copper-doped zinc oxide nanoparticles. Mater. Res. Bull. **43**, 2719–2726 (2008)
44. Zhao, M., Wang, X., Ning, L., Jia, J., Li, X., Cao, L.: Electrospun Cu-doped ZnO nanofibers for H_2S sensing. Sensors Actuators B **156**, 588–592 (2011)
45. Lupan, O., Chow, L., Chai, G., Roldan, B., Naitabdi, A., Schulte, A., Heinrich, H.: Nanofabrication and characterization of ZnO nanorod arrays and branched microrods by aqueous solution route and rapid thermal processing. Mater. Sci. Eng. B **145**, 57–66 (2007)
46. Chow, L., Lupan, O., Heinrich, H., Chai, G.: Self-assembly of densely packed and aligned bilayer ZnO nanorod arrays. Appl. Phys. Lett. **94**, 163105 (2009)
47. Lupan, O., Emelchenko, G.A., Ursaki, V.V., Chai, G., Redkin, A.N., Gruzintsev, A.N., Tiginyanu, I.M., Chow, L., Ono, L.K., Roldan Cuenya, B., Heinrich, H., Yakimov, E.E.: Synthesis and characterization of ZnO nanowires for nanosensor applications. Mater. Res. Bull. **45**, 1026–1032 (2010)
48. Redkin, A.N., Makovei, Z.I., Gruzintsev, A.N., Dubonos, S.V., Yakimov, E.E.: Inorg. Mater. **43**, 253 (2007)
49. Pauporté, T., Lupan, O., Viana, B.: Tunable electroluminescence from low-threshold voltage LED structure based on electrodeposited $Zn_{1-x}Cd_xO$-nanorods/p-GaN heterojunction. Phys. Status Solidi A **209**, 359–369 (2012)
50. Ursaki, V.V., Lupan, O.I., Chow, L., Tiginyanu, I.M., Zalamai, V.V.: Rapid thermal annealing induced change of the mechanism of multiphonon resonant Raman scattering from ZnO nanorods. Solid State Commun. **143**, 437 (2007)
51. Tiginyanu, I.M., Lupan, O., Ursaki, V.V., Chow, L., Enachi, M.: Nanostructures of metal oxides. Comprehensive Semicond. Sci. Tech. **3**, 396–479 (2011)
52. Al-Salman, H.S., Abdullah, M.J.: Hydrogen gas sensing based on ZnO nanostructure prepared by RF-sputtering on quartz and PET substrates. Sensors Actuators B Chem. **181**, 259 (2013)
53. Chow, L., Lupan, O., Chai, G., Khallaf, H., Ono, L.K., Roldan Cuenya, B., Tiginyanu, I.M., Ursaki, V.V., Sontea, V., Schulte, A.: Synthesis and characterization of Cu-doped ZnO one-dimensional structures for miniaturized sensor applications with faster response. Sensors Actuators A **189**, 399–408 (2013)
54. Chai, G., Lupan, O., Chow, L., Heinrich, H.: Crossed zinc oxide nanorods for ultraviolet radiation detection. Sensors Actuators A: Phys. **150**, 184 (2009)
55. Chow, L., Lupan, O., Chai, G.: Fabrication of ZnO nanorod-based hydrogen gas nanosensor. US Patent 8,263,002, 11 Sept 2012
56. Lupan, O., Chow, L., Chai, G., Chernyak, L., Lopatiuk-Tirpak, O., Heinrich, H.: Phys. Status Solidi A **205**, 2673 (2008)
57. Burlacu, A., Ursaki, V.V., Skuratov, V.A., Lincot, D., Pauporté, T., Elbelghiti, H., Rusu, E., Tiginyanu, I.M.: The impact of morphology upon the radiation hardness of ZnO layer. Nanotechnology **19**, 215714 (2008)
58. Peale, R.E., Flitsiyan, E.S., Swartz, C., Lupan, O., Chernyak, L., Chow, L., Vernetson, W.G., Dashevsky, Z.: Neutron transmutation doping and radiation hardness for solution-grown bulk and nano-structured ZnO. In: Mastro, M., LaRoche, J., Ren, F., Chyi, J.-I., Kim, J. (eds.) Performance and Reliability of Semiconductor Devices. Materials Research Society Symposium Proceedings, vol. 1108, pp. A05–03. Materials Research Society, Warrendale, PA (2009)

59. Look, D.C., Reynolds, D.C., Hemsky, J.W., Jones, R.L., Sizelove, J.R.: Production and annealing of electron irradiation damage in ZnO. Appl. Phys. Lett. **75**, 811–813 (1999)
60. Kucheyev, S.O., Jagadish, C., Williams, J.S., Deenapanray, P.N.K., Yano, M., Koike, K., Sasa, S., Inoue, M., Ogata, K.: Implant isolation of ZnO. J. Appl. Phys. **93**, 2972–2976 (2003)
61. Auret, F.D., Goodman, S.A., Hayes, M., Legodi, M.J., van Laarhoven, H.A., Look, D.C.: Electrical characterization of 1.8 MeV proton-bombarded ZnO. Appl. Phys. Lett. **79**, 3074–3076 (2001)
62. Reyntjens, S., Puers, R.: A review of focused ion beam applications in microsystem technology. J. Micromech. Microeng. **11**, 287–300 (2001)
63. Dhara, S., Datta, A., Wu, C.T., Lan, Z.H., Chen, K.H., Wang, Y.L., Chen, L.C., Hsu, C.W., Lin, H.M., Chen, C.C.: Enhanced dynamic annealing in Ga^{+} ion-implanted GaN nanowires. Appl. Phys. Lett. **82**, 451 (2003)
64. Steckla, A.J., Chyr, I.: Focused ion beam micromilling of GaN and related substrate materials (sapphire, SiC, and Si). J. Vac. Sci. Technol. B **17**(2), 362–365 (1999)
65. Rubanov, S., Munroe, P.R.: FIB-induced damage in silicon. J. Microsc. **214**(Pt 3), 213–221 (2004)
66. Chai, G., Lupan, O., Chow, L.: Focused ion beam fabrication of carbon nanotube and ZnO nanodevices. In: Russell, P., Utke, I., Moshkalev, S. (eds.) Nanofabrication Using Focused Ion and Electron Beams: Principles and Applications, pp. 503–514. Oxford University Press, New York (2012)
67. Weisenberg, D., Dürrschnabel, M., Gerthsen, D., Pérez-Willard, F., Reiser, A., Prinz, G.M., Feneberg, M., Thonke, K., Sauer, R.: Conductivity of single ZnO nanorods after Ga implantation in a focused-ion-beam system. Appl. Phys. Lett. **91**, 132110 (2007)
68. Lupan, O., Chai, G., Chow, L., Emelchenko, G.A., Heinrich, H., Ursaki, V.V., Gruzintsev, A.N., Tiginyanu, I.M., Redkin, A.N.: Ultraviolet photoconductive sensor based on single ZnO nanowire. Phys. Status Solidi A **207**, 1735–1740 (2010)
69. Lupan, O., Chow, L., Chai, G.: A single ZnO tetrapod-based sensor. Sensors Actuators B Chem. **141**(2), 511–517 (2009)
70. Kolmakov, A., Moskovits, M.: Chemical sensing and catalysis by one-dimensional metal–oxide nanostructures. Annu. Rev. Mater. Res. **34**, 151–180 (2004)
71. Wang, Z.L.: Zinc oxide nanostructures: growth, properties and applications. J. Phys. Condens. Matter **16**, R829–R858 (2004)
72. Shishiyanu, S.T., Shishiyanu, T.S., Lupan, O.I.: Sensing characteristics of tin-doped ZnO thin films as NO_2 gas sensor. Sensors Actuators B Chem. **107**, 379–386 (2005)
73. Ahn, M.W., Park, K.S., Heo, J.H., Park, J.G., Kim, D.W., Choi, K.J., Lee, J.H., Hong, S.H.: Gas sensing properties of defect-controlled ZnO-nanowire gas sensor. Appl. Phys. Lett. **93**, 263103 (2008)
74. Ferro, R., Podriguez, J.A., Bertrand, P.: Development and characterization of a sprayed ZnO thin film-based NO_2 sensor. Phys. Status Solidi C **2**, 3754–3757 (2005)
75. Lin, H.M., Tzeng, S.J., Hsiau, P.J., Tsai, W.L.: Electrode effects on gas sensing properties of nanocrystalline zinc oxide. Nanostruct. Mater. **10**, 465–477 (1998)
76. Barsan, N., Weimar, U.: Conduction model of metal oxide gas sensors. J. Electroceram. **7**, 143–167 (2001)
77. Arnold, M.S., Avouris, P., Pan, Z.W., Wang, Z.L.: Field-effect transistors based on single semiconducting oxide nanobelts. J. Phys. Chem. B **107**(3), 659–663 (2003)
78. Hassan, J.J., Mahdi, M.A., Chin, C.W., Abu-Hassan, H., Hassan, Z.: A high-sensitivity room-temperature hydrogen gas sensor based on oblique and vertical ZnO nanorod arrays. Sensors Actuators B Chem. **176**, 360–367 (2013)
79. Prades, J.D., Jimenez-Diaz, R., Hernandez-Ramirez, F., Cirera, A., Romano-Rodriguez, A., Morante, J.R.: Harnessing self-heating in nanowires for energy efficient, fully autonomous and ultra-fast gas sensors. Sensors Actuators B **144**, 1–5 (2010)
80. Prades, J.D., Jimenez-Diaz, R., Hernandez-Ramirez, F., Cirera, A., Romano-Rodriguez, A., Morante, J.R.: An experimental method to estimate the temperature of individual nanowires. Int. J. Nanotechnol. **2009**, 783675 (2009)

81. Chai, G.Y., Lupan, O., Rusu, E.V., Stratan, G.I., Ursaki, V.V., Şontea, V., Khallaf, H., Chow, L.: Functionalized individual ZnO microwire for natural gas detection. Sensors Actuators A Phys. **176**, 64–71 (2012)
82. Koerts, T., van Santen, R.A.: The reaction path for recombination of surface CH_x species. J. Mol. Catal. **70**, 119–127 (1991)
83. Kim, S.S., Na, H.G., Choi, S.W., Kwak, D.S., Kim, H.W.: Decoration of Pd nanoparticles on ZnO-branched nanowires and their application to chemical sensors. Microelectron. Eng. **105**, 1–7 (2013)
84. Riu, J., Maroto, A., Rius, F.X.: Nanosensors in environmental analysis. Talanta **69**, 288–301 (2006)
85. Roldan Cuenya, B., Kolmakov, A.: Nanostructures: sensor and catalytic properties. In: Sudipta, S. (ed.) Functional Nanostructures. Nanostructure Science and Technology, pp. 305–344. Springer, New York (2008). doi: 10.1007/978-0-387-48805-9_6
86. Fan, Z., Wang, D., Chang, P.C., Tseng, W.Y., Lu, J.G.: ZnO nanowire field-effect transistor and oxygen sensing property. Appl. Phys. Lett. **85**, 5923 (2004)
87. Fan, Z., Lu, J.G.: Electrical properties of ZnO nanowire field effect transistors characterized with scanning probes. Appl. Phys. Lett. **86**, 032111 (2005)
88. Zhang, Y., Kolmakov, A., Lilach, Y., Moskovits, M.: Electronic control of chemistry and catalysis at the surface of an individual tin oxide nanowire. J. Phys. Chem. B **109**, 1923–1929 (2005)
89. Wei, T.Y., Yeh, P.H., Lu, S.Y., Wang, Z.L.: Gigantic enhancement in sensitivity using schottky contacted nanowire nanosensor. J. Am. Chem. Soc. **131**, 17690–17695 (2009)
90. Lenaerts, S., Roggen, J., Maes, G.: FT-IR characterization of tin dioxide gas sensor materials under working conditions. Spectrochimica Acta Part A: Molecular Spectroscopy **51**, 883 (1995)
91. Yamazoe, N., Fuchigami, J., Kishikawa, M., Seiyama, T.: Interactions of tin oxide surface with O_2, H_2O and H_2. Surf. Sci. **86**, 335 (1979)
92. Chang, S.C.: Oxygen chemisorption on tin oxide: correlation between electrical conductivity and E.P.R. measurements. J. Vac. Sci. Technol. **17**, 366 (1980)
93. Law, J.B.K., Thang, J.T.L.: Improving the NH_3 gas sensitivity of ZnO nanowire sensors by reducing the carrier concentration. Nanotechnology **19**, 205502 (2008)
94. Fan, Z.Y., Lu, J.G.: Gate-refreshable nanowire chemical sensors. Appl. Phys. Lett. **86**, 123510 (2005)
95. Chen, P.C., Shen, G., Zhou, C.: Chemical sensors and electronic noses based on 1-D metal oxide nanostructures. IEEE Trans. Nanotechnol. **7**(6), 668–682 (2008)
96. Basu, S., Dutta, A.: Room-temperature hydrogen sensors based on ZnO. Mater. Chem. Phys. **47**, 93–96 (1997)
97. Zhang, D.H.: Adsorption and photodesorption of oxygen on the surface and crystallite interfaces of sputtered ZnO films. Mater. Chem. Phys. **45**, 248–252 (1996)
98. Li, Q.H., Gao, T., Wang, Y.G., Wang, T.H.: Adsorption and desorption of oxygen probed from ZnO nanowire films by photocurrent measurements. Appl. Phys. Lett. **86**, 123117–123119 (2005)
99. Liao, L., Lu, H.B., Li, J.C., Liu, C., Fu, D.J., Liu, Y.L.: The sensitivity of gas sensor based on single ZnO nanowire modulated by helium ion radiation. Appl. Phys. Lett. **91**, 173110 (2007)
100. Eischens, R.P., Pliskin, W.A., Low, M.J.D.: The infrared spectrum of hydrogen chemisorbed on zinc oxide. J. Catal. **1**, 180–191 (1962)
101. Wang, C., Zhou, G., Li, J., Yan, B., Duan, W.: Hydrogen-induced metallization of zinc oxide $\left(2\bar{1}\bar{1}0\right)$ surface and nanowires: the effect of curvature. Phys. Rev. B **77**, 245303 (2008)
102. Prades, J.D., Jimenez-Diaz, R., Hernandez-Ramirez, F., Barth, S., Cirera, A., Romano-Rodriguez, A., Mathur, S., Morante, J.R.: Ultralow power consumption gas sensors based on self-heated individual nanowires. Appl. Phys. Lett. **93**, 123110 (2008)
103. Fangohr, H., Chernyshenko, D.S., Franchin, M., Fischbacher, T., Meier, G.: Joule heating in nanowires. Phys. Rev. B **84**, 054437 (2011)

Chapter 6
Recent Advances in the Design of Photodetectors Based on Thin-Film and Nanostructured ZnO

Yuri M. Strzhemechny

Abstract This chapter provides a synopsis of the most recent developments in the field of research and design of optoelectronic sensors and related devices based on nano- and microscale ZnO. It is well understood that in addition to advantages of ZnO as a versatile material, its various nanomorphologies introduce additional benefits for sensor applications in optical communications, chemical and biological analysis, and energy conversion materials. Different types of photodetector device structures and architectures are discussed herein, with a brief review of the growth/synthesis processes to generate them, especially those methods that can be scaled up for mass production and be performed at relatively low temperatures and/or low cost. Most essential device properties and figures of merit—responsivity, spectral selectivity, speed, etc.—are reviewed in relation to physical and chemical phenomena, particularly those occurring at nanoscale surfaces and interfaces. The role of native and extrinsic defects in such processes is discussed in detail. Potential solutions for design challenges and functionality roadblocks are also analyzed in the chapter.

6.1 Introduction

The general principle of operation of a semiconductor photodetector is based on creation of electron–hole pairs under illumination when the light-induced separation of electrons and holes produces a photocurrent collected by the electrodes. Most of the time, conductivity changes proportionally to the optical generation rate.

Today there are numerous applications of photodetecting devices. Among most common are automatic light controllers, burglar alarms, optical communications,

Y.M. Strzhemechny (✉)
Department of Physics and Astronomy, Texas Christian University, TCU Box 298840, Fort Worth, TX 76129, USA
e-mail: y.strzhemechny@tcu.edu

S. Li et al. (eds.), *Nanoscale Sensors*, Lecture Notes in Nanoscale Science and Technology 19, DOI 10.1007/978-3-319-02772-2_6,

illumination level sensors, object counters, etc. The choice of a photodetector type for a specific application is determined mainly by the required sensitivity, response speed, and the wavelength range. The latter in most cases is controlled by the band gap of the semiconductor material. To attain fast response times, high responsivities, and low dark currents, it is necessary to clearly elucidate the predominant mechanisms of carrier transport and recombination, which in turn depend on the microscopic phenomena related to the quality and morphology of materials and their surfaces/interfaces. Surface and bulk defects are especially important in this regard.

One can distinguish between several different types of semiconductor photodetectors depending on the arrangement for the separation of the photogenerated electron–hole pairs [1–3]. On the one hand, the simplest arrangement yields a photoconductor, for which the bias is applied across the semiconductor via two metal Ohmic contacts. On the other hand, photovoltaic detectors use electric field inside a *p-n* or a Schottky junction to generate election-hole pairs. Photovoltaic detectors come in several different configurations, such as *p-n* junction detectors, *p-i-n* detectors, Schottky junction detectors, and metal-semiconductor-metal (MSM) detectors. Each of these designs possesses advantages and drawbacks. For example, photoconductors exhibit high gain but slow response. Schottky junctions operate at high switch speeds and low dark currents; however interfacial defects may hamper their performance. The MSM photodetectors exhibit high speeds, high breakdown voltages, and large barrier heights, whereas the *p-i-n* photodiodes produce high current gain and low noise, although a high-quality *p*-type material is essential. Some of these detectors are frequently operated in reverse bias to optimize the width of the depletion region and thus reduce the probability of electron–hole recombination. In recent years novel device designs have emerged based on essentially nanoscale architectures.

Among detection in other spectral ranges, ultraviolet (UV) detection plays an increasingly important role. UV photodetectors have a broad range of applications in chemical, environmental and biological analysis, optical communications, and integrated photonic/optoelectronic devices. Visible-blind UV photodetectors can be used for astronomical studies, astronautical exploration and space communications, as well as environmental (ozone layer, pollution, water sterilization) monitoring. UV detection is desirable for flame sensing and early detection of missile launch plumes. It can be also utilized in vacuum microelectronics, energy conversion, microwave amplifiers, engine control, etc. One has to mention another important aspect of calibration and monitoring of the UV light intensity necessary to minimize exposure of human body to the UV radiation of the Sun and thus to reduce chances of skin cancer. Hence, it is desirable to develop functional and efficient UV photodetectors.

In the past, most of the UV detection employed photomultiplier tubes (PMT). PMT have been ubiquitous because of the mature technology of their production. However, they need high voltages to operate and are not suitable for miniaturization. Moreover, to make them operate in a solar-blind regime, one has to use additional external filters. Therefore, design and production of semiconductor UV photodetectors became predominant in recent decades.

Silicon-based UV photodiodes gained acceptance due to well-established growth processing and assembly protocols. They can be printed on the microchip and require much lower operating voltages. As mentioned above, semiconductor photodetectors are most sensitive to radiation with energy equal or somewhat greater than the band gap. Therefore Si-based photodetectors are not optimized to operate in the UV range. Because of the incomplete visible spectrum rejection rate, they tend to have low detection levels due to substantial dark current. A viable alternative is to employ materials with a direct wide band gap, such as SiC, GaN, InGaN, and AlGaN. Nitride-based detectors gained a widespread acceptance today with numerous mature device designs and production techniques. They exhibit excellent figures of merit and operate well in harsh environments. Nevertheless, their production remains rather expensive due to natural scarcity of gallium. In addition, novel nanoscale device structures are somewhat limited due to complex technological procedures involved. These complications can be successfully addressed by introducing zinc oxide as a basis for a UV detecting device.

ZnO, a wurtzite compound semiconductor, has found extensive use in varistors, sensors, and solar cells. Recently, ZnO has attracted considerable attention as a promising material for new important applications in optoelectronics and spintronics. Studies of physical and chemical properties of ZnO represent a thriving field with a broad domain of topics. ZnO is a material with a unique combination of valuable traits. ZnO demonstrates remarkable excitonic properties (exciton binding energy of 60 meV at room temperature, more than three times greater than that for GaN) and outstanding chemical and environmental stability (high damage threshold due to the high Zn-O bond strength). In fact, it outperforms GaN in radiation resistance and high temperature stability. Both constituent elements are abundant; the fabrication cost is low, including production of high-quality native substrates; the synthesis methods are well established; and the product is a naturally robust *n*-type material. Latest advances in crystal growth allowing bulk production of high-quality ZnO single crystals with reproducible characteristics render this material a good candidate for such applications as room temperature excitonic lasers, UV emitters and detectors, as well as high-temperature/high-power microelectronics. ZnO doped with magnetic impurities demonstrates characteristics of a dilute magnetic semiconductor at room temperature. For the purposes of UV detection, ZnO has an appropriate band gap energy of 3.37 eV, exhibits a strong UV photoresponse, reveals high transparency in the visible range, and is thermally and chemically stable. The saturation velocity and the breakdown electric field are much higher than those for SiC and GaN. ZnO is biocompatible, piezoelectric, and allows wet chemical etching, all of which may be beneficial for specific UV sensing applications.

In modern technology, implementation of nanoscale structures achieves ever increasing role. It becomes a conventional wisdom that ZnO is one of those materials exceptionally suitable for production and engineering of nanostructures (see, e.g., [4]). It is easy to synthesize nanoscale ZnO employing a broad and constantly growing variety of techniques for such synthesis. These fabrication techniques are mostly inexpensive, with ease of control, and allowing production of relatively large quantities of nanoscale ZnO materials. In many instances such growth yields high-quality ZnO nanocrystals. Moreover, ZnO nanostructures can be produced in a wide range of different morphologies and dimensionalities, such as nanorods, nanowires, nanobelts, nanourchins, and nanotetrapods. Currently, research in the field of ZnO nanostructures is one of the major areas in materials science and it is rapidly expanding. In the past few years, numerous emerging applications of nano-ZnO have been suggested and tested: LEDs, field-effect transistors, bio- and chemical sensors, transducers, nanocantilevers, thermal transport elements, and, central to our topic, UV photodetectors. The fact that the objects of interest are nanosize crystals brings about distinctly new phenomena. Moreover, since these ZnO nanostructures come in different sizes, shapes, and dimensionalities, each specific case of a nanostructure often demonstrates its own set of properties. Generally, the most important feature of a nanostructure, which affects its physical and chemical properties and frequently yields a qualitatively dissimilar behavior, is the high surface-to-volume ratio. It manifests itself in the substantially modified density of states, especially in the difference of surface vs. bulk states. Because of that, in nanostructures one deals with surface-enhanced phenomena (e.g., increased efficiency of radiative recombination), surface-induced strain, confinement, etc. The latter is brought mostly due to a discrete spectrum of eigenstates of elementary excitations (phonons, excitons, plasmons) in the crystal lattice and has a strong dependency on the shape, size, etc.

Major thrust in the studies of defects in bulk ZnO crystals has been recently focused primarily on overcoming the difficulties with producing a high-quality reliably *p*-type material. Such material is sought because of a great potential for spintronic and high-performance optoelectronic applications. For the former, a hole-mediated exchange between magnetic dopants is required to achieve high Curie temperatures, whereas for the latter, *p*-*n* homojunctions are needed for efficient lasing. As of today, efforts to obtain a device-grade *p*-type ZnO have not resulted in a material suitable for mass production.

Remarkably, nanoscale ZnO structures are offering numerous applications that do not require resolving the challenge of *p*-type conductivity. The recently reported studies demonstrated that nanosize ZnO can be employed in such systems as nanotransducers [5], random lasers [6], dye-sensitized solar cells [7], nanosensors [8], etc. This diversity of the range of applications regardless of the conductivity type is a significant argument in favor of great potential of nanocrystalline ZnO.

In particular, all the factors listed above make it feasible and relatively easy to incorporate ZnO nanostructures into new photodetector architectures, as will be discussed below in this chapter. For the purposes of UV sensing applications, ZnO nanostructures possess many advantages essential for adequate performance, such as high sensitivity due to large surface-to-volume ratio and low-dimensional carrier confinement, optical transmittance, electrical conductivity, negative electron affinity, a good electron field emission and field enhancement factor, as well as tunable electronic properties. A variety of ZnO nanostructures are readily available through numerous cheap and simple preparation techniques, which is not the case for most other semiconductor materials used for UV detection. Due to this, key device parameters can be tuned by adjusting sizes and shapes of the nanostructures.

Photodetector designs employing ZnO continuous thin films are also gaining popularity. Their growth techniques (to be discussed below) are also numerous and are implemented in many traditional and emerging applications, such as transistors, sensors, photodetectors, and memory devices. In particular, most of the ZnO thin-film implementations in the photodetectors are employed mostly in the MSM or Schottky junction configurations. Advantages of using ZnO thin films in photodetectors stem from such factors as, e.g., existence of well-established growth procedures, low price of the precursors, ability to scale up the production, as well as thermal and environmental stability of the obtained films.

Once again, in many of those ZnO systems, surface properties play a critical role. Surface is, effectively, the largest defect of a crystal lattice. Moreover, in the process of termination and reconstruction that can involve chemisorbed and physisorbed contamination, microchemical imperfections, and generated stresses, the surface develops elevated concentration of defects. These defects may not be necessarily confined within the immediate vicinity of the surface but could permeate inward driven by diffusion, stress, cation/anion imbalance, etc. It is quite natural to assume that such defective layer, with distinct stoichiometric, optoelectronic and structural characteristics, is a generic feature weakly affected by the growth conditions. Results for bulk ZnO crystals (e.g., [9, 10]) demonstrate that indeed there likely exists a defective nanoscale layer with distinct optoelectronic and structural characteristics. It is quite natural to presume that the same statements apply to a free surface of a nanosize ZnO object. In this case the volume ratio of the defective layer vs. relatively defect-free core becomes very significant and thus contribution of the surface and near-surface states may dramatically increase. Moreover, in many photodetector devices nanoscale interfaces are essential for effective device operation, and thus not only the free surface defects but also the interfacial defects play a crucial role.

It should be noted that understanding of the character and effects of the surface/interface defects in ZnO is largely incomplete. Not only elucidation, but the entailing modification and control of the surface defects are required for performance improvement of ZnO-based sensors. This aspect will be emphasized below.

Despite all the advantages ZnO offers for UV detector applications and several decades of research on the photoresponse of ZnO [11], its vigorous device implementation began in earnest only at the turn of the twenty-first century. This trend started when, almost simultaneously, research teams from the Rutgers University [12], University of Maryland [13], and University of California [14] reported on successful device implementation for novel ZnO-based UV detectors. While the former two produced thin-film MSM structures, the latter described operation of a photodetector based on a single ZnO nanowire. Lu's group investigated properties of a Schottky detector made using ZnO proper, whereas Venkatesan's group employed a $Mg_xZn_{1-x}O$ alloy. The authors obtained the following performance parameters for the MSM structures: photoresponsivities of 1.5 A/W [12] and 1,200 A/W [13], the leakage current of 1 nA at 5 V [12], and rise and fall times of 12 ns [12]/8 ns [13] and 50 ns [12]/1,400 ns [13] correspondingly. Rejection of the visible signal was reported to be more than four orders of magnitude [13].

During the next few years, the number of publications on ZnO-based UV photodetectors began to rapidly grow, with a variety of new approaches emerging for both thin-film and nanostructured geometries. For example, in reference [15], lateral $Mg_xZn_{1-x}O$ continuous composition spreads were implemented in a compact broadband photodetector. Epitaxial film MSM structures were shown to improve performance by employing ruthenium [16] and palladium [17, 18] electrodes. Many other groups reported on UV photodetection employing single ZnO nanowires [19–23], with demonstrated polarized photodetection [24] and high internal gain [25]. In addition, devices based on arrays of ZnO nanostructures have been designed: Kumar et al. [26] synthesized ZnO nano-needle multilayer UV detectors, in [27] multiple *n*-type ZnO nanowire-formed heterojunction with *p*-type silicon, while Hsu et al. [28] reported on a device based on structurally uniform vertical ZnO nanowires grown on a ZnO:Ga/glass template. All these, as well as other studies, laid a solid foundation for the continuously expanding field of ZnO-based photodetectors. Several reviews appeared in press reflecting progress in these devices based on nanoscale [29, 30] and thin-film ZnO [2]. The most recent few years clearly demonstrate significant and accelerating progress achieved in this domain, which will be discussed in detail below.

6.2 Types of High-Performance ZnO Photodetectors

Even though the latest publications on high-performance ZnO photodetectors were dealing with the designs based on both thin films and nanostructures, there has been a significant increase in the share of reports on devices formed primarily by nanostructures, such as nanowires, nanorods, nanobelts, tetrapods, nanoflakes, nanosheets, nanoparticles, nanotips, nanotubes, and nanohelixes. Common trends gravitate toward just a few major groups of device architectures based on nanoscale ZnO. The most prevalent approaches offered and reported recently for the efficient photodetection include two major groups. The first one represents ordered arrays of vertical nanorods/nanowires either in the form of a standard interdigitated MSM structure, a *p-n* junction, or a Schottky contact, for which enhanced performance is attributed to the very large surface-to-volume ratio of the ZnO nanostructures. A good illustration of the first type can be seen in Fig. 6.1 [31]. An example of a ZnO nanowire/n-Si heterojunction is presented in Fig. 6.2 [32]. The second group covers structures containing Schottky barriers between a metal and quasi-1D ZnO nanostructures, such as stand-alone nanowires, nanorods, and nanobelts (see, e.g., [33], Fig. 6.3a) or lateral networks thereof. However, many emerging nanoscale-based designs do not fall into these two classes, and they will be discussed below as well. Architecturally, both the thin-film and nanoscale-based detectors fall primarily into two categories—MSM and non-MSM, as illustrated in the following two subsections.

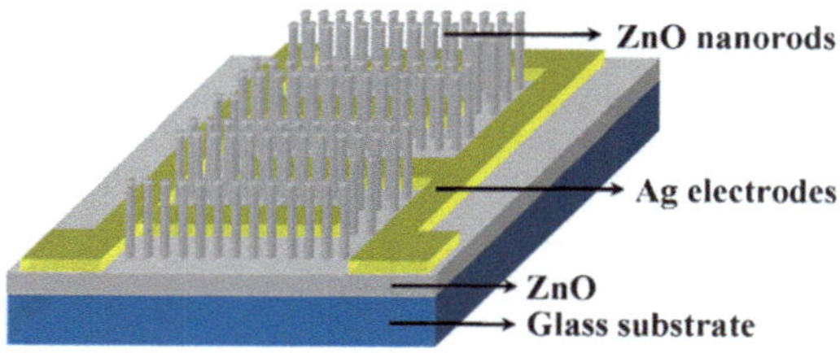

Fig. 6.1 Schematic for UV MSM photodetectors with ZnO nanorod arrays [31]. Reprinted with permission from Ji, L.W., Peng, S.M., Su, Y.K., Young, S.J., Wu, C.Z., Cheng, W.B., Appl. Phys. Lett. **94**, 203106, (2009). Copyright 2009, AIP Publishing LLC

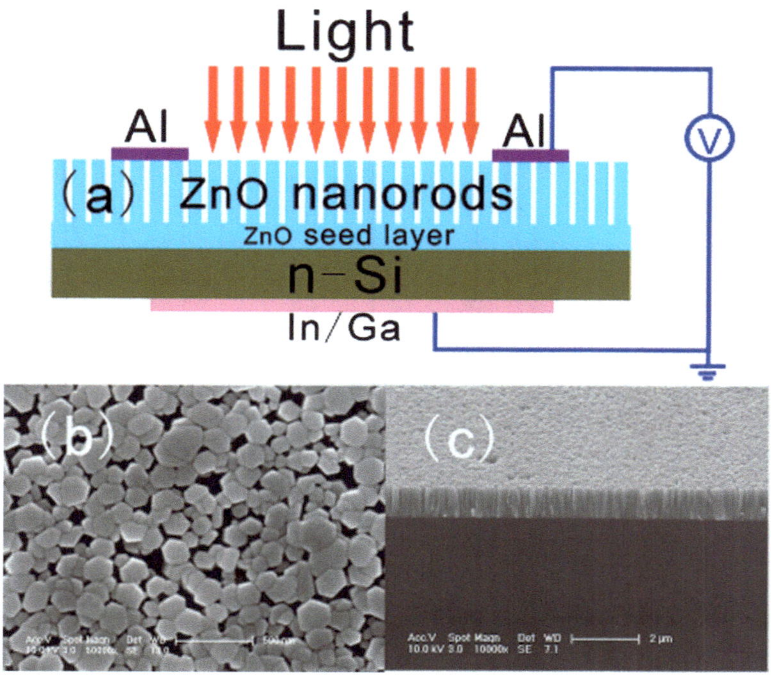

Fig. 6.2 (**a**) The diagram of the ZnO nanorods/n-Si heterojunction. (**b**) Top view of the SEM photograph of the as-prepared ZnO nanorods. (**c**) A cross-sectional SEM image of the as-prepared ZnO nanorods [32]. Reprinted with permission from Huang, H., Fang, G., Mo, X., Yuan, L., Zhou, H., Wang, M., Xiao, H., Zhao, X., Appl. Phys. Lett. **94**, 063512, (2009). Copyright 2009, AIP Publishing LLC

6.2.1 MSM Architectures

In the MSM configurations based on both continuous thin films and nanostructures, performance of metals forming rectifying contacts (cf., e.g., [31, 34–41]) has been addressed more often than of those forming Ohmic contacts (cf., e.g., [42–44]).

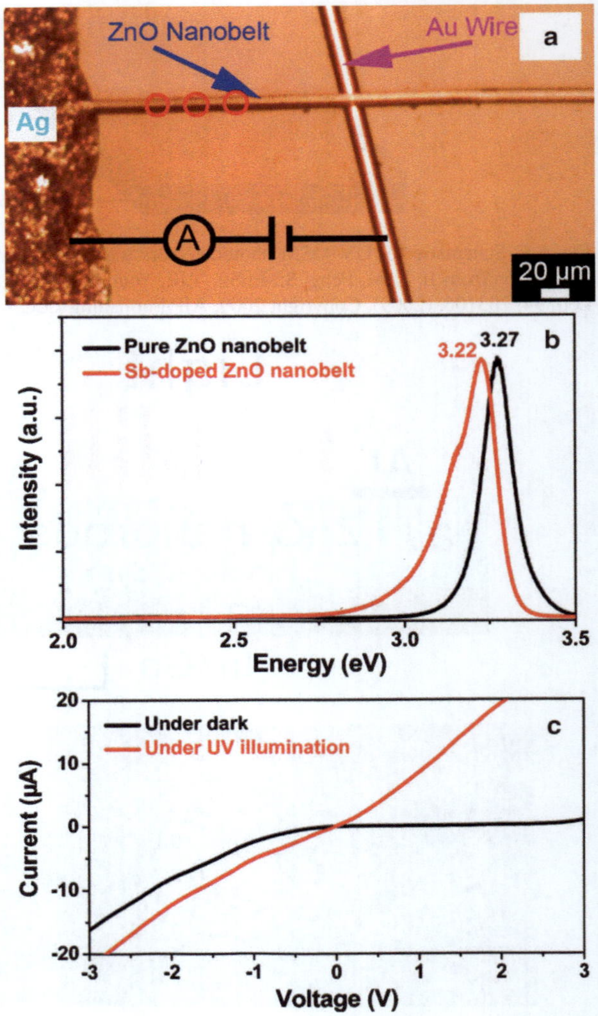

Fig. 6.3 (**a**) Optical image of the device based on a single ZnO nanobelt (**b**) PL spectra of single pure and Sb-doped ZnO nanobelts on a glass substrate. (**c**) I–V characteristics of the device under dark and UV light illumination at the Schottky contact [33]. Reprinted with permission from Yang, Y., Guo, W., Qi J., Zhao, J., Zhang, Y., Appl. Phys. Lett. **97**, 223113 (2010). Copyright 2010, AIP Publishing LLC

Overall, it appears that good performance of the Schottky MSM UV photodiodes strongly depends on obtaining a large Schottky barrier height. For thin-film MSM structures, the most common metals employed to produce Schottky contacts were Au (e.g., [34]), Ti/Au for the lateral interdigitated back-to-back structure in [36], as well as aluminum (cf. [41]), whereas for Ohmic contacts, mostly aluminum (e.g., [42, 43]). There were numerous architectural improvements reported for the MSM thin-film detectors. One of them is growing a thick active layer on a ZnO substrate to alleviate the decrease of the electric field with depth within the MSM structure [37]. For the Ohmic MSM contacts, decrease of the interelectrode spacing was demonstrated to improve electron mobility [42]. By the same token, double-layer

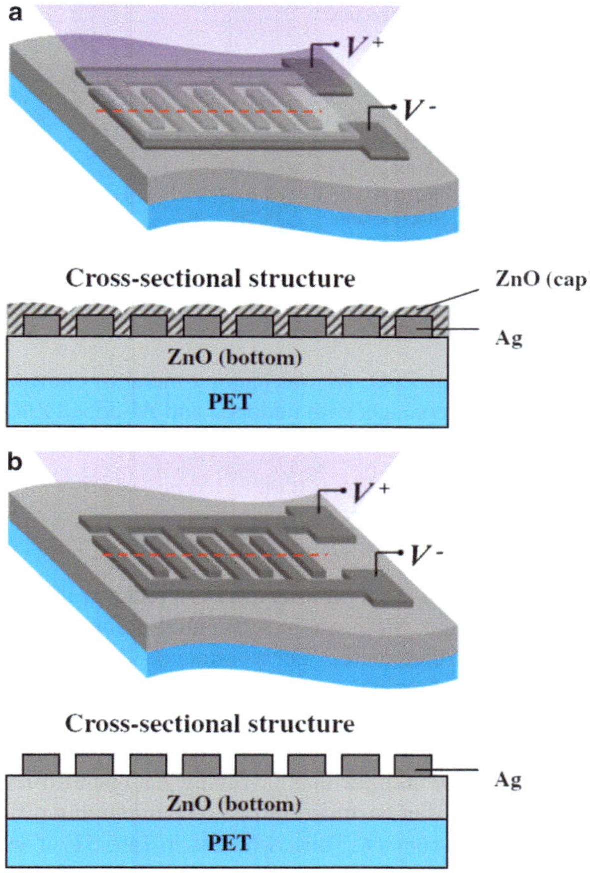

Fig. 6.4 (**a**) Schottky-type MSM photodetector with ZnO cap layer. (**b**) Traditional Schottky-type MSM photodetector [40]. Copyright 2010, The Japan Society of Applied Physics

photodetectors fabricated on flexible substrates demonstrated a better performance compared to the more conventional MSM devices based on a single ZnO layer [40] (Fig. 6.4).

Nanocrystalline ZnO-based Schottky-type devices with the MSM architecture were produced employing well-aligned vertical ZnO nanorods selectively grown in the gap between the interdigitated electrodes on ZnO/glass [31] (Fig. 6.1) and on ZnO/polyimide substrate [38] (Fig. 6.5). Among new approaches within the MSM architectures using nano-ZnO, one can mention a design using Wheatstone bridge-type electrodes exhibiting enhanced carrier collection and substantial miniaturization of the detector [45]. Such metals as Pd [39], Ag [40], and Ag/Au [38] were successfully utilized to produce Schottky contacts for nano-ZnO-based devices with the MSM structures. Nanomorphologies other than nanorods (e.g., tetrapods and hair-like nanowires, cf. [39], Fig. 6.6) have been also tested in Schottky-type MSMs. For Ohmic contacts in MSM detectors based on

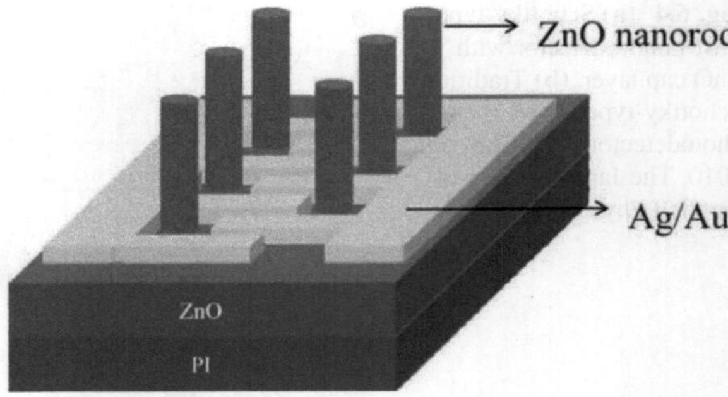

Fig. 6.5 Schematic structure for the fabricated ZnO nanorod-based MSM photodetectors [38]. Reproduced by permission of The Electrochemical Society from Chen, T.P., Young, S.J., Chang, S.J., Hsiao, C.H., Huang, C.S., J. Electrochem. Soc. **159**, J153–J157 (2012)

nanocrystalline ZnO, Al was again a metal of choice. For example, in [44] it was used to form contacts with undoped and Al-doped ZnO nanorod arrays.

6.2.2 Other Architectures

Although MSM configuration has been recently the most popular in the ZnO-based detector design, there was a substantial number of reports on devices employing non-MSM architectures. For continuous ZnO thin films, they were reported in [36, 46–48]. For example, the authors of [48] studied the effects of annealing on the performance of a Schottky barrier Au/Cr/SiO_2/ZnO/Al MIS photodetector. Consistent with the trend, a lot more numerous were recent publications on non-MSM detector architectures involving ZnO nanostructures. Single nanowire-based devices were demonstrated in [49–51] employing different approaches to forming the Schottky contact. Thus, whereas in [49, 51] it was reported on the Schottky contact between a ZnO nanowire and gold, in [50] a nanowire sandwiched between two graphene sheets had Schottky barriers formed at the interfaces with graphene (Fig. 6.7). Another Schottky barrier diode was fabricated using ZnO nanowire/Au contacts in conjunction with polymethyl methacrylate (PMMA) microlens arrays [3]. The PMMA microlens array was employed to improve the light trapping and as a passivation layer to minimize the leakage current. Also, in [52] the authors tested a photodetecting device fabricated using vertically aligned dense ZnO nanowires forming mechanical Schottky contacts with platinum-coated polyimides. Another Schottky UV photodetector was demonstrated in [33] (Fig. 6.3) where a self-powered device operated based on a single Sb-doped ZnO nanobelt in contact with Au (and an Ohmic contact on the other side). Moreover, in [51] it was suggested that when Ag nanoparticles were deposited on the surface of a ZnO nanowire bridging two Au/Cr electrodes, it led to the performance enhancement due to the formation of a thin AgO_x layer. Among non-MSM ZnO Schottky barrier photodetectors, one can further mention hybrid nanostructures based on a reduced graphene oxide/ZnO nanoparticle interfaces [53], self-assembled nanowires on Si with Ti/Pt contacts [33] (Fig. 6.8), and lateral ZnO nanowire-based photodetectors

Fig. 6.6 (**a**1) Low-and (**a**2) high-magnification SEM images of ZnO nanorods on Si substrate; (**b**1) low-and (**b**2) high-magnification SEM images of hair-like ZnO nanowires grown on porous silicon/Si substrate; (**c**1) low- and (**c**2) high-magnification SEM images of tetrapod-like ZnO nanorods grown on quartz substrate [39]. Reprinted with permission from Abdulgafour, H.I., Hassan, Z., Ahmed, N.M., Yam, F.K., J. Appl. Phys. **112**, 074510 (2012). Copyright 2012, AIP Publishing LLC

on glass with gold Schottky contacts [54]. UV detectors employing only Ohmic contacts with nanostructured ZnO were also reported. For example, gold contacts were shown to be Ohmic for closely packed ZnO nanowire arrays on Si [55] (Fig. 6.9) and for spatial networks of tetrapod ZnO nanostructures [56]. Ohmic contacts were formed between high-density ZnO nanoflakes/nanowires on glass and

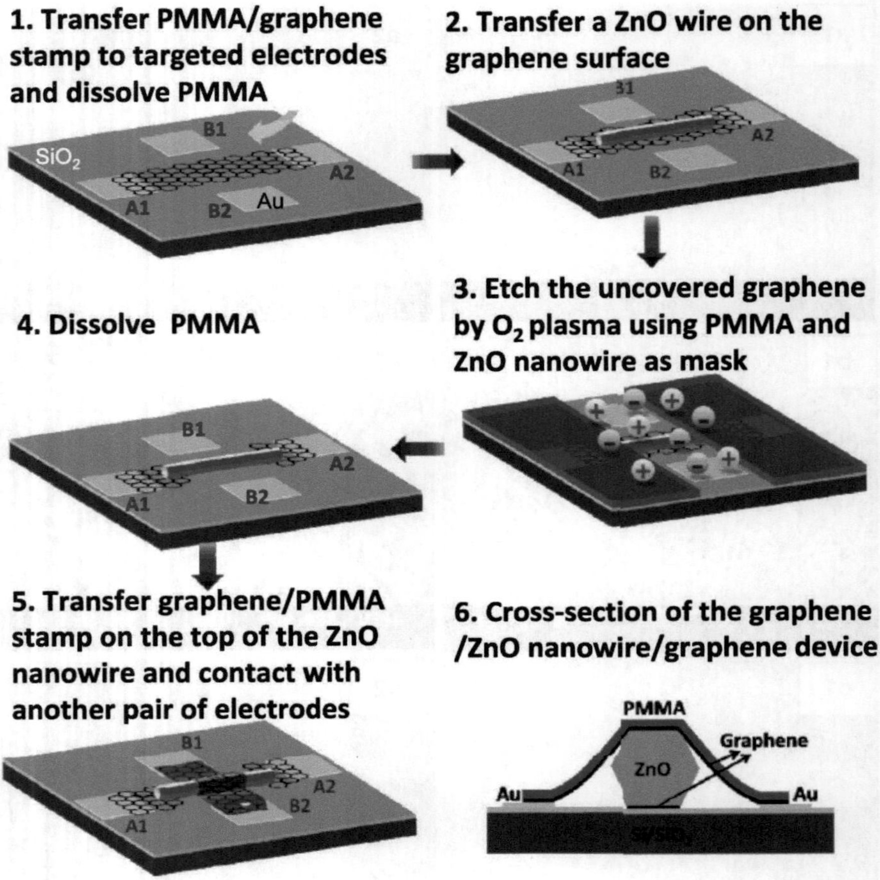

Fig. 6.7 (*Steps 1–5*) Schematic diagrams of the processes to fabricate the graphene/ZnO NW/graphene vertical sandwich device. (*Step 6*) Schematic diagram of the cross section of the graphene/ZnO NW/graphene-based sandwich structure UV detector [50]. Reprinted with permission from Fu X.W., Liao, Z.M., Zhou, Y.B., Wu, H.C., Bie, Y.Q., Xu, J., Yu, D.P., Appl. Phys. Lett. **100**, 223114 (2012). Copyright 2012, AIP Publishing LLC

a Zn pad [57] (Fig. 6.10). Aluminum Ohmic contacts were also obtained for the UV detectors employing unannealed and annealed ZnO nanorods on Si [32] (Fig. 6.2) and nanorod arrays on seed-layer-coated glass [58]. Ohmic indium tin oxide contacts were reported for nanowires on a polymer substrate [59] (Fig. 6.11). Other designs for nanoscale ZnO-containing photodetectors included *p-n* homojunctions employing ZnO nanowire arrays [60] (Fig. 6.12), metal-insulator-semiconductor device employing ZnO/SiO_2 core-shell nanowires [61], ZnO nanocrystal-containing unipolar circuitry (graphite electrodes) pencil-drawn on paper [62], multiple quantum well-based device [63], and photodiodes employing Ga-doped ZnO films on GaN films [64, 65].

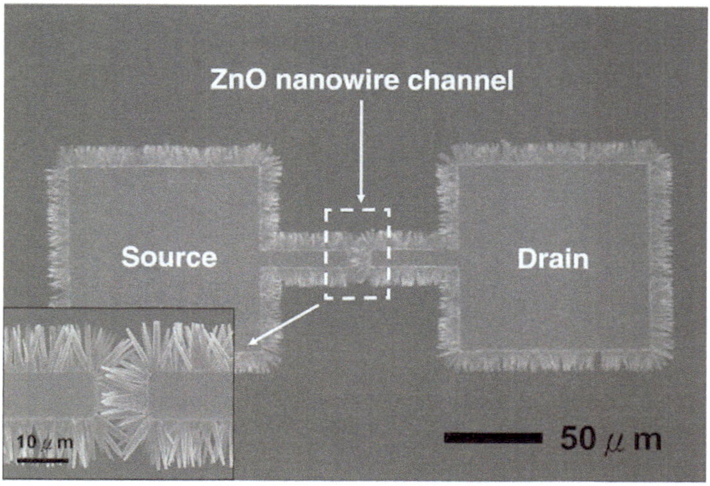

Fig. 6.8 Optical microscope image of the device discussed in [33]. The *inset* is an enlarged image of laterally self-assembled ZnO nanowires with Ti/Pt electrodes. Reprinted with permission from Yang, Y., Guo, W., Qi J., Zhao, J., Zhang, Y., Appl. Phys. Lett. **97**, 223113 (2010). Copyright 2010, AIP Publishing LLC

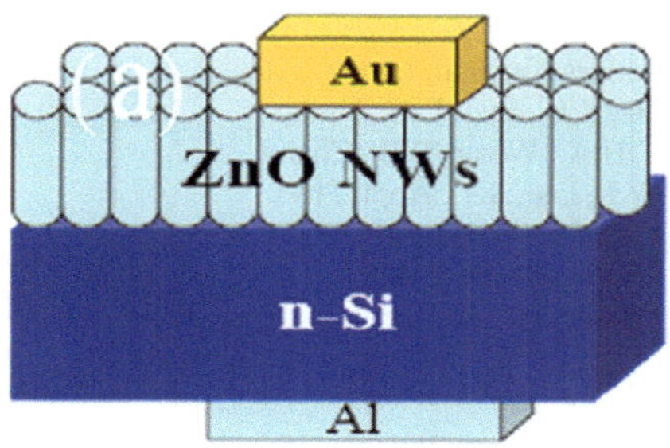

Fig. 6.9 The schematic of the ZnO nanowire array/n-Si heterojunction [55]. Reprinted with permission from Guo, Z., Zhao, D., Liu, Y., Shen, D., Zhang, J., Li, B., Appl. Phys. Lett. **93**, 163501 (2008). Copyright 2008, AIP Publishing LLC

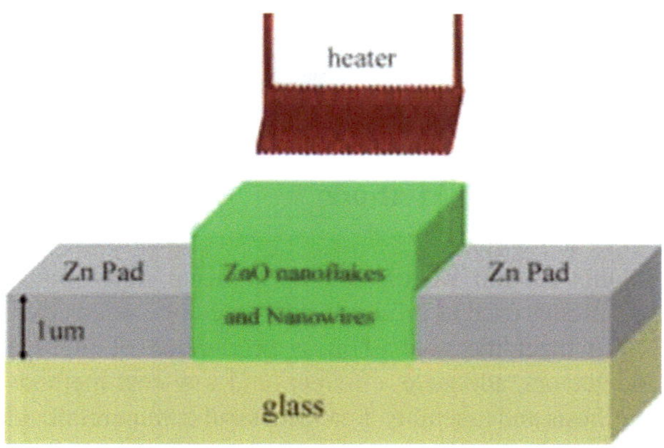

Fig. 6.10 System for synthesizing ZnO nanoflakes and nanowires [57]. Reproduced by permission of The Electrochemical Society from Hsu, C.L., Hsueh, T.J., Chang, S.P., J. Electrochem. Soc. **155**, K59–K62 (2008)

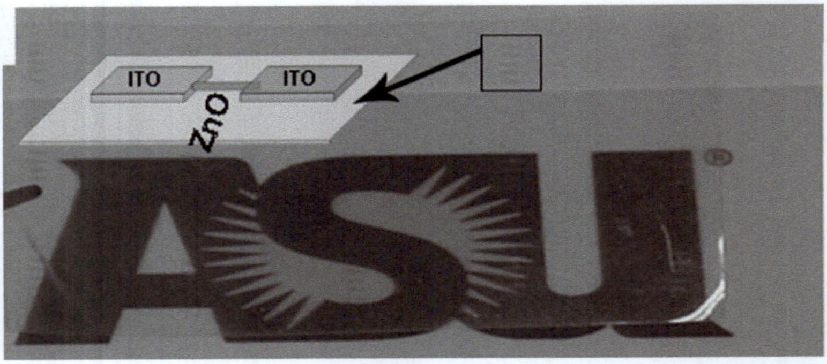

Fig. 6.11 ZnO nanowire-based visible-transparent UV detector on a polyethylene 2,6-naphthalate substrate (*inset*: schematic of the device) [59]. Reprinted with permission from Yu, H., Azhar, E. A., Belagodu, T., Lim, S., Dey, S., J. Appl. Phys. **111**, 102806 (2012). Copyright 2012, AIP Publishing LLC

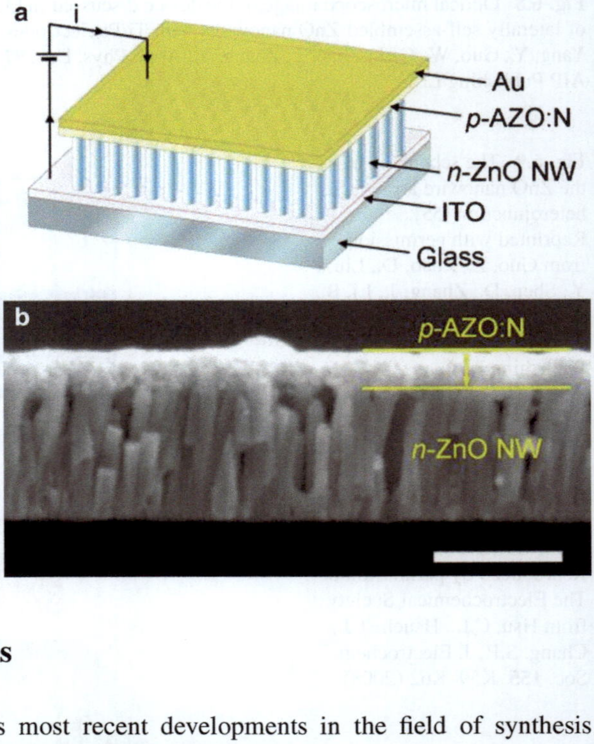

Fig. 6.12 (**a**) Schematic diagram of the ZnO homojunction device. (**b**) An SEM image of the cross section of the homojunction device. The scale bar is 500 nm [60]. Reprinted with permission from Leung, Y.H., He, Z.B., Luo, L.B., Tsang, C.H.A., Wong, N.B., Zhang, W.J., Lee, S.T., Appl. Phys. Lett. **96**, 053102 (2010). Copyright 2010, AIP Publishing LLC

6.3 Growth Methods

In this section we discuss most recent developments in the field of synthesis techniques used to produce ZnO-based photodetectors.

To meet the goal of mass production of high-performance ZnO-based photodetectors, adequate, efficient, and low-cost methods should be implemented for synthesis and assembly. For successful commercialization, obviously, these methods should embrace simplicity of material synthesis and, preferably, low-temperature processes, and they must be easily scaled up through inexpensive bottom-up protocols. In modern applications, complex nanoscale architectures and designs, such as coaxial or branched nanostructures, as well as organic/inorganic hybrid nanosystems gain prominent role, with many challenges still remaining en route of commercialization for these systems. As an example, for the photodetector applications, integration of nanowire/nanorod arrays into an operable device and mass production thereof is significantly hindered by the difficulties with the metallization processes during contact formation.

As was mentioned above, there are scores of growth methods in use to fabricate photodetectors based on both nanoscale and thin-film ZnO.

One of the most common growth techniques to produce ZnO thin films for photodetector applications is sputtering. Radio frequency (rf) magnetron sputtering was used to grow ZnO thin films on such hard substrates as SiO_2/p^+Si [42, 66] or glass with ultrathin metal overlayers [46, 47]. Flexible substrates were also used to sputter ZnO thin films onto, e.g., DC-sputtering on polypropylene carbonate [35] or rf-sputtering on polyethylene terephthalate [40]. Co-sputtering was used to grow films of ZnO alloys with MgO (e.g., [37]) and Ga_2O_3 (e.g., [64]) for photodetector devices. In the latter, a thick MgZnO epilayer was grown on a ZnO substrate to suppress its response. There were also papers that reported using magnetron co-sputtering to grow nanoscale ZnO, e.g., an array of closely packed ZnO nanowires on Si [55].

At the present time, one of the most popular and probably the most promising techniques capable of producing relatively large quantities of nanoscale ZnO is a low-temperature aqueous solution method. Not surprisingly, in recent years it has been utilized to synthesize nanoscale ZnO for photodetector applications. For example, this method was used on silicon substrates to grow nanorods [32] (Fig. 6.2), [45] and lateral self-assembled nanowires with a ZnO seeding layer [33] (Fig. 6.8). The aqueous solution growth was implemented to produce ZnO nanorods on a flexible polyimide substrate to generate an MSM photodetector, while the interdigital structure was fabricated by a standard photolithography/lift-off procedure [38] (Fig. 6.5). Solution growth was also reported in [67] to create thin-film photosensitive transistors based on a blend of indium gallium zinc oxide and single-walled carbon nanotubes.

Other low-cost methods reported lately were the thermal evaporation and the sol-gel growth, the latter also in conjunction with sonication and spin coating. The sol-gel method has been employed to create photodetectors based on both thin-film [48] and nanoscale [44, 58] ZnO. Thermal evaporation yielded devices based on lateral nanowires on glass substrates [54] as well as ZnO/SiO_2 core-shell nanowires on SiO_2/Si [61]. Thermal oxidation of metallic Zn is also a low-cost method that has been shown to produce nanoflakes and nanowires on glass [57] (Fig. 6.10). Hydrothermal solution growth was used in [60] (Fig. 6.12) to grow ZnO nanorod arrays on glass substrates.

On the other hand the growth techniques requiring more complex growth techniques/equipment to produce ZnO-based photosensitive devices were also reported by numerous authors. For example, for the ZnO-based photosensing applications, chemical vapor deposition was applied to synthesize vertical nanowire arrays [52], Sb-doped nanobelts bridging Ohmic and Schottky contacts [33] (Fig. 6.3) [49], and graphene/ZnO nanowire/graphene vertical structures [50] (Fig. 6.7). Atomic layer deposition was used in photodetector designs to grow thin films of ZnO on glass [34] and Al-doped ZnO buffer layers below ZnO nanowire arrays [52]. Pulsed laser deposition was reported as a growth method for ZnO films in [36], whereas molecular beam epitaxy was to synthesize ZnMgO/ZnO multiple quantum wells in [63].

One can mention other techniques to generate ZnO nanostructures for photodetectors: carbothermal method [68] and vapor-liquid-solid process [59] (Fig. 6.11) to grow nanowires, hydrothermal process to grow nanowires and subsequent dielectrophoresis to align them [3], catalyst-free vapor-solid mechanism to produce Pd/ZnO/Pd nanostructures on different substrates [39], chemical solution through photolithography for selective growth of nanorod arrays in a gap between interdigitated electrodes [31] (Fig. 6.1), etc. The method described in [39] is a good example of a low-cost technique: ZnO nanostructures (nanorods, nanotetrapod, hair-like nanowires) can be synthesized on various substrates without using any metal catalyst or a ZnO seed layer.

6.4 Devices on Flexible Substrates and Other Low-Cost Designs

As we mentioned above, flexible substrates have been increasingly utilized in the design of ZnO-based photodetectors. It is well know that these substrates are gaining significant popularity in science and technology due to their potential to fully overhaul the design paradigms for future commercial products. Electronic devices—such as displays, detectors, and sensors—on flexible and transparent substrates present appealing prospects to create integrated and wearable products with multiple functionalities. Such gadgets could be renewable, recyclable, conformable, implantable, with multilayer architectures, and suitable for large-scale production. Importantly, ZnO is fitting very well into this new concept since it can be grown by low-temperature processes (see the preceding section).

Here we will mention several illustrative examples of successful implementation of flexible substrates into different schemes of device operation for ZnO-based photodetectors. For instance, in the abovementioned report [40], the authors employed poly(ethylene terephthalate) as a substrate for ZnO epitaxial films (Figs. 6.4 and 6.13b) to create a detector with a performance superior to that of the traditional MSM ZnO photodetectors. In [38] polyimide was used as a flexible substrate for a photodetector device based on the vertical arrays of ZnO nanorods (Fig. 6.5). A case of a flexible high-performance UV detector operating on single ZnO nanowires was demonstrated in [59]. Polyethylene 2,6-naphthalate was employed to support active channels of ZnO nanowires between Ohmic indium tin oxide contacts as

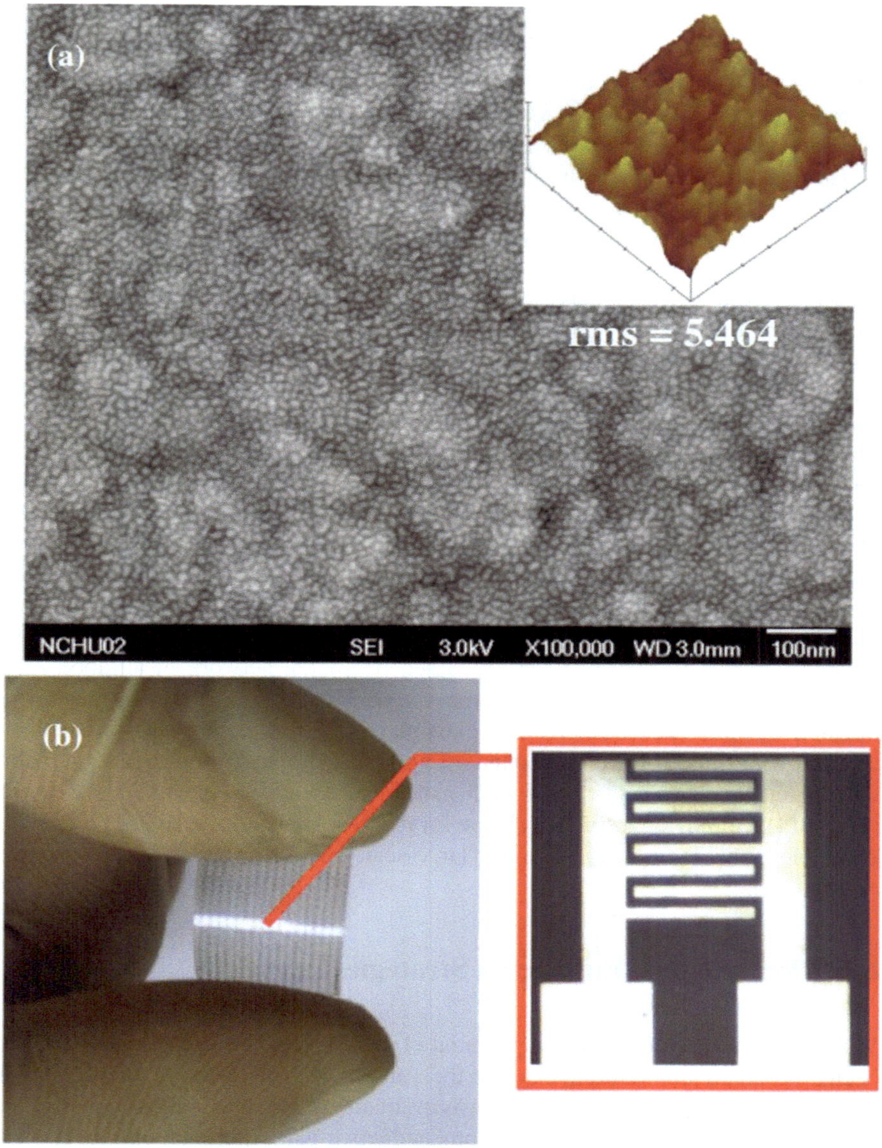

Fig. 6.13 (**a**) SEM image for the ZnO epitaxial film grown on PET substrate. The *inset* shows the corresponding AFM image with a surface roughness rms value of 5 nm. (**b**) Photo of the fabricated flexible MSM photodetectors. The *inset* shows the image of a MSM device obtained by optical microscopy [40]. Copyright 2010, The Japan Society of Applied Physics

shown in Fig. 6.11. A novel approach to flexible detector with high photoresponse was reported in [62]. The UV sensor was produced on paper by pencil drawing of interdigitated graphitic electrodes on screen-printed ZnO nanocrystals. This scheme offers such advantages as light weight, corrosion/heat resistance, and obviously low cost and easiness of production.

Turning our discussion to the related topic of design strategies aimed at lowering the production costs, it is worth mentioning several other examples of novel approaches to economical fabrication of ZnO-based devices. Thus, in [57], the authors reported on UV photodetectors employing high-density nanoflakes/nanowires on glass. Simple and quick synthesis technique involving localized oxidation of rf sputter-deposited Zn particles at a low temperature (see Fig. 6.10) coupled with a simple contact scheme yielded devices suitable for inexpensive mass production. Low-cost large-area photodetector arrays could be also produced by a low-temperature aqueous solution growth used in [33] to synthesize lateral self-assembled nanowires on Si with a ZnO seeding layer. A resulting photodetector (Fig. 6.8) demonstrated good photoresponse characteristics in both time domain and intensity. A cost-effective production of ZnO-based photodetectors was also demonstrated in [39]. The device based on nanoscale ZnO, already discussed in this text, was fabricated employing inexpensive catalyst-free vapor-solid mechanism highly suitable for mass production. The authors of [56] described a photodetector produced by another easy and cost-effective method that could be also scaled up. Based on spatial network of tetrapod ZnO nanostructures grown via thermal evaporation [56], the detector was made by transferring these nanostructures on a glass plate. In [68] a device was produced via the following processing route: standard photolithography/etching was combined with a selective CVD to obtain ZnO nanowire-based photodetector on SiO_2/Si (Fig. 6.14), employing a simple scheme for macroscopic contacts to ZnO nanowire ensembles suitable for integration into large-scale circuits.

6.5 Performance-Enhancing Strategies

With the growing interest in the potential of ZnO-based photodetectors, there has been a sustained joint effort to enhance their many performance parameters. Among these one should mention photoresponsivity, quantum efficiency, signal/noise ratio, response/switching/recovery time, photoconductive gain, illumination power density, low dark currents, photocurrent/dark current contrast, as well as performance stability. In this section we review some recent new ideas and approaches aimed at improving performance of ZnO-based photodetectors.

Significant enhancements in photoresponsivity were reported in, e.g., [31, 39, 44, 67]. Henceforth, the authors of [31] demonstrated an increased responsivity (Fig. 6.15) in MSM detectors on glass substrates when the traditional thin-film semiconductor component was replaced with vertical nanorod arrays (Fig. 6.1). Such increase in photoresponse was attributed to the significantly greater surface-

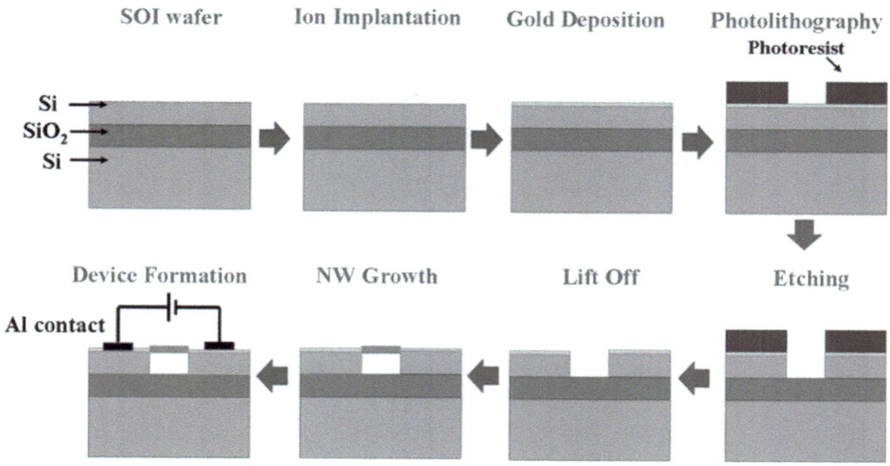

Fig. 6.14 Simplified synthesis schematics for the device described in [68]. Reprinted with permission from Xie, Y., Madel, M., Neuschl, B., Jie, W., Röder, U., Feneberg, M., Thonke, K., J. Vac. Sci. Technol. **B30**, 061801 (2012). Copyright 2012, AVS

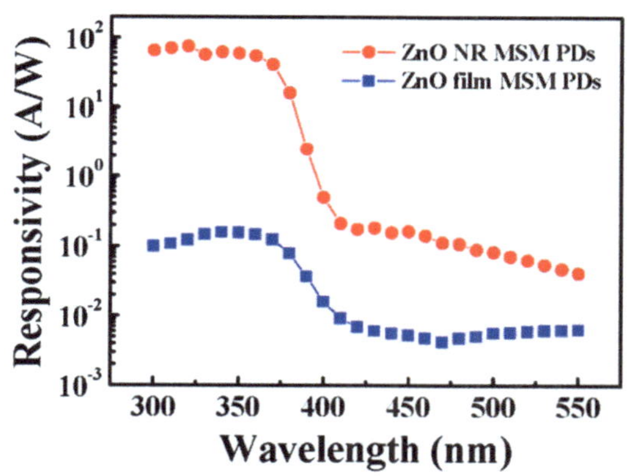

Fig. 6.15 Measured spectral responsivities of the ZnO MSM photodetectors at 5 V applied bias [31]. Reprinted with permission from Ji, L.W., Peng, S.M., Su, Y.K., Young, S.J., Wu, C.Z., Cheng, W.B., Appl. Phys. Lett. **94**, 203106 (2009). Copyright 2009, AIP Publishing LLC

to-volume ratio for the nanorod array system, which in turn leads to an increase in hole trapping, carrier injection, and transport. In a similar MSM nanorod-based system on a polyimide substrate [38] (Fig. 6.5), the high surface-to-volume ratio was also shown to significantly increase photocurrent [38] (Fig. 6.16) as well as response speeds. The authors of [39] investigated the effects of different substrates (Si, porous silicon/Si, and quartz) and the resulting ZnO nanomorphologies (aligned nanorods, tetrapod-like nanorods, and hair-like nanowires) on the improved performance of MSM Schottky photodiodes. The best responsivity (Fig. 6.17) and the fastest photoresponse were obtained for the photodetector based on nanowires grown on the porous silicon/Si substrate. Again it was suggested that the higher surface-to-volume ratio in the nanowire networks (Fig. 6.6) was the main reason behind this enhancement.

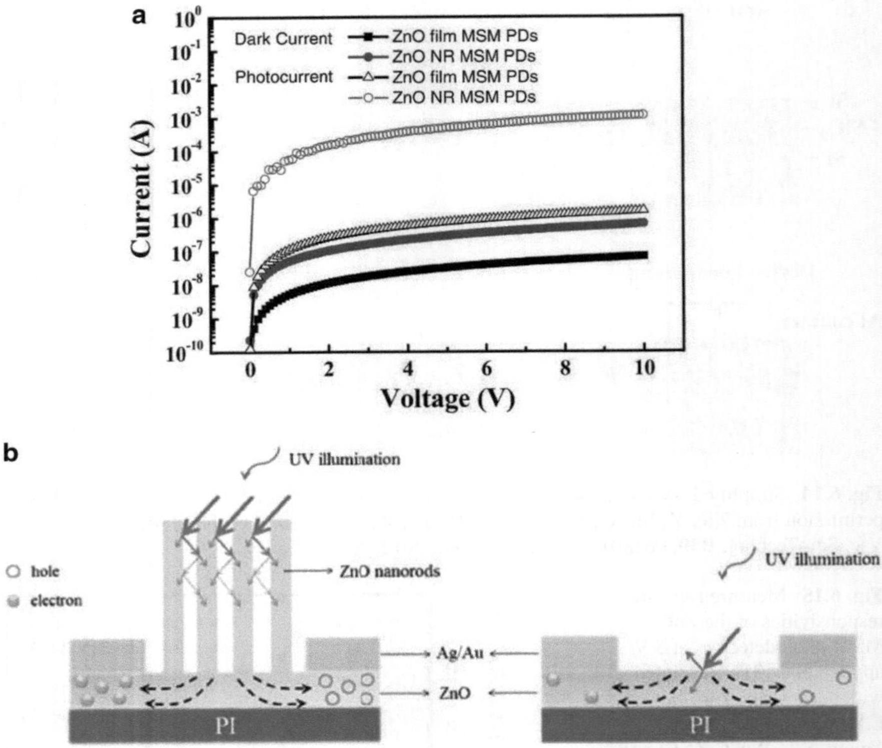

Fig. 6.16 (**a**) The I–V curves of ZnO film and nanorods photodetectors measured in the dark and under UV illumination. (**b**) Schematic diagram of light-trapping mechanism of nanorods [38]. Reproduced by permission of The Electrochemical Society from Chen, T.P., Young, S.J., Chang, S.J., Hsiao, C.H., Huang, C.S., J. Electrochem. Soc. **159**, J153–J157 (2012)

The effects of the variation of an ultrathin metal overlayer were reported in [46, 47] for bilayer thin-film MSM detectors [47] (Fig. 6.18). Photoresponse was strongly affected by the interface conductivity, with the best performance found for the Sn overlayers—see [46] (Fig. 6.19).

An alternative approach for improved functionality oftentimes could be based on a suitable doping procedure. For example, Mamat et al. [44] investigated the effects of Al doping of ZnO vertical nanorod arrays in a photoconductive sensor on its performance. Al is a good choice for a dopant because of easiness of its incorporation into the host ZnO lattice, as well as the resulting enhancement of optical transmittance and electrical conductivity. This, in turn, leads to the improvements in the sensing performance (Fig. 6.20).

Annealing was also shown to improve performance of ZnO-based photodetectors. For example, in [42] fast response of the MSM thin-film device was reported as

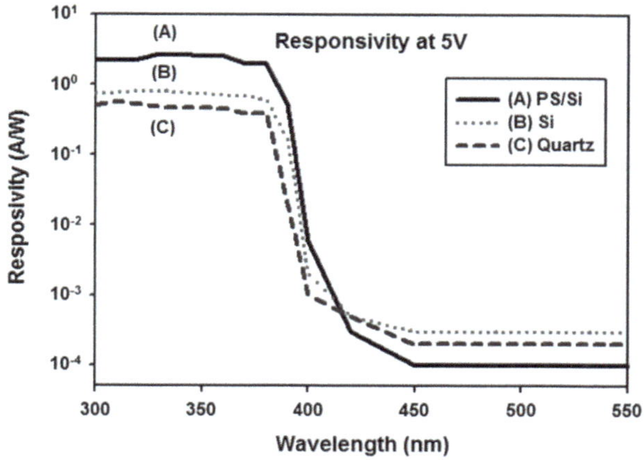

Fig. 6.17 Responsivity spectra of MSM photodetectors based on (**a**) ZnO hair-like nanowires grown on PS/Si substrate, (**b**) ZnO nanorods on Si substrate, and (**c**) ZnO tetrapod-like nanorods grown on quartz substrates at room temperature under applied bias of 5 V. Reprinted [39] with permission from Abdulgafour, H.I., Hassan, Z., Ahmed, N.M., Yam, F.K., J. Appl. Phys. **112**, 074510 (2012). Copyright 2012, AIP Publishing LLC

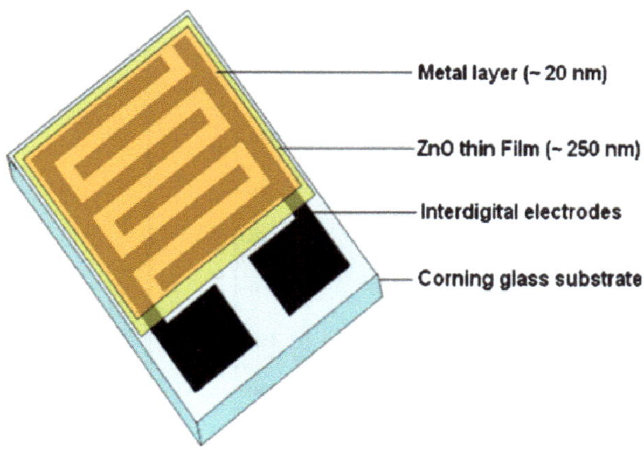

Fig. 6.18 Schematic of the metal-/ZnO-based UV photodetector [47]. Reprinted with permission from Yadav, H.K., Sreenivas, K., Gupta, V., Appl. Phys. Lett. **96**, 223507 (2010). Copyright 2010, AIP Publishing LLC

a result of annealing in pure oxygen atmosphere and thus a subsequent improvement in the quality of the ZnO material (also, see below).

Photosensitivity enhancements were reported in other detector systems, with more complex structures and compositions. For example, blending single-walled carbon nanotubes with indium gallium zinc oxide to produce thin-film transistors [67] improves the photosensitivity (Fig. 6.21) due to higher photocurrents in these oxide/nanotube blend-based devices. Another example of improved performance by nanocarbon incorporation was reported in [50] for a graphene/ZnO nanowire/graphene photodetector (Fig. 6.7). Time response characteristics of this

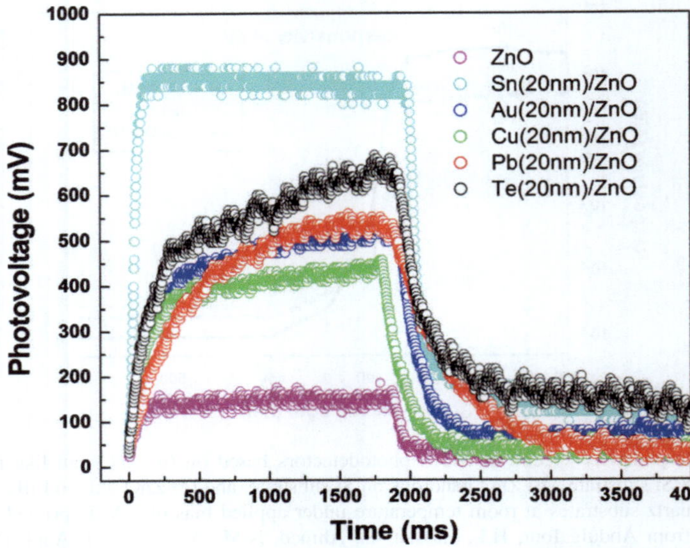

Fig. 6.19 UV photoresponse of the metal/ZnO bilayer detectors [46]. Reprinted with permission from Yadav, H.K., Sreenivas, K., Gupta, V., Appl. Phys. Lett. **107**, 044507 (2010). Copyright 2010, AIP Publishing LLC

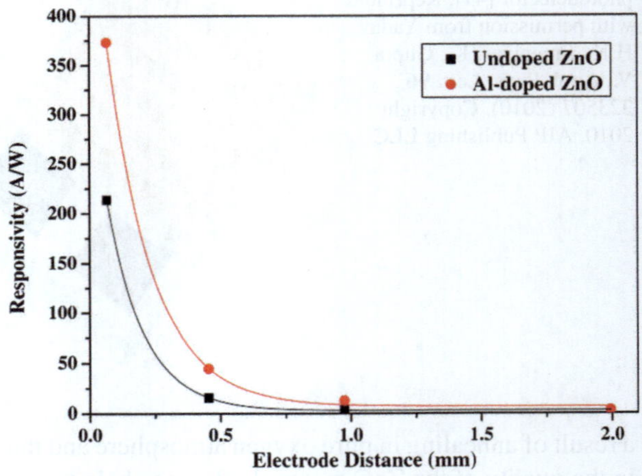

Fig. 6.20 Plot of a responsivity as a function of the electrode distance of the undoped and Al-doped ZnO nanorod array-based UV photoconductive sensors [44]. Copyright 2012, The Japan Society of Applied Physics

device were shown to be superior to those of more conventional detectors based on ZnO nanowires due to the exceptional quality of the Schottky barriers between ZnO and graphene.

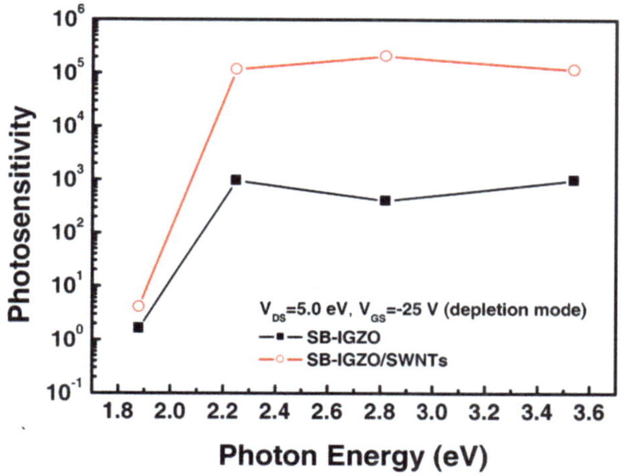

Fig. 6.21 Photosensitivity vs. the energy of incident photons for devices described in [67]. Reprinted with permission from Lee, K.W., Heo, K.Y., Kim, H.J., Appl. Phys. Lett. **94**, 102112 (2009). Copyright 2009, AIP Publishing LLC

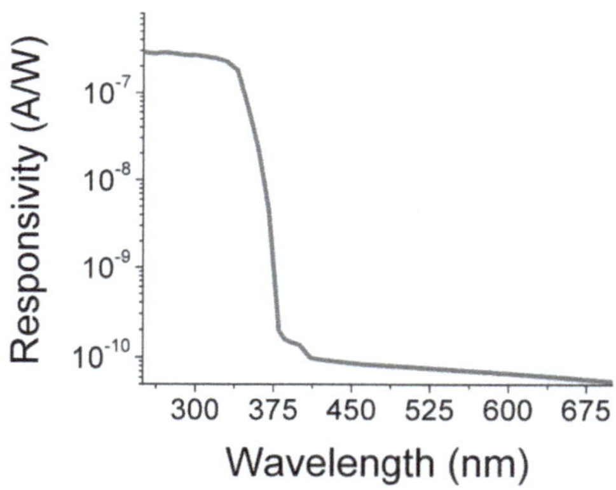

Fig. 6.22 Measured spectral responsivity of a ZnO nanowire detector at 1.5 V bias [59]. Reprinted with permission from Yu, H., Azhar, E.A., Belagodu, T., Lim, S., Dey, S., J. Appl. Phys. **111**, 102806 (2012). Copyright 2012, AIP Publishing LLC

6.6 Spectral Selectivity, Self-Powered Designs

In ZnO-based photodetecting devices, different aspects of wavelength selectivity, such as transparency range, levels of transmittance, rejection efficiency, and absorption spectrum, would be important in exploitation. Most of ZnO-based devices are purposely designed to operate exclusively in the UV spectral region by transmitting most of the visible and IR radiation and increasing the UV-to-visible rejection ratio. Examples of such performance can be found in the reports mentioned above, such as [31] (Fig. 6.15) or [59] (Fig. 6.22). In [60] (Fig. 6.23), a visible-blind UV photodetector was fabricated employing a *p*-*n* homojunction consisting of an *n*-type ZnO nanowire array grown by a hydrothermal method covered with a *p*-type sol-gel grown Al,N:ZnO thin film [60] (Fig. 6.12). Many other examples of high UV rejections can be provided in this context.

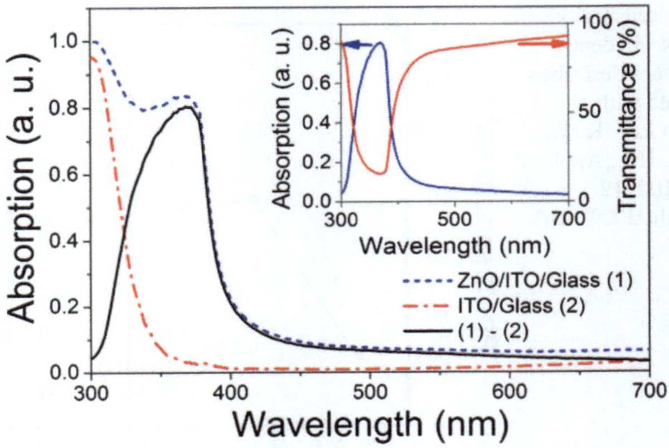

Fig. 6.23 Absorption spectra of the whole device (*1*), only the ITO/glass substrate (*2*), and the measured absorption of the ZnO layers (*1–2*). The inset shows the absorption and the transmittance spectra of the ZnO layers [60]. Reprinted with permission from Leung, Y.H., He, Z.B., Luo, L.B., Tsang, C.H.A., Wong, N.B., Zhang, W.J., Lee, S.T., Appl. Phys. Lett. **96**, 053102 (2010). Copyright 2010, AIP Publishing LLC

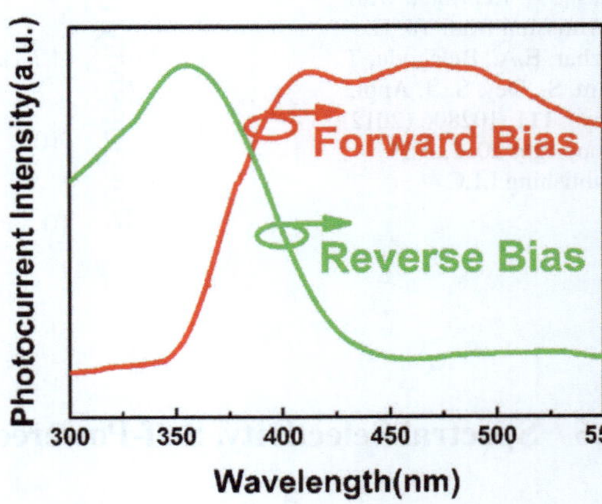

Fig. 6.24 Photocurrent spectra of the ZnO nanowire array/n-Si heterojunction for a forward bias (*red color*) and a reverse bias (*green color*), respectively [55]. Reprinted with permission from Guo, Z., Zhao, D., Liu, Y., Shen, D., Zhang, J., Li, B., Appl. Phys. Lett. **93**, 163501 (2008). Copyright 2008, AIP Publishing LLC

Conversely, there were successful attempts reported recently to manufacture ZnO-based photodetectors designed for an expanded absorption spectrum into the visible range. For example, in [32] and [55] the authors reported on devices based on similar ZnO/Si heterojunctions [32] (Fig. 6.2)/[55] (Fig. 6.9) in which the photoresponse to UV and visible was engaged for the reverse and forward bias, respectively (Figs. 6.24 and 6.25). A different concept was employed in [53], where a UV/visible photodetector was produced based on a hybrid nanostructure of ZnO and graphene oxide by a solvothermal method. It was shown that when the graphene component is thermally reduced, it significantly expands the absorption spectrum and thus the operational wavelength range of the device.

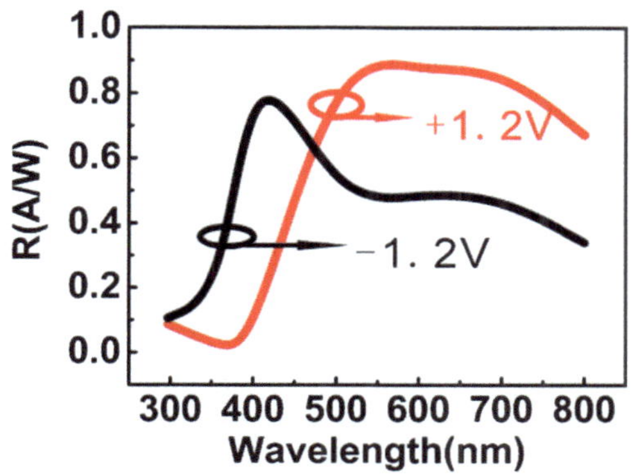

Fig. 6.25 Photoresponsivity spectra of the annealed sample for a forward and a reverse bias [32]. Reprinted with permission from Huang, H., Fang, G., Mo, X., Yuan, L., Zhou, H., Wang, M., Xiao, H., Zhao, X., Appl. Phys. Lett. **94**, 063512 (2009). Copyright 2009, AIP Publishing LLC

Another important factor in designing photodetectors for cost-effective operation crucial in many mature and emerging applications is the ability of the device to operate at low voltages or to self-power. Such applications are especially important in circumstances where energy consumption is limited (e.g., space exploration). An increasing number of efforts were made in recent years to achieve low- or zero-voltage operation of ZnO-based photosensors. For example, UV photodetector operation at low voltages was demonstrated in [66] (Fig. 6.26). The device had an architecture of a thin-film transistor where isolation of the active ZnO channel was achieved by boron implantation (Fig. 6.27). On the other hand the ZnO/graphene oxide UV/visible photodetector, described above [53], demonstrated ability to efficiently operate at zero bias. In another report discussed here [32], annealing of the samples significantly increased the responsivity of the device at zero bias. Similarly, the authors of [33] showed that a detector based on a single Ohmic/Schottky-bridging Sb-doped ZnO nanobelt can detect UV light as zero bias with a high degree of selectivity (Fig. 6.28a). Many other low-voltage and self-powered ZnO photodetectors have been reported recently.

6.7 Influence of Defect Properties

As we mentioned above, defects have significant impact on the operation of photodetectors based on both thin-film and nanoscale ZnO. Specifically, one can point out surface states at the free surfaces of nanostructures as well as interface defects at ZnO/metal contacts (both Schottky and Ohmic: we refer readers to a recent comprehensive review on the topic of metal contacts on ZnO [69]).

High performance of the photosensitive devices based on nanoscale ZnO is shown to be in good correlation with the low concentration of defects, corroborated, e.g., by photoluminescence (PL) spectroscopy. In single nanobelt-based photodetector devices described above [33], pure and Sb-doped ZnO had an excellent

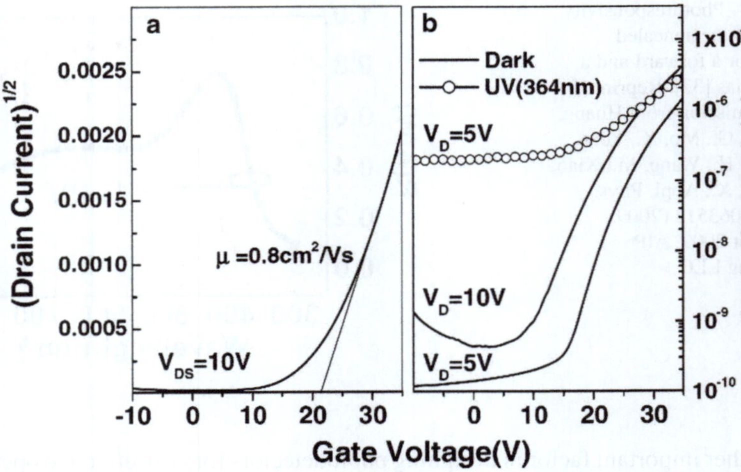

Fig. 6.26 Performance parameters for ZnO-based thin-film transistors described in [66]. Copyright 2008, The Japan Society of Applied Physics

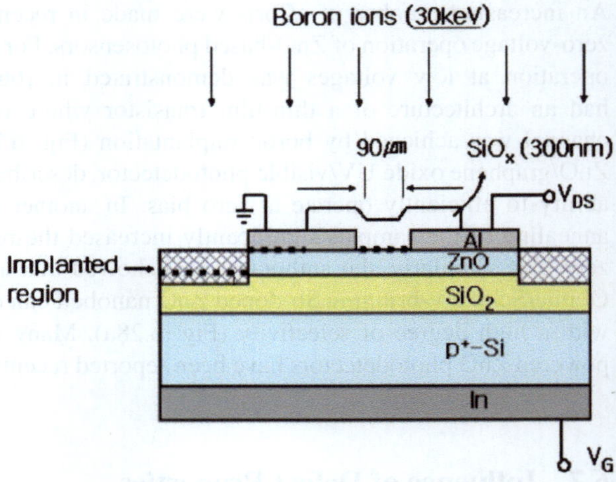

Fig. 6.27 Schematic cross section of the boron-implanted ZnO-based thin-film transistor [66]. Copyright 2008, The Japan Society of Applied Physics

crystal quality, with no defect luminescence detected (Fig. 6.3b). In this case surface states were trapping holes under illumination and releasing them in the dark, thus contributing to the dark current. No detectable defect luminescence was reported in [68] (Fig. 6.29) for the ZnO nanowires, also in a detector with remarkable performance parameters. However, in this study photoconductivity spectra revealed some nonradiative defect states (Fig. 6.30). Similarly, in [59], the high-performance device was based on ZnO nanowires with low concentration of deep-level defects (as confirmed by the PL spectra) and linear or planar structural defects (as detected by the transmission electron microscopy). Relatively low dark current was attributed

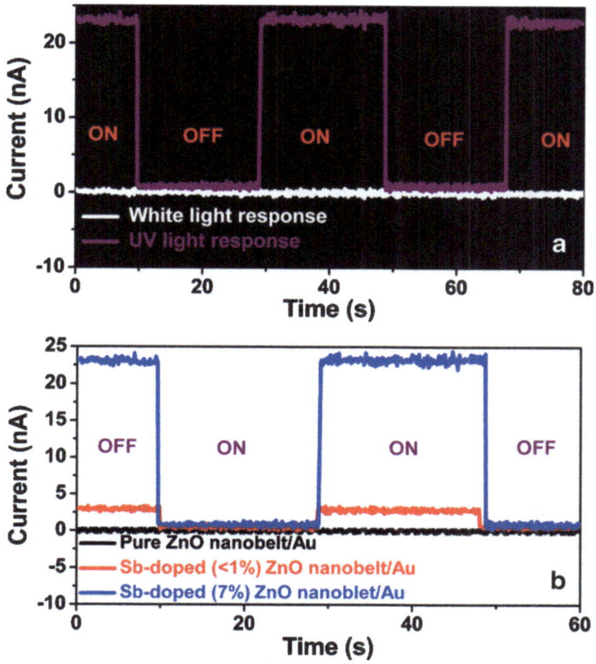

Fig. 6.28 (**a**) The current response curves of the device under white light and UV light illumination. (**b**) The current response curves of the devices based on single ZnO nanobelts with different Sb-doping concentrations [33]. Reprinted with permission from Yang, Y., Guo, W., Qi J., Zhao, J., Zhang, Y., Appl. Phys. Lett. **97**, 223113 (2010). Copyright 2010, AIP Publishing LLC

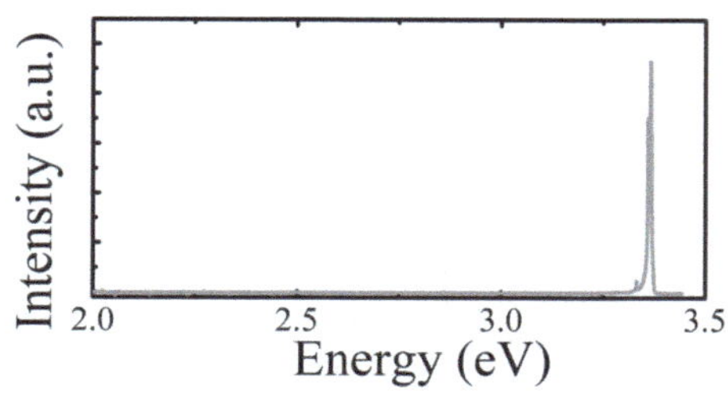

Fig. 6.29 Low-temperature ($T = 8$K) PL spectra of the high-quality ZnO nanowires in a broad spectral range. No defect-related green or yellow luminescence band is observed [68]. Reprinted with permission from Xie, Y., Madel, M., Neuschl, B., Jie, W., Röder, U., Feneberg, M., Thonke, K., J. Vac. Sci. Technol. **B 30**, 061801 (2012). Copyright 2012, AVS

to such donors as Zn interstitials and O vacancies. In the tetrapod network-based detector with high-level sensitivity [56], the lower dark current was also ascribed to low defect density and good crystalline quality.

In view of this, it is obvious that in order to have an efficient control of the device performance, one has to have good understanding of the defect-related phenomena and ability to tailor the defect properties. Numerous recent reports have demonstrated various successful approaches to achieve such control. For example, the authors of [46, 47] by means of integrating a thin-film device with an ultrathin tin overlayer enhanced photoresponse due to compensation of the surface traps

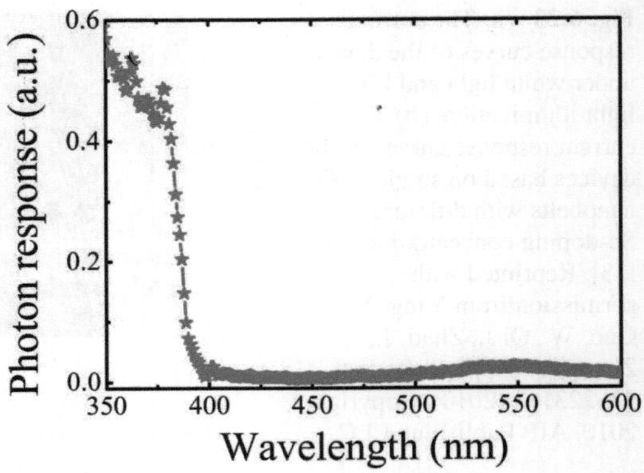

Fig. 6.30 Spectral responsivity of the silicon-on-insulator-based ZnO photodetector [68]. Reprinted with permission from Xie, Y., Madel, M., Neuschl, B., Jie, W., Röder, U., Feneberg, M., Thonke, K., J. Vac. Sci. Technol. **B 30**, 061801 (2012). Copyright 2012, AVS

by the metal overlayer. A different type of surface modification leading to the passivation of surface states was demonstrated in [3] where PMMA was used to cap the ZnO nanowire/Au Schottky diodes thus reducing the leakage currents by more than three orders of magnitude. Such reduction was well correlated with the PL results, which revealed elimination of the visible luminescence (arguably due to Zn and O vacancies) after PMMA encapsulation. The authors proposed that the acetate groups of the PMMA macromolecules are annihilating the surface defects in ZnO. Alternatively, in the ZnO:Al nanorod array-based UV photosensors [58], the relative intensity of the luminescence related to the surface and lattice defects (e.g., oxygen vacancies, zinc interstitials, hydroxyl groups) was controlled by the time of a sonicated sol-gel immersion and the resulting length/thickness ratio of the nanorods. Possibly, such change in the morphology of the nanorods leads to improvements in crystal quality. Subsequently, decrease of the nanorod array thickness leads to improved responsivities and response times (Fig. 6.31). In a related work on similar devices [44], it was demonstrated that Al doping improved transmittance of the ZnO nanorod arrays and reduced the unwanted visible light absorption. The authors suggested that aluminum reduces intrinsic defects (oxygen vacancies, zinc interstitials), as confirmed by the PL experiments. Al doping was also shown to improve responsivity (Fig. 6.20). Generally, benefits of doping on device performance were reported by many other authors. For example, in [66] B ion implantation was used to isolate ZnO channel in a thin-film transistor, which led to high mobility and operation at a zero gate bias. Annealing-driven phosphorus diffusion from the *p*-doped Si substrate into the ZnO nanorods and subsequent formation of a *p*-*n* ZnO homojunction were suggested in [32]. It resulted in a substantial improvement in the responsivity measured at zero bias. Another ZnO homojunction obtained by an Al/N co-doping was reported in [60] for a ZnO nanowire-based UV photodetector. Sb-doping of single ZnO contact-bridging nanobelts reported in [33] was shown to decrease the width of the Schottky barrier leading to a higher field in the barrier and thus a significant increase in the UV photocurrent and device sensitivity.

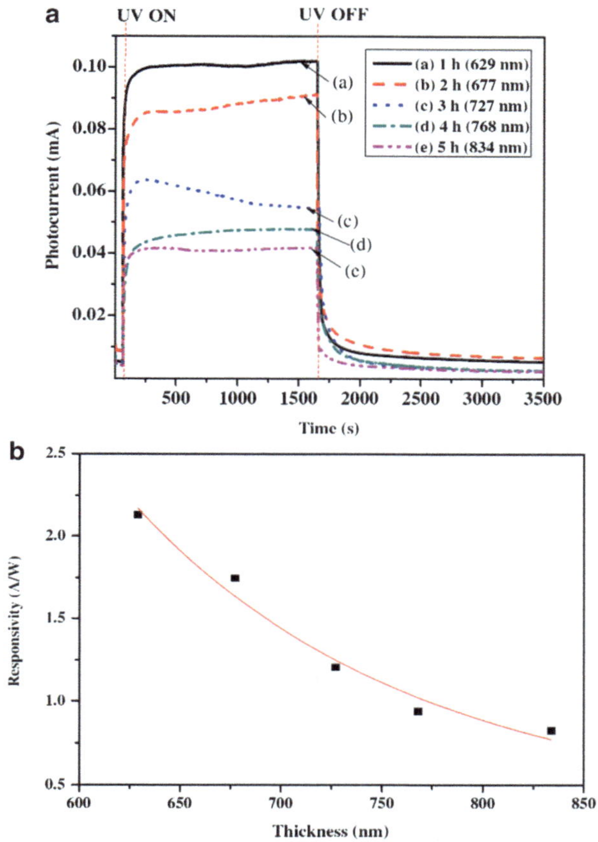

Fig. 6.31 (**a**) Photocurrent properties of the Al-doped ZnO nanorod array-based UV photoconductive sensors of various nanorod array thicknesses under UV illumination (365 nm) and a 10 V bias. (**b**) Plot of the UV sensor responsivity as a function of Al-doped ZnO nanorod array thickness [58]. Copyright 2012, The Japan Society of Applied Physics

On the other hand, there were a number of recent papers discussing the impact of deep states on the performance of ZnO-based photodetectors. Bian et al. [42] observed a substantial difference in the performance of ZnO thin films grown on SiO_2/Si substrates before and after annealing in pure oxygen at high temperature. It was especially apparent in the response times, which became much shorter for the annealed films. The authors explained this effect in terms of reduction of oxygen vacancies and migrating zinc interstitial into the crystal resulting in a diminished intergranular defective phase. This, in turn, improved the quality of the material, reduced electron scattering, and increased electron mobility. The importance of deep-level defects was also pointed out in the performance of a ZnO homojunction-based device, discussed above [60], where deep-level defects were likely to affect response in the red part of the spectrum.

Undoubtedly, one of the most important factors in the defect-related phenomena in ZnO-based detectors is the impact of interface and surface states. It is hard to overestimate the role of the metal contacts—both Ohmic and Schottky—in modern-time photodetectors. Quality and nature of these interfaces should be carefully

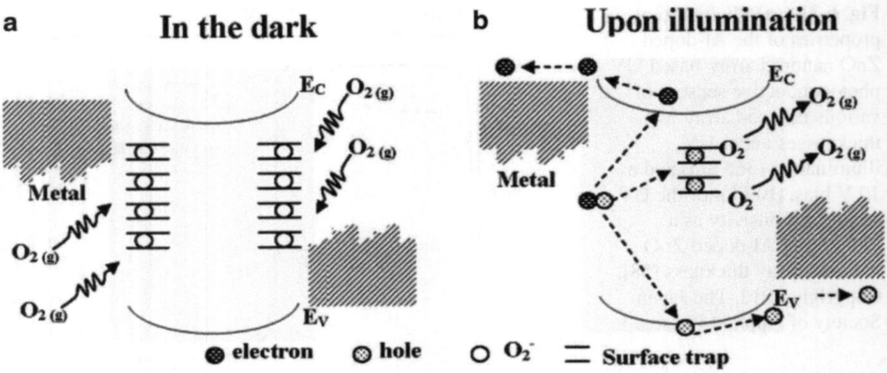

Fig. 6.32 Energy band illustration for photoconduction mechanism discussed in [31]: (**a**) In the dark, oxygen molecules are adsorbed onto the ZnO surfaces by capturing free electrons, which creates a low-conductivity depletion layer near the surface. (**b**) Under UV illumination at photon energy greater than the band gap, electron–hole pairs are photogenerated. The holes migrate along the potential gradient and discharge the oxygen ions; consequently, oxygen is photodesorbed from the surface. Reprinted with permission from Ji, L.W., Peng, S.M., Su, Y.K., Young, S.J., Wu, C.Z., Cheng, W.B., Appl. Phys. Lett. **94**, 203106 (2009). Copyright 2009, AIP Publishing LLC

tailored and interfacial states as well as interfacial phases should be well controlled. As an example of the interfacial phases' impact, one can mention stability testing for a Au/Cr/SiO_2/ZnO/Al MIS device performed in [48]. It was demonstrated that annealing may be responsible for the formation of interfacial phases (e.g., chromium oxide) with different work functions and different potential barriers.

Surface states are gaining recognition as a dominating factor in the performance of nanoscale ZnO-based photodetectors, due to the huge surface-to-volume ratio and especially regarding their role in the charge carrier trapping/release as well as oxygen absorption. Ji and coauthors [31] addressed the role of surface traps in the MSM detectors based on ZnO nanorods grown on ZnO thin films (Fig. 6.1). They proposed that the photoresponse is determined by the adsorption/desorption of oxygen assisted by the trap states at the surface of ZnO nanorods as follows (Fig. 6.32): in the dark, oxygen is adsorbed by the ZnO surface leading to a low-conductivity surface depletion layer, whereas during UV illumination electron–hole pairs are produced, with holes subsequently neutralizing oxygen ions and desorbing them from the surface. Oxygen adsorption/desorption at a nanoscale surface was also discussed in [49] for ZnO single nanowire-based devices and was shown to be responsible for the UV detection. Moreover, ZnO/Au interface states were also shown to affect performance of the detector in reverse bias, by trapping holes and facilitating electron tunneling through the effectively thinner depletion region. Incidentally, it was revealed that low concentration of deep-level defects, as measured by PL, makes this device blind to visible light. Similar effects of both the surface states as oxygen ion traps and the interface states as hole traps were considered for the MSM photodetectors based on ZnO nanostructures on Si and

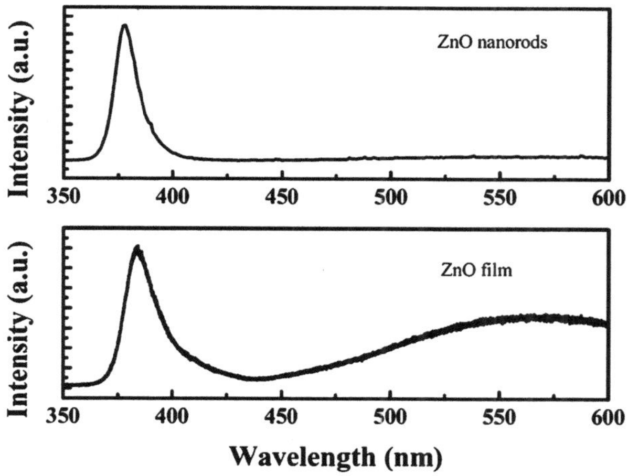

Fig. 6.33 PL spectra of ZnO nanorods and thin films at room temperature [38]. Reproduced by permission of The Electrochemical Society from Chen, T.P., Young, S.J., Chang, S.J., Hsiao, C.H., Huang, C.S., J. Electrochem. Soc. **159**, J153–J157 (2012)

quartz substrates [39] as well as for the MSM photodetectors with ZnO nanorods grown on ZnO thin films and flexible polymer substrates [38]. In the latter reference, it was also demonstrated that the superior performance of the ZnO nanorods in comparison with that of the ZnO thin films is related to the substantial difference in the concentration of deep-level defects, with nanorods being practically defect-free (Fig. 6.33). Control of the charge trapping related to oxygen absorption/desorption at ZnO nanoscale surfaces was recently successfully demonstrated in a number of works. For example, in [52] for the device based on vertically aligned ZnO nanowires, the concentration of oxygen deficiency was controlled by oxygen flow rate during growth. As a result, nanowires with higher concentration of oxygen vacancies were more conductive, which was linked to the oxygen interaction with the surface.

By the same token, Tzeng et al. [51] obtained improved performance of the detectors based on single ZnO nanowires by decorating the nanowire surfaces with Ag nanoparticles. Such modification was claimed to increase recombination of photogenerated holes, since it was occurring not only via the ZnO surface-absorbed oxygen ions but also at the Ag/ZnO interfaces (Fig. 6.34). Yet another type of the surface modification aimed to control the surface oxygen-mediated charge transfer was demonstrated in [53] for a self-powered visible light detector based on reduced graphene oxide/ZnO hybrid nanostructure. In this work it was shown that carbon doping of ZnO leads to creation of additional donor states contributing to a continuous charge turnover involving surface adsorption and desorption of oxygen ions, trapping of photogenerated holes with a simultaneous transfer of electrons to the electrode via the reduced graphene oxide [53].

Impact on the light detection capabilities of the oxygen trapping by surface defects was demonstrated not only in ZnO nanostructures but also in thin films (e.g., [36, 40]).

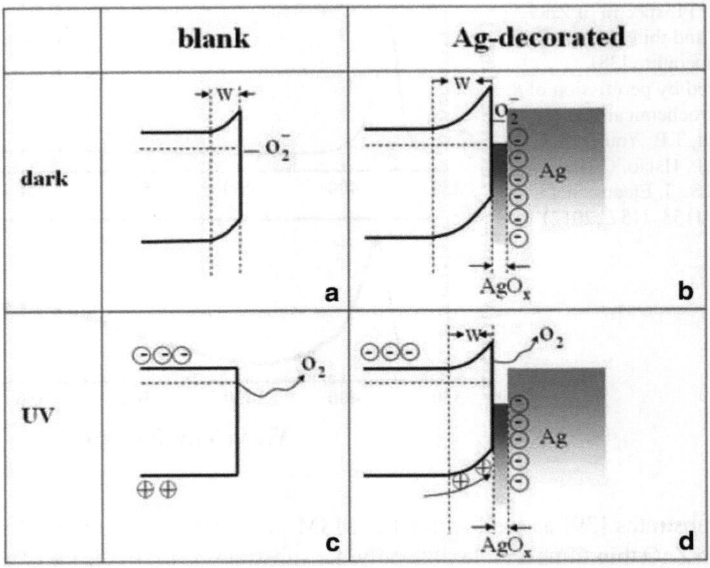

Fig. 6.34 Schematic energy band diagrams for blank and Ag-decorated ZnO nanowire-based photodetectors in the dark and under UV light illumination [51]. Reproduced by permission of The Electrochemical Society from Tzeng S.K., Hon, M.H., Leu, I.C., J. Electrochem. Soc. **159**, H440–H443 (2012)

6.8 Conclusions

In summary, we have outlined the most recent trends in the design and development of ZnO-based photodetectors. This chapter is by no means a complete review of all the publications in this area within the past several years, but rather a compilation of the most representative reports dealing with the topics discussed here. The body of publications on the subject is rapidly increasing and many novel ideas in the area suggested recently are gaining wide acceptance, both scholarly and practical. This is obviously aided by the common recognition of ZnO as a very versatile material, especially in its many nanoscale morphologies. We believe that the interest in ZnO-based photosensing devices is not going to wane anytime soon, with more exciting new physics, chemistry, and design ideas down the line.

Acknowledgments The author is grateful to his many colleagues and collaborators for helpful discussions and encouragement.

References

1. Decoster, D., Harari, J. (eds): Optoelectronic sensors. Wiley-ISTE, London, UK (2010)
2. Liu, K., Sakurai, M., Aono, M.: ZnO-based ultraviolet photodetectors. Sensors **10**, 8604–8634 (2010). and references therein
3. Tzeng, S.K., Hon, M.H., Leu, I.C.: Improvement in the photocurrent to dark current contrast ratio of ZnO/Au Schottky photodetector with PMMA microlens arrays. J. Electrochem. Soc. **157**(10), H919–H923 (2010)
4. Wang, Z.L.: Zinc oxide nanostructures: growth, properties and applications. J. Phys. Condens. Matter **16**, R829 (2004). and references therein
5. Gao, P.X., Song, J., Liu, J., Wang, Z.L.: Nanowire piezoelectric nanogenerators on plastic substrates as flexible power sources for nanodevices. Adv. Mater. **19**, 67–72 (2007)
6. Noh, H., Scharrer, M., Anderson, M.A., Chang, R.P.H., Cao, H.: Photoluminescence modification by a high-order photonic band with abnormal dispersion in ZnO inverse opal. Phys. Rev. B **77**, 115136 (2008)
7. Zhang, Q., Chou, T.P., Russo, B., Jenekhe, S.A., Cao, G.: Aggregation of ZnO nanocrystallites for high conversion efficiency in dye-sensitized solar cells. Angew. Chem. Int. Ed. **47**, 2402–2406 (2008)
8. Wang, Z.L.: Piezoelectric nanostructures: from growth phenomena to electric nanogenerators. MRS Bull. **32**, 109–116 (2007)
9. Mosbacker, H.L., Strzhemechny, Y.M., White, B.D., Smith, P.E., Look, D.C., Reynolds, D.C., Litton, C.W., Brillson, L.J.: Role of near-surface states in ohmic-Schottky conversion of Au contacts to ZnO. Appl. Phys. Lett. **87**, 012102 (2005)
10. Strzhemechny, Y.M., Mosbacker, H.L., Look, D.C., Reynolds, D.C., Litton, C.W., Garces, N.Y., Giles, N.C., Halliburton, L.E., Niki, S., Brillson, L.J.: Remote hydrogen plasma doping of single crystal ZnO. Appl. Phys. Lett. **84**, 2545 (2004)
11. Fabricius, H., Skettrup, T., Bisgaard, P.: Ultraviolet detectors in thin sputtered ZnO films. Appl. Opt. **25**, 2764–2767 (1986)
12. Liang, S., Sheng, H., Liu, Y., Huo, Z., Lu, Y., Shen, H.: ZnO Schottky ultraviolet photodetectors. J. Cryst. Growth **225**, 110–113 (2001)
13. Yang, W., Vispute, R.D., Choopun, S., Sharma, R.P., Venkatesan, T., Shen, H.: Ultraviolet photoconductive detector based on epitaxial $Mg_{0.34}Zn_{0.66}O$ thin films. Appl. Phys. Lett. **78**, 2787 (2001)
14. Kind, H., Yan, H., Messer, B., Law, M., Yang, P.: Nanowire ultraviolet photodetectors and optical switches. Adv. Mater. **14**, 158–160 (2002)
15. Takeuchi, I., Yang, W., Chang, K.-S., Aronova, M.A., Venkatesan, T., Vispute, R.D., Bendersky, L.A.: Monolithic multichannel ultraviolet detector arrays and continuous phase evolution in $Mg_xZn_{1-x}O$ composition spreads. J. Appl. Phys. **94**, 7336 (2003)
16. Lin, T.K., Chang, S.J., Su, Y.K., Huang, B.R., Fujita, M., Horikoshi, Y.: ZnO MSM photodetectors with Ru contact electrodes. J. Cryst. Growth **281**, 513–517 (2005)
17. Young, S.J., Ji, L.W., Chang, S.J., Su, Y.K.: ZnO metal–semiconductor–metal ultraviolet sensors with various contact electrodes. J. Cryst. Growth **293**, 43–47 (2006)
18. Young, S.J., Ji, L.W., Chuang, R.W., Chang, S.J., Du, X.L.: Characterization of ZnO metal–semiconductor–metal ultraviolet photodiodes with palladium contact electrodes. Semicond. Sci. Technol. **21**, 1507–1511 (2006)
19. Keem, K., Kim, H., Kim, G.-T., Lee, J.S., Min, B., Cho, K., Sung, M.-Y., Kim, S.: Photocurrent in ZnO nanowires grown from Au electrodes. Appl. Phys. Lett. **84**, 4376 (2004)
20. Heo, Y.W., Kang, B.S., Tien, L.C., Norton, D.P., Ren, F., La Roche, J.R., Pearton, S.J.: UV photoresponse of single ZnO nanowires. Appl. Phys. A **80**, 497–499 (2005)
21. Heo, Y.W., Tien, L.C., Norton, D.P., Kang, B.S., Ren, F., Gila, B.P., Pearton, S.J.: Electrical transport properties of single ZnO nanorods. Appl. Phys. Lett. **85**, 2002 (2004)
22. Suehiro, J., Nakagawa, N., Hidaka, S.-I., Ueda, M., Imasaka, K., Higashihata, M., Okada, T., Hara, M.: Dielectrophoretic fabrication and characterization of a ZnO nanowire-based UV photosensor. Nanotechnology **17**, 2567–2573 (2006)

23. Law, J.B.K., Thong, J.T.L.: Simple fabrication of a ZnO nanowire photodetector with a fast photoresponse time. Appl. Phys. Lett. **88**, 133114 (2006)
24. Fan, Z., Chang, P.-C., Lu, J.G., Walter, E.C., Penner, R.M., Lin, C.-H., Lee, H.P.: Photoluminescence and polarized photodetection of single ZnO nanowires. Appl. Phys. Lett. **85**, 6128 (2004)
25. Soci, C., Zhang, A., Xiang, B., Dayeh, S.A., Aplin, D.P.R., Park, J., Bao, X.Y., Lo, Y.H., Wang, D.: ZnO nanowire UV photodetectors with high internal gain. Nano Lett. **7**, 1003–1009 (2007)
26. Kumar, S., Gupta, V., Sreenivas, K.: Synthesis of photoconducting ZnO nano-needles using an unbalanced magnetron sputtered ZnO/Zn/ZnO multilayer structure. Nanotechnology **16**, 1167–1171 (2005)
27. Luo, L., Zhang, Y., Mao, S.S., Lin, L.: Fabrication and characterization of ZnO nanowires based UV photodiodes. Sens. Actuator A Phys. **127**, 201–206 (2006)
28. Hsu, C.L., Chang, S.J., Lin, Y.R., Li, P.C., Lin, T.S., Tsai, S.Y., Lu, T.H., Chen, I.C.: Ultraviolet photodetectors with low temperature synthesized vertical ZnO nanowires. Chem. Phys. Lett. **416**, 75–78 (2005)
29. Zhai, T., Fang, X., Liao, M., Xu, X., Zeng, H., Yoshio, B., Golberg, D.: A comprehensive review of one-dimensional metal-oxide nanostructure photodetectors. Sensors **9**, 6504–6529 (2009). and references therein
30. Shen, G., Chen, D.: One-dimensional nanostructures for photodetectors. Recent. Pat. Nanotechnol. **4**, 20–31 (2010). and references therein
31. Ji, L.W., Peng, S.M., Su, Y.K., Young, S.J., Wu, C.Z., Cheng, W.B.: Ultraviolet photodetectors based on selectively grown ZnO nanorod arrays. Appl. Phys. Lett. **94**, 203106 (2009)
32. Huang, H., Fang, G., Mo, X., Yuan, L., Zhou, H., Wang, M., Xiao, H., Zhao, X.: Zero-biased near-ultraviolet and visible photodetector based on ZnO nanorods/n-Si heterojunction. Appl. Phys. Lett. **94**, 063512 (2009)
33. Yang, Y., Guo, W., Qi, J., Zhao, J., Zhang, Y.: Self-powered ultraviolet photodetector based on a single Sb-doped ZnO nanobelt. Appl. Phys. Lett. **97**, 223113 (2010)
34. Shan, C.X., Zhang, J.Y., Yao, B., Shen, D.Z., Fan, X.W.: Ultraviolet photodetector fabricated from atomic-layer-deposited ZnO films. J. Vac. Sci. Technol. B **27**, 1766–1768 (2009)
35. Jandow, N.N., Ibrahim, K., Hassan, H.A.: I-V characteristic for ZnO MSM photodetector with Pd contact electrodes on PPC plastic. AIP. Conf. Proc. **1250**, 424–427 (2009)
36. Li, L., Ryu, Y., White, H.W., Yu, P.: Optical properties of metal-semiconductor-metal ZnO UV photodetectors. Proc. SPIE **7603**, 76031A (2010)
37. Zheng, Q., Huang, F., Ding, K., Huang, J., Chen, D., Zhan, Z., Lin, Z.: MgZnO-based metal-semiconductor-metal solar-blind photodetectors on ZnO substrates. Appl. Phys. Lett. **98**, 221112 (2011)
38. Chen, T.P., Young, S.J., Chang, S.J., Hsiao, C.H., Huang, C.S.: Field-emission and photoelectrical characteristics of ZnO nanorods photodetectors prepared on flexible substrate. J. Electrochem. Soc. **159**, J153–J157 (2012)
39. Abdulgafour, H.I., Hassan, Z., Ahmed, N.M., Yam, F.K.: Comparative study of ultraviolet detectors based on ZnO nanostructures grown on different substrates. J. Appl. Phys. **112**, 074510 (2012)
40. Ji, L.-W., Wu, C.-Z., Lin, C.-M., Meen, T.-H., Lam, K.-T., Peng, S.-M., Young, S.-J., Liu, C.-H.: Characteristic improvements of ZnO-based metal-semiconductor-metal photodetector on flexible substrate with ZnO cap layer. Jpn. J. Appl. Phys. **49**, 052201 (2010)
41. Hassan, N.K., Sulieman, K.M., Hashim, M.R.: ZnO thin film prepared by simple technique as UV photodetector. AIP. Conf. Proc. **1328**, 217–219 (2011)
42. Bian, X., Zhang, J., Bi, Z., Wang, D., Zhang, X., Hou, X.: Fast-response photoconductive metal-semiconductor-metal ultraviolet detector based on ZnO film grown by radio-frequency magnetron sputtering. Opt. Eng. **47**, 064001 (2008)
43. Moore, J.C., Covington, L.R., Foster, R.L., Gee, E.J., Jones, M.R., Morris, S.A.: ZnO ultraviolet photodetectors grown via thermal oxidation of Zn-metal on glass and sapphire substrates. Proc. SPIE **7940**, 79401L (2011)

44. Mamat, M.H., Khusaimi, Z., Zahidi, M.M., Bakar, S.A., Yusop, M.Z., Tanemura, M., Mahmood, M.R.: Effects of aluminium doping and electrode distance on the performance of aligned zinc oxide. Jpn. J. Appl. Phys. **51**, 06FE04 (2012)
45. Vasudevan, A., Jung, S., Ji, T.: UV detector from ZnO nanorods with electrodes resembling a wheatstone bridge pattern. Proc. SPIE **8102**, 81020G (2011)
46. Yadav, H.K., Sreenivas, K., Gupta, V.: Study of metal/ZnO based thin film ultraviolet photodetectors: the effect of induced charges on the dynamics of photoconductivity relaxation. Appl. Phys. Lett. **107**, 044507 (2010)
47. Yadav, H.K., Sreenivas, K., Gupta, V.: Persistent photoconductivity due to trapping of induced charges in Sn/ZnO thin film based UV photodetector. Appl. Phys. Lett. **96**, 223507 (2010)
48. Ali, G.M., Chakrabarti, P.: Effect of thermal treatment on the performance of ZnO based metal-insulator-semiconductor ultraviolet photodetectors. Appl. Phys. Lett. **97**, 031116 (2010)
49. Das, S.N., Moon, K.-J., Kar, J.P., Choi, J.-H., Xiong, J., Lee, T., Myoung, J.-M.: ZnO single nanowire-based UV detectors. Appl. Phys. Lett. **97**, 022103 (2010)
50. Fu, X.-W., Liao, Z.-M., Zhou, Y.-B., Wu, H.-C., Bie, Y.-Q., Xu, J., Yu, D.-P.: Graphene/ZnO nanowire/graphene vertical structure based fast-response ultraviolet photodetector. Appl. Phys. Lett. **100**, 223114 (2012)
51. Tzeng, S.-K., Hon, M.-H., Leu, I.-C.: Improving the performance of a zinc oxide nanowire ultraviolet photodetector by adding silver nanoparticles. J. Electrochem. Soc. **159**, H440–H443 (2012)
52. Kim, D.C., Jung, B.O., Kwon, Y.H., Cho, H.K.: Highly sensible ZnO nanowire ultraviolet photodetectors based on mechanical schottky contact. J. Electrochem. Soc. **159**, K10–K14 (2012)
53. Zhan, Z., Zheng, L., Pan, Y., Sun, G., Li, L.: Self-powered, visible-light photodetector based on thermally reduced grapheme oxide–ZnO (rGO–ZnO) hybrid nanostructure. J. Mater. Chem. **22**, 2589 (2012)
54. Weng, W.Y., Chang, S.J., Hsu, C.L., Hsueh, T.J., Chang, S.P.: A lateral ZnO nanowire photodetector prepared on glass substrate. J. Electrochem. Soc. **157**, K30–K33 (2010)
55. Guo, Z., Zhao, D., Liu, Y., Shen, D., Zhang, J., Li, B.: Visible and ultraviolet light alternative photodetector based on ZnO nanowire/*n*-Si heterojunction. Appl. Phys. Lett. **93**, 163501 (2008)
56. Hao, Y., Zhao, J., Qin, L., Guo, Q., Feng, X., Wang, P.: Facile fabrication of UV photodetector based on spatial network of tetrapod ZnO nanostructures. Micro. Nano. Lett. **7**, 200–203 (2012)
57. Hsu, C.-L., Hsueh, T.-J., Chang, S.-P.: Preparation of ZnO nanoflakes and a nanowire-based photodetector by localized oxidation at low temperature. J. Electrochem. Soc. **155**, K59–K62 (2008)
58. Mamat, M.H., Ishak, N.I., Khusaimi, Z., Zahidi, M.M., Abdullah, M.H., Muhamad, S., Sin, N.D.M., Mahmood, R.: Thickness-dependent characteristics of aluminium-doped zinc oxide nanorod-array-based, ultraviolet photoconductive sensors. Jpn. J. Appl. Phys. **51**, 06FF03 (2012)
59. Yu, H., Azhar, E.A., Belagodu, T., Lim, S., Dey, S.: ZnO nanowire based visible-transparent ultraviolet detectors on polymer substrates. J. Appl. Phys. **111**, 102806 (2012)
60. Leung, Y.H., He, Z.B., Luo, L.B., Tsang, C.H.A., Wong, N.B., Zhang, W.J., Lee, S.T.: ZnO nanowires array p-n homojunction and its application as a visible-blind ultraviolet photodetector. Appl. Phys. Lett. **96**, 053102 (2010)
61. Afsal, M., Wang, C.-Y., Chu, L.-W., Ouyang, H., Chen, L.-J.: Highly sensitive metal-insulator-semiconductor UV photodetectors based on ZnO/SiO2 core-shell nanowires. J. Mater. Chem. **22**, 8420 (2012)
62. Hasan, K.U., Nur, O., Willander, M.: Screen printed ZnO ultraviolet photoconductive sensor on pencil drawn circuitry over paper. Appl. Phys. Lett. **100**, 211104 (2012)
63. Tabares, G., Hierro, A., Vinter, B., Chauveau, J.-M.: Polarization-sensitive Schottky photodiodes based on a-plane ZnO/ZnMgO multiple quantum-wells. Appl. Phys. Lett. **99**, 071108 (2011)

64. Sheu, J.K., Chang, K.H., Lee, M.L.: Ultraviolet band-pass photodetectors formed by Ga-doped ZnO contacts to n-GaN. Appl. Phys. Lett. **92**, 113512 (2008)
65. Chang, K.-H., Sheu, J.-K., Lee, M.-L., Kang, K.-S., Huang, J.-F., Wang, W.-L., Lai, W.-C.: Ga-Doped ZnO/GaN Schottky barrier UV band-pass photodetector with a low-temperature-grown GaN cap layer. Jpn. J. Appl. Phys. **49**, 04DF12 (2010)
66. Bae, H., Im, S., Song, J.: Low-voltage ultraviolet detectors using ZnO thin-film transistor isolated by B ion implantation. Jpn. J. Appl. Phys. **47**, 5362–5364 (2008)
67. Lee, K.W., Heo, K.Y., Kim, H.J.: Photosensitivity of solution-based indium gallium zinc oxide single-walled carbon nanotubes blend thin film transistors. Appl. Phys. Lett. **94**, 102112 (2009)
68. Xie, Y., Madel, M., Neuschl, B., Jie, W., Röder, U., Feneberg, M., Thonke, K.: Silicon-on-insulator based ZnO nanowire photodetector. J. Vac. Sci. Technol. B **30**, 061801 (2012)
69. Brillson, L.J., Lu, Y.: ZnO Schottky barriers and Ohmic contacts. J. Appl. Phys. **109**, 121301 (2011)

Chapter 7
Thin Film Gas Sensors Based on Nanocarbon Materials

L.B. Kong and H. Huang

Abstract Gas sensors have found various applications in human daily lives and industries. Semiconductive oxides, including SnO_2, ZnO, Fe_2O_3, and In_2O_3, have been studied as promising candidates for gas sensor applications for many years. Recent studies indicated that nanocarbons, such as carbon nanotubes (CNTs) and nanosheets of graphene or graphene oxide, as well as their hybrids with nanosized noble metals and semiconducting oxides, have been emerged as new types of thin film gas sensors with advanced sensing performances. This chapter was aimed to summarize the recent progress in processing and characterization of nanosized hybrids based on nanocarbons and various nanoparticles as thin film gas sensors.

7.1 Introduction

A gas sensor is a device which detects the presence of various gases within an area, usually as part of a safety system. This type of equipment is used to detect a gas leak and interface with a control system so that a process can be automatically shut down or a function can be switched on. A gas detector can also sound an alarm to operators in the area where the leak is occurring, warning them to leave the area. This type of device is important because there are many gases that can be harmful to organic lives, such as humans or animals. Gas detectors can be used to detect

L.B. Kong (✉)
School of Materials Science and Engineering, Nanyang Technological University, 50 Nanyang Avenue, Singapore 639798, Singapore
e-mail: elbkong@ntu.edu.sg

H. Huang
Singapore Institute of Manufacturing Technology (SIMTech), A*STAR, 71 Nanyang Drive, Singapore 638075, Singapore
e-mail: hhuang@SIMTech.a-star.edu.sg

S. Li et al. (eds.), *Nanoscale Sensors*, Lecture Notes in Nanoscale Science and Technology 19, DOI 10.1007/978-3-319-02772-2_7,

combustible, flammable, and toxic gases, as well as oxygen depletion. This type of device is used widely in industry and can be found in a variety of locations such as on oil rigs, to monitor manufacture processes and emerging technologies such as photovoltaic. They may also be used in firefighting. Development of gas sensors with high sensitivity, high selectivity, fast responses, and recovery, for different gases, is always a challenge.

There are various types of gas sensors. One group of the conventional gas sensors is semiconducting oxide gas sensors, made with SnO_2, Fe_2O_3, ZnO, In_2O_3, WO_3, and so on. Some of the gas sensors have been used in our daily life and industries. Recent studies indicated that nanocarbon materials, including carbon nanotubes (CNTs) and graphene (or reduced graphene oxide), possessed very promising gas sensing performances [1–4]. Furthermore, the performances of the gas sensors made with these nanocarbon materials could be enhanced by incorporating various nanoparticles (NPs), such as noble NPs and oxide NPs. Depending on the number of carbon atomic layers, CNTs can be classified into three groups: single-walled (SWCNTs), double-walled (DWCNTs), and multiwalled (MWCNTs). Sometimes, DWCNTs are also included in MWCNTs. In this case, there are two types of groups. Properties of CNTs could be very different due to the presence of defects, chirality, and so on, which are not within the scope of this article.

Review articles on progress in CNTs [1–3] and graphene [4] as gas sensing materials were published recently. For example, Kauffman and Star [1] summarized the progress in CNT-based gas sensors, with emphasis on their uses in biomedicine and defense applications. Another review paper on CNT-based gas sensors was presented by Zhang et al. [2]. They also mentioned the enhancement in sensing performances with noble metal nanoparticle decoration. However, oxide-CNT hybrids were not included in their review. The review of Bondavalli et al. [3] focused only on CNT-based transistor gas sensors. Graphene-based sensors were reviewed by Yavari and Koratkar [4], but without the mention of graphene hybrid. More notably, the three reviews on CNT gas sensors were published 3–4 years ago. It is timely to present an update on the progress in this special research area.

With this background, this chapter aims to present an overview on the progress in nanocarbon materials that have been used as thin film gas sensors. The materials include various CNTs (SW, DW, and MW), graphene and reduced graphene oxides, hybrids of the two nanocarbons and noble metal, and oxide NPs. No attempt has been made to provide with an exhausted coverage of the literatures; it is more likely to bring out an outline of the latest development of this new group of thin film gas sensors.

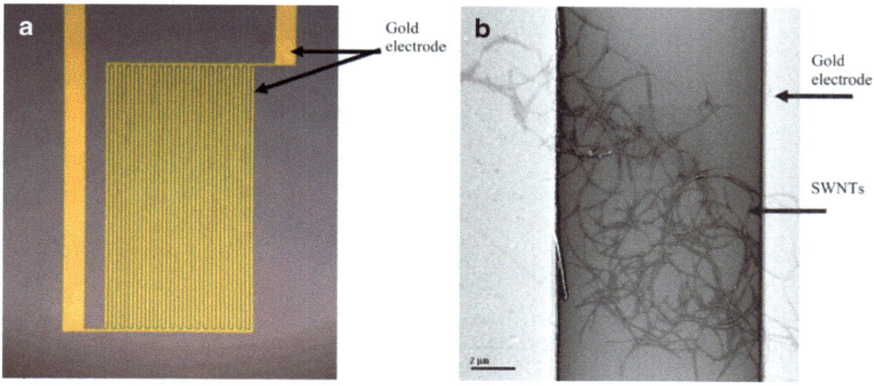

Fig. 7.1 Interdigitated electrode (IDE) (**a**) and SEM image of the SWCNT network across two neighboring gold electrodes (**b**). Reproduced with permission from ref. [5]. Copyright 2003, American Chemical Society

7.2 Gas Sensors Based on Nanocarbons

7.2.1 Carbon Nanotubes (CNTs)

7.2.1.1 SWCNTs

A simple method to fabricate gas sensor with SWCNTs was reported by Li et al. [5]. In that study, SWCNTs were casted on substrates with an interdigitated electrode (IDE) from a nanotube suspension in dimethylformamide (DMF). Gas sensors with SWCNT networks were formed after the evaporation of DMF. Figure 7.1 shows a photograph of the IDE deposited on silicon substrates and a SEM image of the SWCNT network across two gold electrodes. The SWCNTs were synthesized by the high-pressure carbon monoxide disproportionate (HiPCo) process and purified with acid to remove the residuals of the metallic catalyst, followed by air oxidation at high temperature to remove the graphitic carbon. The final purified SWNTs had a purity of >99.6 % and a surface area of 1,587 m^2/g. The high purity ensured that the sensing behavior was the intrinsic property of the SWCNTs, not from impurities.

Electrical conductivity of the SWCNT networks was controlled by the density of the nanotubes across the neighboring electrodes. The density of nanotubes was determined by either the concentration of the DMF suspensions or the amount of the suspension used. As shown in Fig. 7.2, three groups of sensors were fabricated by using three, two, and one drop of the SWCNT-DMF suspension. Conductivity was increased with increasing number of droplets. The sensors exhibited high sensitivity to NO_2, acetone, benzene, and nitrotoluene, with sensitivity being almost linearly proportional to the concentrations of the gases tested. Figure 7.3 shows sensing performance of the group 3 sensor to nitrotoluene.

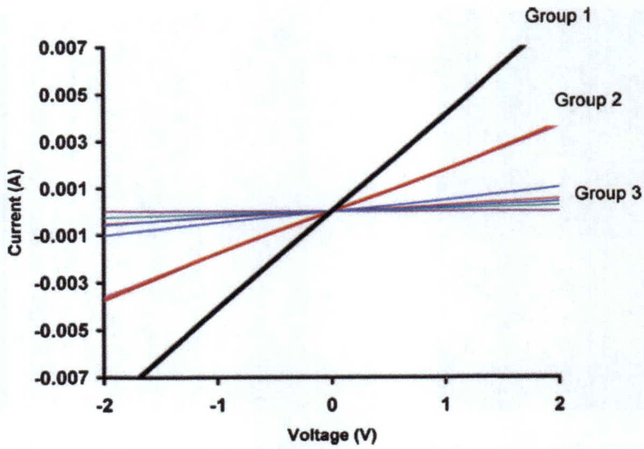

Fig. 7.2 I–V curves of different groups of sensors. Group 1: three droplets of 0.05 μL SWCNT-DMF solution were deposited on the IDE; each droplet was added after the previous one dried. Group 2: two droplets of 0.05 μL SWCNT-DMF solution were deposited on the IDE; the second droplet was added after the first one dried. Group 3: one droplet of 0.05 μL SWCNT-DMF solution was deposited on the IDE. The initial sensor resistance values were 244, 568, and 5.4 ± 2.2 kΩ, for groups 1–3, respectively. Reproduced with permission from ref. [5]. Copyright 2003, American Chemical Society

A similar method to prepare SWCNT gas sensor was reported by Ndiaye et al. [6]. They used a commercially available SWCNT powder, with an average external diameter of about 1.3 nm, purity of >90 %, and specific surface area of 300–600 m^2/g. The sensors were fabricated on alumina substrates, with IDEs made of screen-printed Pt and a Pt heating resistor on the back side of the substrates. Two sensors (A and B) were made from two different types of nanotube suspensions (A and B). Suspension A was prepared with 2 mg SWCNTs mixed with 10 ml chloroform ($CHCl_3$), while suspension B was made by dispersing the SWCNTs in 10 ml 0.7 wt% NaDDBS (sodium dodecylbenzene sulfonate). The use of the surfactant NaDDBS allowed for better dispersion of the SWCNTs.

In both cases, the resistance decreased with increasing temperature, implying that the SWCNTs were semiconductive and their semiconductivity was not affected by the dispersion method. However, the way of dispersion had a significant influence on resistivity of the sensor. Sensor A had a lower resistivity than sensor B. As explained by the authors, the use of surfactant led to sensing layers with the SWCNTs being homogeneously distributed on the sensor surface and thus giving rise to higher resistance. The absence of surfactant resulted in a homogeneous distribution and thus a lower resistivity due to the presence of CNT bundles.

The SWCNT sensors were demonstrated to show high sensitivity to NO_2. The sensor resistance was decreased in NO_2. SWCNTs were *p*-type semiconductor and NO_2 was a strong oxidant. When the sensors were exposed in NO_2, NO_2 molecules were adsorbed on surfaces of the SWCNTs. Their interaction led to the depletion of electrons in the SECNTs and thus an increase in concentration of holes. The increase in the concentration of holes was responsible for the decrease in resistance of

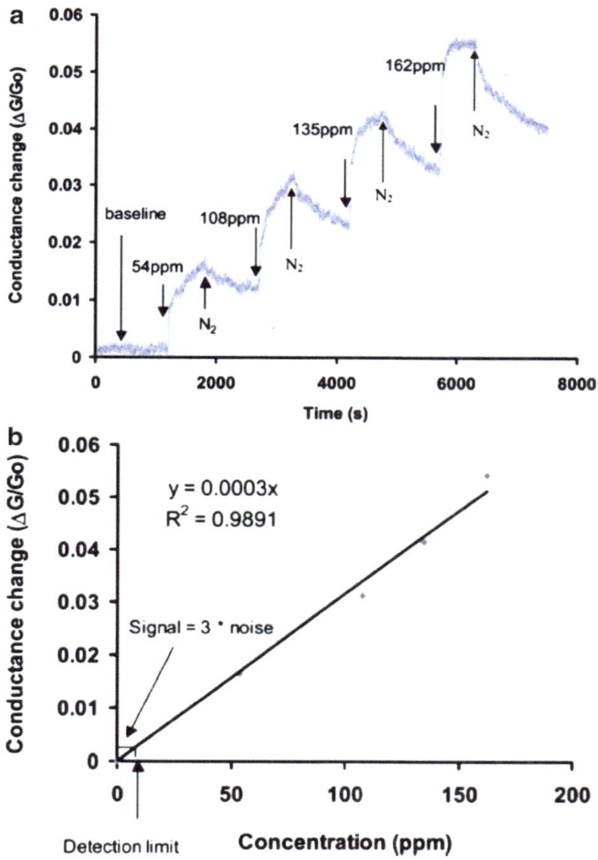

Fig. 7.3 Sensor response of group 3 to nitrotoluene. (**a**) Sensor response is a step function of concentration. (**b**) Calibration curve from (**a**). The nitrotoluene vapor was evaporated using a bubbler with a 100 cm^3/min flow rate of ultrapure nitrogen at room temperature, and this vapor stream was further diluted by nitrogen to a total flow rate of 400 cm^3/min. The purge gas was nitrogen as well. Reproduced with permission from ref. [5]. Copyright 2003, American Chemical Society

the *p*-type SWCNT sensors. A sensitivity, $\triangle R/R_0$ ($\triangle R = R_G - R_0$, with R_G and R_0 being the resistances of the sensors in gas NO_2 and in air, respectively), was used to characterize the sensing performance of the sensors. It was found that the sensitivity of sensor A is higher than sensor B. The difference in electrical conductivity and sensing performance between the two SWCNT sensors was explained in terms of the difference in dispersion state of the SWCNTs.

An alternative method to use SWCNT thin film as gas sensor was reported by Nguyen et al. [7]. The SWNTs were directly synthesized and deposited on substrates in situ in an arc-discharge chamber. Porous SWCNT thin film was formed due to the steric hindrance during the deposition. The SWCNT thin film sensor was sensitive to NO and NH_3 with concentrations down to 2 ppm. The effect of the sensor porosity was examined and compared with that of a nonporous sensor structure. Figure 7.4 shows representative SEM images of the as-deposited and posttreated SWCNT thin films. The dense film was obtained by immersing the as-deposited sample in methanol for 2 min, followed by drying at 100 °C for 20 min.

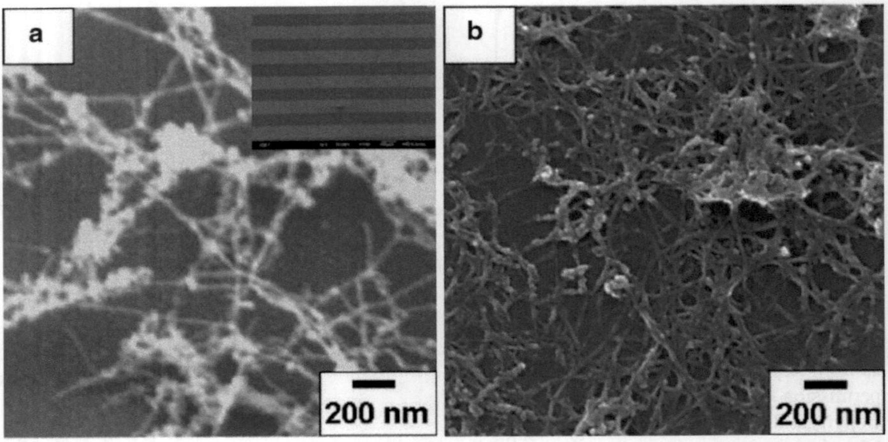

Fig. 7.4 SEM images of the as-synthesized SCWNTs (**a**) and methanol-treated SCWNTs (**b**). Reproduced with permission from ref. [7]. Copyright 2009, Elsevier

As shown in Fig. 7.5, the as-deposited and the dense SWCNT thin film sensors exhibited similar resistances as a function of temperature [7]. They had a "V"-shaped resistance–temperature curve as the temperature was increased from room temperature to 400 °C, while they demonstrated a semiconductor behavior when they were cooled down from 400 °C. In ambient condition, the SWCNTs absorbed nitrogen and oxygen on their surface, so that there was electron transfer from the nanotubes to oxygen molecules. When the SWCNTs were heated in vacuum, the majority of the physisorbed molecules were desorbed, and the intrinsic properties, i.e., semiconducting behaviors, of the SWCNTs were demonstrated. As a result, temperature-dependent resistivity curves were observed during cooling. In other words, the "V"-shaped *R–T* curves of non-annealed SWCNT were not the intrinsic properties of the carbon nanotubes. It resulted from two competitive factors: the increase in resistance with desorption of the physisorbed water molecules and the resistance decrease due to the semiconducting behavior of the nanotubes. In addition, the as-deposited SWCNT thin film had a much higher resistance than the methanol-treated one, which was attributed to the fact that the treated sample exhibited better inter-contact among the individual nanotubes.

Figure 7.6 shows sensing behaviors of the as-deposited SWCNT thin film sensor to NO and NH_3 at room temperature. The resistance of the sensor increased in NH_3 and decreased in NO. As an active electron dopant, polar NH_3 molecule donated electrons and neutralized the holes of the *p*-type SWCNTs, resulting in an increase in resistivity. Therefore, the resistance of the SWCNT sensor increased upon exposure to NH_3. NO is an acceptor dopant to the *p*-type SWCNTs, thus increasing the resistance of the SWCNT sensor. Sensitivity of the as-deposited sensor was 17 and 46 % to 40 ppm of NH_3 and NO, respectively. The sensor responded to NO was about three times higher than to NH_3. However, the sensitivity

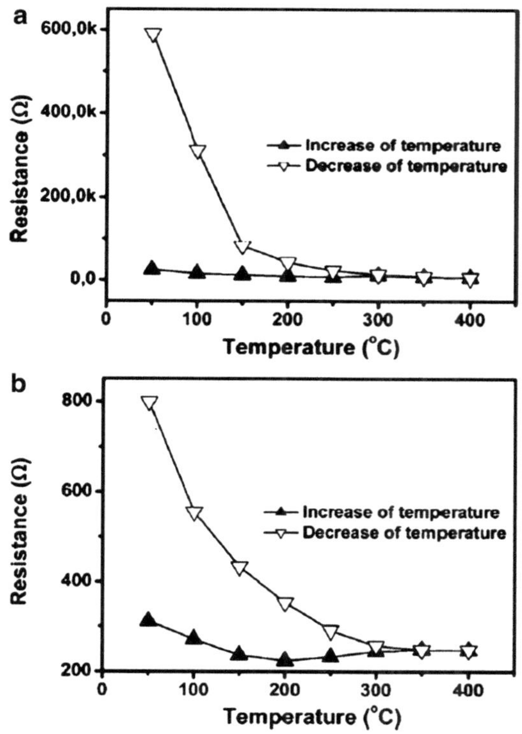

Fig. 7.5 Resistances of the sensors as a function of temperature for (**a**) as-synthesized and (**b**) methanol-treated SWCNTs. The dependence was measured firstly with increasing temperature and successively with decreasing temperature in vacuum. Reproduced with permission from ref. [7]. Copyright 2009, Elsevier

to NO started to saturate above 20 ppm, while it increased linearly up to 40 ppm for NH_3. This difference in sensitivity between the two gases was tentatively attributed to the differences in the adsorption efficiency, electron transfer efficiency from the gases to the CNTs, and the binding energies between the gases and the CNTs. One problem was that this SWCNT sensor had a poor recovering capability at room temperature.

The recovery performance of the SWCNT thin film sensor increased with increasing temperature, as shown in Fig. 7.7. For example, at 300 °C, the sensor recovered immediately once it was surged with N_2. However, the sensitivity decreased with increasing temperature. This is simply because the adsorption efficiency of the molecules sensed by the sensor was decreased with increasing temperature. Therefore, there was a trade-off between sensitivity and recovering rate when SWCNT sensors are used for real applications.

SWCNT thin film gas sensors have also reported to be deposited by using chemical vapor deposition (CVD) method [7, 8]. For instance, Wongwiriyapan et al. [8] deposited SWCNT network on alumina substrates with Pt ITEs by using a CVD. The SWCNT sensor demonstrated high sensitivity to NO_2 at concentration down to 50 ppb. Han et al. [9] used CVD to deposit SWCNT thin layer on silicon substrate. They made an attempt to control the morphology and electrical connectivity of

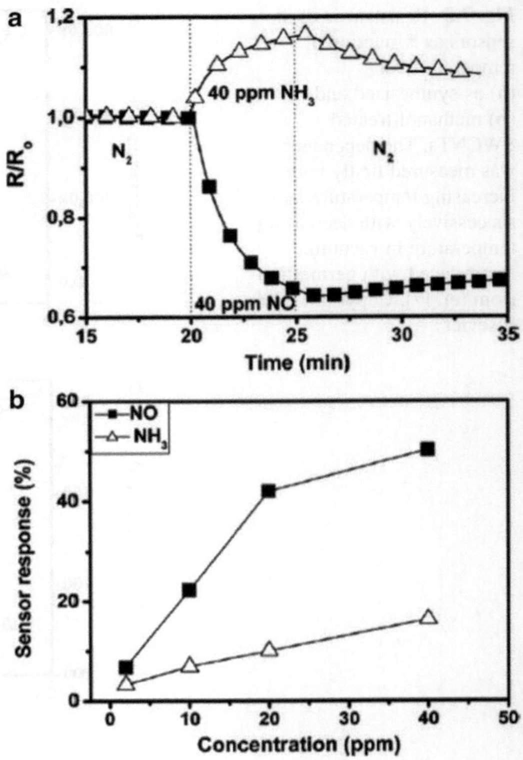

Fig. 7.6 Sensing behaviors of the as-deposited sensor: (**a**) the responses to 40 ppm NO and NH_3 gases and (**b**) sensor response as a function of gas concentration. Reproduced with permission from ref. [7]. Copyright 2009, Elsevier

SWCNT by using a rough nanoporous silica support of Fe catalyst particles. The SWCNTs produced in this way possessed very high quality, whose resistivity was almost three orders of magnitude lower than that of the SWCNTs prepared by wet chemical methods. The synthesized SWCNT thin network was integrated into Si-based nanodevices, which showed high sensitivity to NH_2 and NO_2.

By using Schottky barrier contacts between Pd and semiconducting SWCNTs, Ganzhorn et al. [10] developed a gas sensor that was very sensitive to H_2. The devices made with the SWCNTs of diameter of 1–1.6 nm exhibited a 100-fold change in the on-state conductance at 100 ppm H_2 as compared to in air. The sensors had a strong dependence on the diameter of the SWCNTs. For devices of intermediate diameter, CNTs achieved a detection limit of 1 ppm to H_2.

Lee et al. [11] developed a planar SWCNT sensor fabricated on paper substrate to detect small concentrations of NH_3 gas, using the shift in resonance frequency of a patch antenna as the discriminator. Gas sensitivity of the antenna sensor was enhanced by using a polymer, poly(m-aminobenzene sulfonic acid) (or PABS), to functionalize the SWCNTs. This new type of antenna-based wireless gas sensor had several advantages for practical applications, such as small size, light weight, and very low power consumption. A similar strategy of coating SWCNTs with

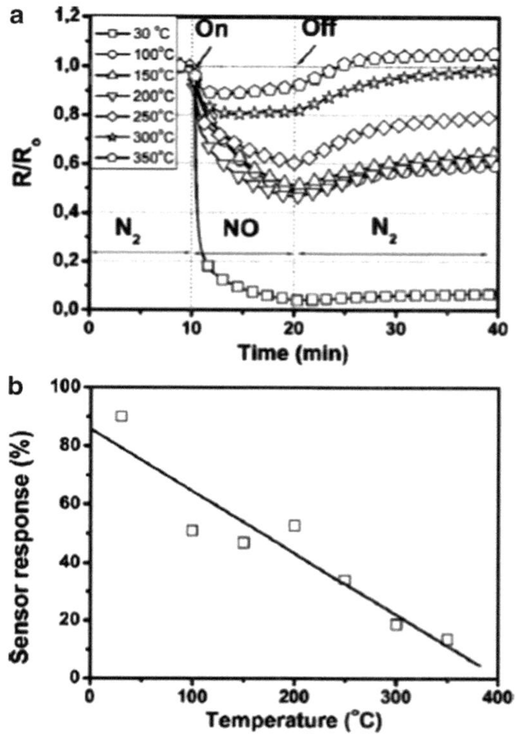

Fig. 7.7 Sensing behaviors of the as-deposited sensor: (**a**) change in electrical resistance ratio of the sensor upon exposure of 60 ppm NO at different temperatures and (**b**) temperature dependence of the sensor response. Reproduced with permission from ref. [7]. Copyright 2009, Elsevier

polypyrrole (PPy) to fabricate gas sensor was reported by An et al. [12]. The coating of PPy had no significant effect on electrical properties of the SWCNTs and sensitivity of the sensors, but improved greatly the reproducibility of the sensors.

7.2.1.2 MWCNTs

Compared to SWCNTs, MWCNTs are easier to synthesize and thus much cheaper and popular. Valentini et al. [13, 14] used a plasma-enhanced CVD method to deposit MWCNT films on Si_3N_4/Si substrates, as resistive gas sensors to NO_2. Resistance of the sensors decreased upon exposure to NO_2. The sensors were found to show the highest sensitivity at about 165 °C. The sensing concentration limit to NO_2 was as low as 10 ppb, with fast response and good selectivity. Furthermore, the performances of the MWCNT sensors could be improved by proper heat treatment and oxygen annealing [14].

A similar CVD process to deposit MWCNT thin films for sensor application was demonstrated by Cho et al. [15]. As shown in Fig. 7.8, the authors adopted a different sensor configuration. The sensor consisted of a CNT thin film, a Cr-resistive heater, a pair of electrodes, and a passivation layer (SiO_2) on a thermally insulated dielectric

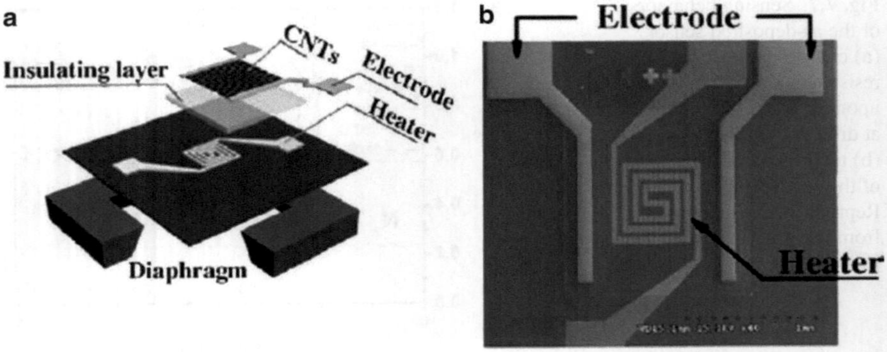

Fig. 7.8 Photograph (**a**) and FE-SEM photomicrograph (**b**) of the MWCNT-based gas sensor. Reproduced with permission from ref. [15]. Copyright 2006, Elsevier

diaphragm used to reduce the thermal gradient of the $Si/SiO_2/Si_3N_4$ structure. The diaphragm with a square shape was formed by etching the back side of the Si substrates through the anisotropic etching with KOH. The sensors were constructed by the conventional silicon process, as shown in Fig. 7.9. Detailed fabrication process could be found in the original literature. Figure 7.10 shows representative SEM images of the functional parts of the MWCNT thin film sensors. The MWCNT films were grown with either a mesh shape (Fig. 7.10a) or a serpentine shape (Fig. 7.10b). The MWCNTs were vertically aligned to the substrate with a uniform distribution (Fig. 7.10c). The diameter and height of the MWCNTs were 30–40 nm and 3–4 μm, respectively. The dimension and thickness of the CNT films could be readily controlled by the properties of catalyst and deposition conditions. An overlapped part of the catalyst metal and the electrode metal was used to ensure good electrical contact (Fig. 7.10d). The MWCNT-based sensors showed high sensitivity to ppm level NO_2 at room temperature.

Another way to use MWCNTs for sensor application was dispersion method [16–20], similar to that for SWCNTs as discussed above [5–7]. For example, Feng et al. [16] used commercially available MWCNT powders to fabricate humidity (vapor gas of water) sensor by using spray coating. The CNT powders were dispersed in *N*-methyl-2-pyrrolidone (NMP). The dispersion was sprayed on PPE substrates with pre-deposited IDEs for resistance measurement. The performance of the MWCNT sensors could be further improved through functionalization with acid (H_2SO_4 and HNO_3) treatment. Resistance of the sensor increased with increasing humidity. Various assumptions were proposed to explain the humidity sensing mechanism, such as electron donation to the *p*-type semiconducting MWCNTs from water molecules, increase in tunneling barriers due to the water absorption on the tube-to-tube interfaces, weak bonding between the H atoms of water molecules and C atoms on the surface of the MWCNTs, and hydrogen bonding between the polar water molecules with the oxygen-containing defects of the MWCNTs [16]. Further studies should be conducted to elucidate the underlying mechanisms.

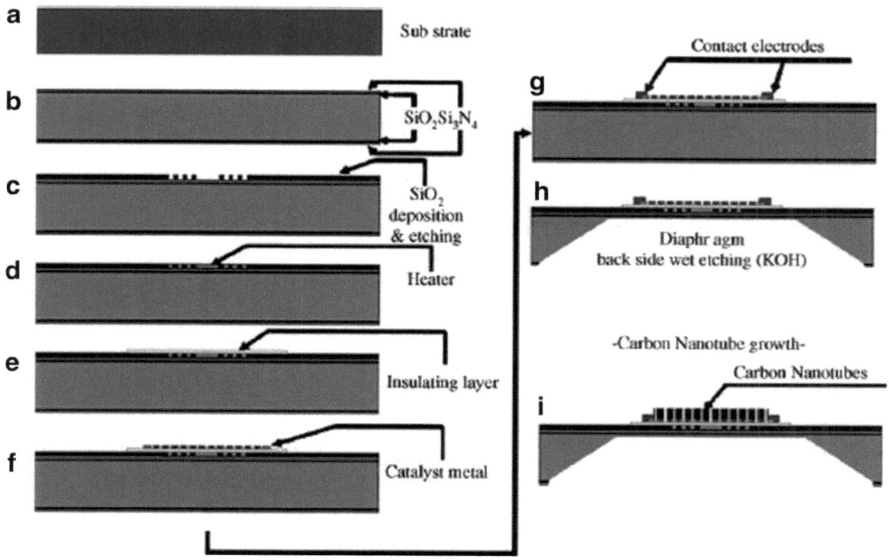

Fig. 7.9 Processing flow to fabricate selectively grown vertical MWCNT gas sensor. Reproduced with permission from ref. [15]. Copyright 2006, Elsevier

A comparison study on O_2 sensing properties of MWCNT thin film sensors made with two methods was reported by Cava et al. [17]. Before the deposition of thin films of substrates with ITEs, commercial MWCNT powder was dispersed in toluene. Figure 7.11 shows a schematic diagram of the two CNT film processing methods. The first method was to drop the CNT suspension directly onto the substrates (Fig. 7.11a). Thickness of the MWCNT film was simply controlled by the amount of the suspension used. The other method was called self-assembly deposition (Fig. 7.11b). To do this, the MWCNT dispersion in toluene was first mixed in a container with water. The mixture was left for the occurrence of phase separation, which led to the formation of a grey and transparent film at the interface of the two liquids. MWCNT thin film was transferred to the substrates after the liquids evaporated. Microstructure characterization indicated that the film made by the self-assembly had a much uniform morphology than that deposited by dropping. The self-assembly film exhibited higher sensitivity to O_2. Sensitivity of the SWCNT thin film sensor was found to be affected by operating temperature. As shown in Fig. 7.12, the optimized operating temperature was between 140 and 160 °C. The operation temperature was related to the intrinsic property of the MWCNT devices. At this temperature, all processes of interactions between O_2 and CNTs were reinforced. Because O_2 acted as a dopant of hole, conductivity of the *p*-type MWCNTs would increase. As a result, resistance of the MWCNT thin film sensor device decreased with increasing concentration of O_2, as shown in Fig. 7.13.

Fig. 7.10 FE-SEM images of the as-grown MWCNTs: (**a**) top view of the serpentine-shaped MWCNT films, (**b**) top view of the mesh-shaped MWCNT films on the stacked structured substrates with a Cr heater and a diaphragm, (**c**) details of the MWCNT film on the stacked structured substrate, and (**d**) cross-sectional view of the electrode part. Reproduced with permission from ref. [15]. Copyright 2006, Elsevier

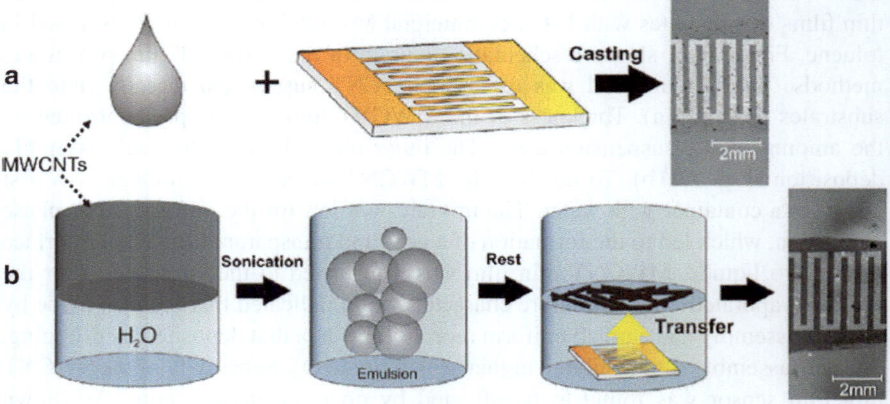

Fig. 7.11 Schematic representation of different deposition systems and the photograph of the respective IDE device: (**a**) casting method where the MWCNT dispersion was casted onto the IDE and (**b**) self-assembly method; after an emulsion process and resting time, a thin film of MWCNTs was formed and transferred to an IDE. Reproduced with permission from ref. [17]. Copyright 2012, Elsevier

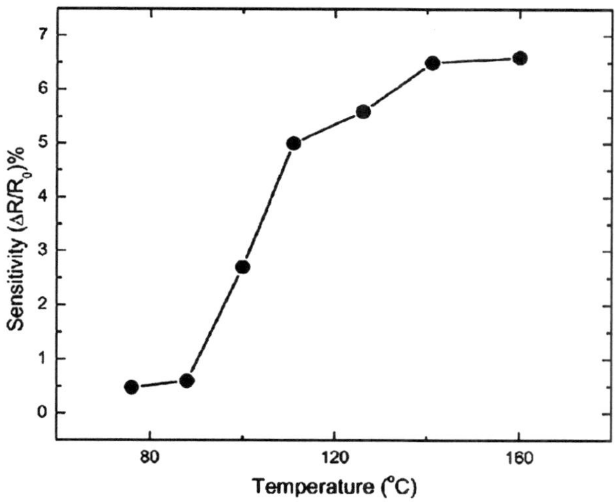

Fig. 7.12 Sensitivity as function of temperature for the self-assembly IDE device. For each temperature increase, the rate of change in the electrical resistance—when an amount of oxygen (10 %) is added to a dried nitrogen flow—was measured. The standard uncertainty of this measurement was 0.03 %. Reproduced with permission from ref. [17]. Copyright 2012, Elsevier

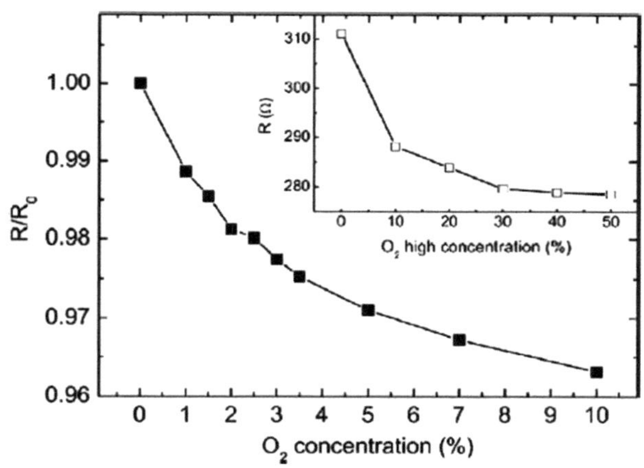

Fig. 7.13 Resistance versus oxygen concentration. The *curve* shows the self-assembly CNT IDE device at operation temperature of 160 °C. The electrical resistivity was monitored as function of concentration of the O_2 at the flow of N_2. The *inset* figure refers to the O_2 high concentrations. The standard uncertainty of this measurement was 0.03 %. Reproduced with permission from ref. [17]. Copyright 2012, Elsevier

Different from the CNT sensors that functioned due to their electrical properties, a new type of gas sensor, surface acoustic wave (SAW) sensor, was developed by Hsu et al. [20]. Surface acoustic wave sensors are a class of devices which rely on the modulation of surface acoustic waves to sense a physical phenomenon. The sensor transduces an input electrical signal into a mechanical wave which, unlike an electrical signal, can be easily influenced by the variation in physical properties. The device then transduces this wave back into an electrical signal. Changes in amplitude, phase, frequency, or time delay between the input and output electrical signals can be used to measure the presence of the desired phenomenon. The SAW device reported by Hsu et al. [20] was a quartz crystal. Commercial MWCNT powder was coated onto the quartz crystal. The MWCNT layer acted as an absorbent to absorb gases to be sensed. The absorption of gases meant increase in mass of the quartz crystal, which in turn would decrease the frequency of the device and thus could be used as sensors. To improve the sensitivity of the sensor, the MWCNT powder was treated with acid to attach –COOH groups and then was mixed with poly(*n*,*n*-dimethylamino propylsilsesquioxane) (SXNR) in THF before being sprayed. The SAW quartz crystal coated with the MWCNT-COOH/SXNR possessed very promising sensitivity and quick response to ethanol vapor. It is believed that more and more types of gas sensors based on CNTs will be developed in the future.

7.2.2 Graphene

The use of graphene as gas sensors has become an increasingly hot topic of research in the area. Graphene is a single layer of carbon atoms formed in honeycomb lattice, with very unique thermal, mechanical, electronic, and optical properties. Graphene is expected to be very sensitive to changes in its environment, due to its extremely high electron mobility at room temperature, great surface area, and low electrical noise. A relatively thorough review on the progress in chemical sensors based on graphene and graphene-related materials appeared recently [4]. There have also been theoretical studies to predict the possible applications of graphene-based materials as chemical sensors [21]. This section was aimed to present thin film gas sensors based on graphene prepared by using different methods, with representative examples instead of an exhausted literature review.

Graphene can be produced by using micromechanical cleavage, graphite oxide exfoliation, andchemical vapor deposition (CVD). An example of using micromechanical cleavage to prepare graphene thin layer used for gas sensor application was presented by Hwang et al. [22]. High-quality graphene layers were produced from highly oriented pyrolytic graphite (HOPG) first and then transferred onto suitable substrates with pre-deposited electrodes, as shown in Fig. 7.14a. Figure 7.14b shows a representative AFM image of the graphene layer. The graphene device was connected to a fabricated chip holder by Al wires to have good contact. The whole assembly was then used gas sensor to be tested in a chamber. Figure 7.15 illustrates photographs of the chip holder with graphene device and a schematic of the graphene device on SiO_2/Si substrate that was connected to the Au electrodes on the chip holder. Sensing responses were monitored by measuring resistance of the device. The resistance was constant in N_2, which increased and decreased when exposed to NH_3 and NO_2, respectively, as shown in Fig. 7.16a. In this case, NH_3 acted as an electron donor, shifting the valence band of graphene away from the Fermi level, leading to a depletion of hole and thus increase in resistance of the device. In contrast, NO_2 took electrons away from the graphene, which generated

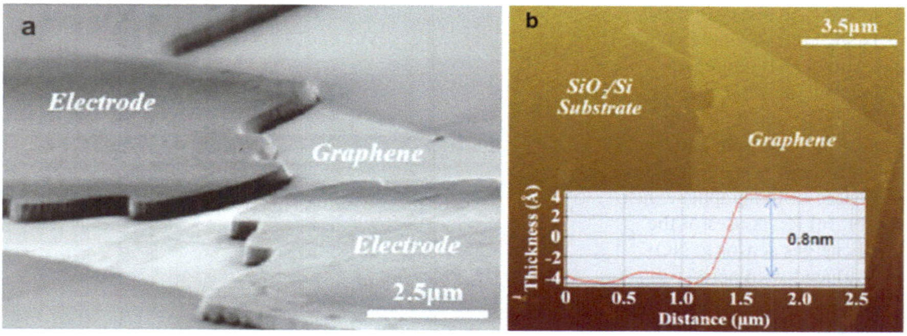

Fig. 7.14 SEM image of the graphene layer embedded on a silicon wafer with Au electrodes (**a**) and AFM image of graphene before electrode deposition, with the *inset* showing the graphene thickness (**b**). Reproduced with permission from ref. [22]. Copyright 2012, Elsevier

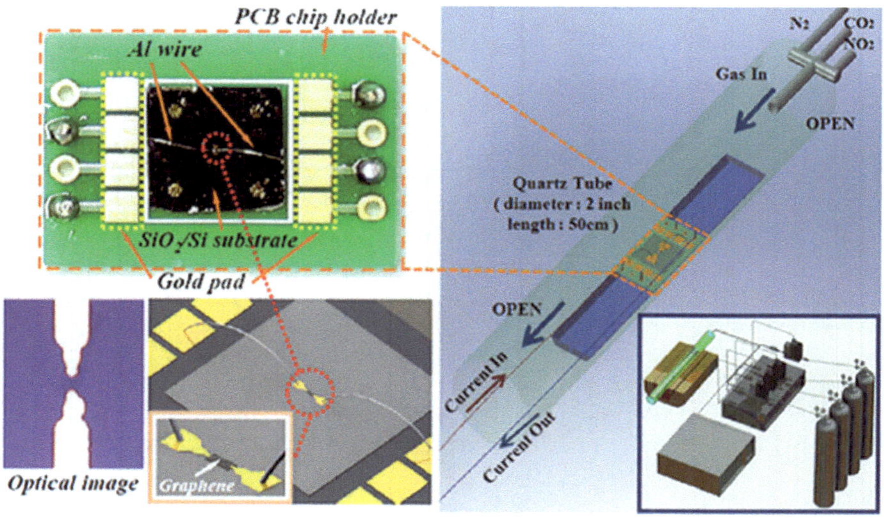

Fig. 7.15 Photographs of a chip holder with graphene, a schematic of the SiO_2/Si substrate, an optical image of the graphene device, and a schematic of the measurement system. Reproduced with permission from ref. [22]. Copyright 2012, Elsevier

holes and thus decreased the resistance of the graphene device. In this study, the effects of geometry of the graphene layer, layer thickness (number of graphene), and thermal treatment on sensing performances of the sensors were systematically investigated [22], providing useful guidance for commercial applications of this kind of graphene gas sensors.

Several chemical exfoliation methods have been developed to synthesize graphene, among which the Hummers method is most widely used in the open literatures. This method involves the use of strong acid and oxidants. The derived

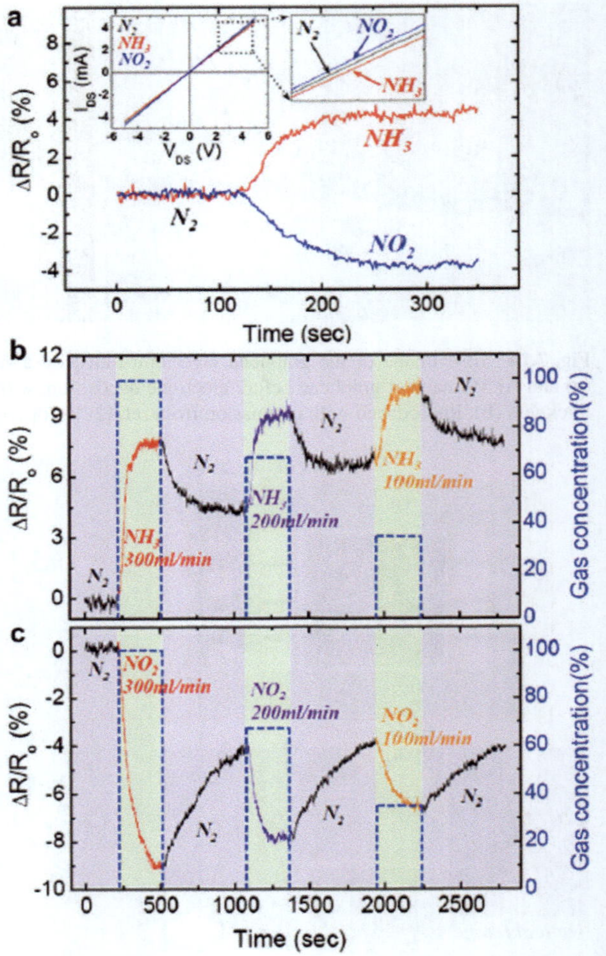

Fig. 7.16 (**a**) Responses in resistance of the sensor upon exposure to NH_3 and NO_2, with *inset* being *I–V* curves of N_2, NH_3, and NO_2 after joining the graphene and the chip. (**b**) NH_3 sensing responses of the graphene gas sensor. (**c**) Responses of the graphene gas sensor to NO_2. Reproduced with permission from ref. [22]. Copyright 2012, Elsevier

graphene has large amount of OH group, which is usually called graphene oxide (GO). Due to the presence of the OH group, GO is soluble in water. GO can be reduced to form graphene, which is also named as reduced graphene oxide (rGO). Wang et al. [23] reported rGO thin films which was sensitive to H_2. Different conditions were tried to reduce the GO thin films. Optimized condition was identified through the sensing measurement.

A simple chemical sensor was developed by using randomly stacked graphene nanoflakes [24]. The graphene nanoflakes were synthesized by a surfactant exfoliation method. Graphene thin sheets were prepared by using vacuum filtration, which were then transferred onto Si substrates through a stamp printing technique. The graphene sensors exhibited high sensitivity to toluene and 1,2-dichlorobenzene molecules. By using hydrothermal method, Huang et al. [25] synthesized a defect

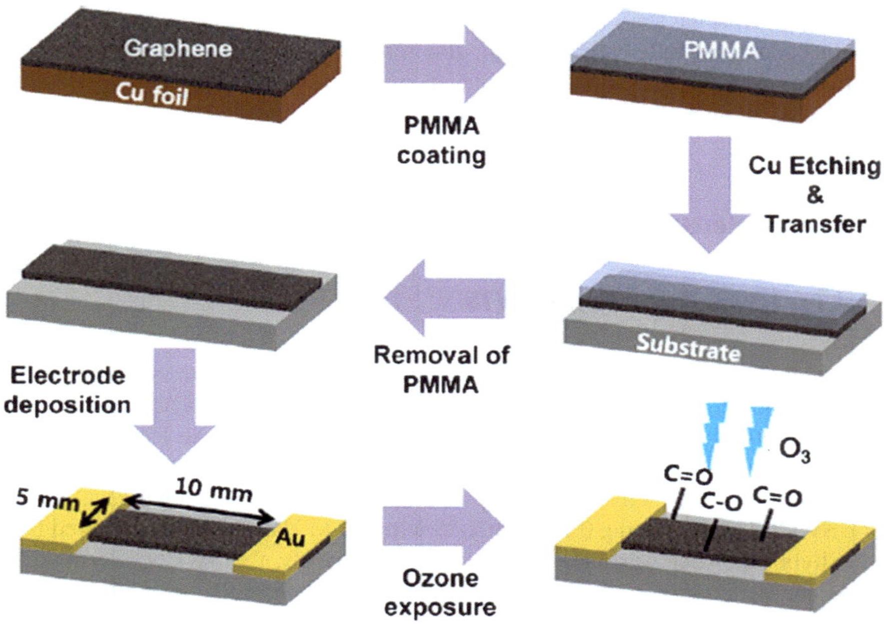

Fig. 7.17 Fabrication process of the ozone-treated graphene gas sensor. Reproduced with permission from ref. [27]. Copyright 2012, Elsevier

graphene that was sensitive to humidity. The synthesis was conducted by hydrothermally treating mixture of sodium, carbohydrate sugar, and ethanol at 200 °C. The treated samples were then rapidly pyrolyzed at 600 °C before vacuum filtration and drying. The derived graphene powder was mixed with ethanol. Graphene thin film sensors were fabricated by dropping the mixture suspension onto alumina substrates with ITEs.

Different from micromechanical cleavage and chemical exfoliation which started from graphite, CVD method to produce graphene uses gaseous precursors as the carbon sources. The most striking advantage of CVD is its capability of depositing graphene layer with large area, uniform, and controllable thickness and no need to use hazard chemicals [26–29]. Yavari et al. [26] used CVD to deposit high-quality graphene thin film on copper (Cu) foil using hexane, a liquid precursor. The self-limiting process of precipitation of carbon on Cu allowed for very thin films of graphene. The deposited graphene thin films were transferred onto Si substrates to fabricate sensor devices. The graphene thin film sensors were sensitive to NH_3 and NO_2 at sub-ppm level in air at room temperature and atmospheric pressure. The sensitivity was higher than that of commercial sensors. A similar process was reported by Chung et al. [27]. They used ozone to treat the graphene thin films to improve sensor performances. Figure 7.17 shows the fabricationprocess of the ozone-treated graphene thin film sensors. Graphene thin films could be used to develop flexible gas sensors, which might extend the application regime [28]. Recent studies indicated that graphene thin film sensors could detect gas molecules at extremely low concentrations, for instance, hundreds of ppq (parts-per-quadrillion) level [29], showing their potential applications in some special areas.

7.3 Gas Sensors Based on Nanosized Hybrids

7.3.1 Hybrids Based on CNTs

Nobel metal and general metal nanoparticles have been used to improve the sensing performances of CNT gas sensors [30–34]. A systematic study was reported by Ju et al. [30] on sensing properties of Pd-SWCNT gas sensors to H_2. Commercial SWCNT powder was oxidized with nitric acid and sulfuric acid, which was then reacted to yield terminal amine groups. Pd nanoparticles were grown on surface of the SWCNTs by reducing 0.1 M $PdCl_4^{2-}$ with 1.0 M $NaBH_4$. Sensor derived from this hybrid powder was called Sample I. Alternatively, the SWCNT surface was modified with hyper-branched PAMAM dendrimers before the decoration with Pd NPs. Such kind of powder was Sample II, which Sample III was obtained by heating Sample II at 200 °C in Ar for 12 h. Fabrication process of the Pd-SWCNT hybrid gas sensor is shown in Fig. 7.18. All Samples were prepared by ultrasonic spreading of the precursor powders on Si substrates with Au electrodes.

Figure 7.19 shows high-resolution transmission electron microscopy (HR-TEM) images (Fig. 7.19a, c, e) and real-time electrical resistance responses (Fig. 7.19b, d, f) to H_2 gas for Sample I, Sample II, and Sample III, respectively [30]. The density of Pd NPs in Sample I was lower than those in Sample II and Sample III. This was attributed to the fact that the 3-dimensional structure of the dendrimers in Samples II and III was beneficial in hosting Pd NPs. The lone pair of electrons of the amine groups reacted easily with Pd-containing cations, leading to high contents of Pd NPs. The particle size of Sample III was slightly larger than that of Sample II, due to the agglomeration of Pd NPs caused by the thermal annealing.

Representative electrical sensitivities of the three sensors to 10,000 ppm H_2 gas at room temperature are shown in Fig. 7.19b, d, f, respectively. Sample I had a response of 25 % with a response time of 324 s. After it was exposed to air, it experienced a very long recovering time. In contrast, the response time of Sample II was only 3 s, much faster than that of Sample I, as seen in Fig. 7.19d. The sensitivity of Sample II was only 8.6 %, much lower than that of Sample I. The faster response time and lower sensitivity of Sample II were explained by the attraction of dipole moments formed in the SWCNTs with the dendrimers, which was 2–3 nm thick, so that the Pd NPs were on the dendrimer ends [30]. This problem was addressed by the thermal annealing. As shown in Fig. 7.19f, Sample III exhibited a sensitivity of 25 % and a response time of 7 s to 10,000 ppm H_2 gas. The thermal annealing reduced the distance between the SWCNTs and the Pd NPs.

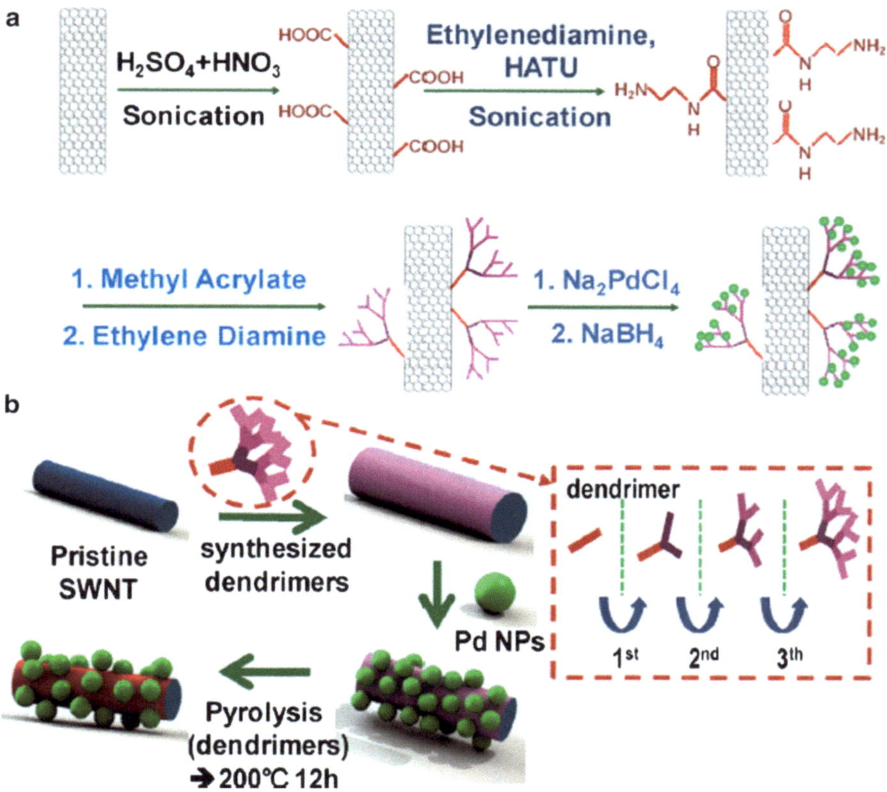

Fig. 7.18 The schematic illustrations of the modification of (**a**) SWCNTs with NH_2-terminated PAMAM dendrimers and Pd NPs and (**b**) fabrication procedure for Sample III. Reproduced with permission from ref. [35]. Copyright 2010, Elsevier

Figure 7.20a shows the real-time electrical resistance response to H_2 gas for Sample III at various concentrations of H_2 gas at room temperature. Variation of sensitivity with the square root of concentration of H_2 gas and response time as a function of H_2 gas concentration are shown in Fig. 7.20b, c. The results indicated that Sample III was able to detect an ultralow concentration of H_2 gas down to 10 ppm. As claimed by the authors, the concentration was the lowest detection limit reported for CNT-based H_2 gas sensors. The correlation between the sensitivity and H_2 gas concentration was in a good agreement with the law in the range of 10–1,000 ppm. The fitted sensitivity was consistent with the experimental value. Response time was inversely proportional to H_2 gas concentration. All these experimental observations could be well explained in terms of the functionalization of the SWCNTs by the Pd NPs.

Liu et al. [31] reported an aerosol jet printing method to fabricate Pt-decorated SWCNT H_2 gas sensor. The ink for the jet printing was a solution of hybrid of

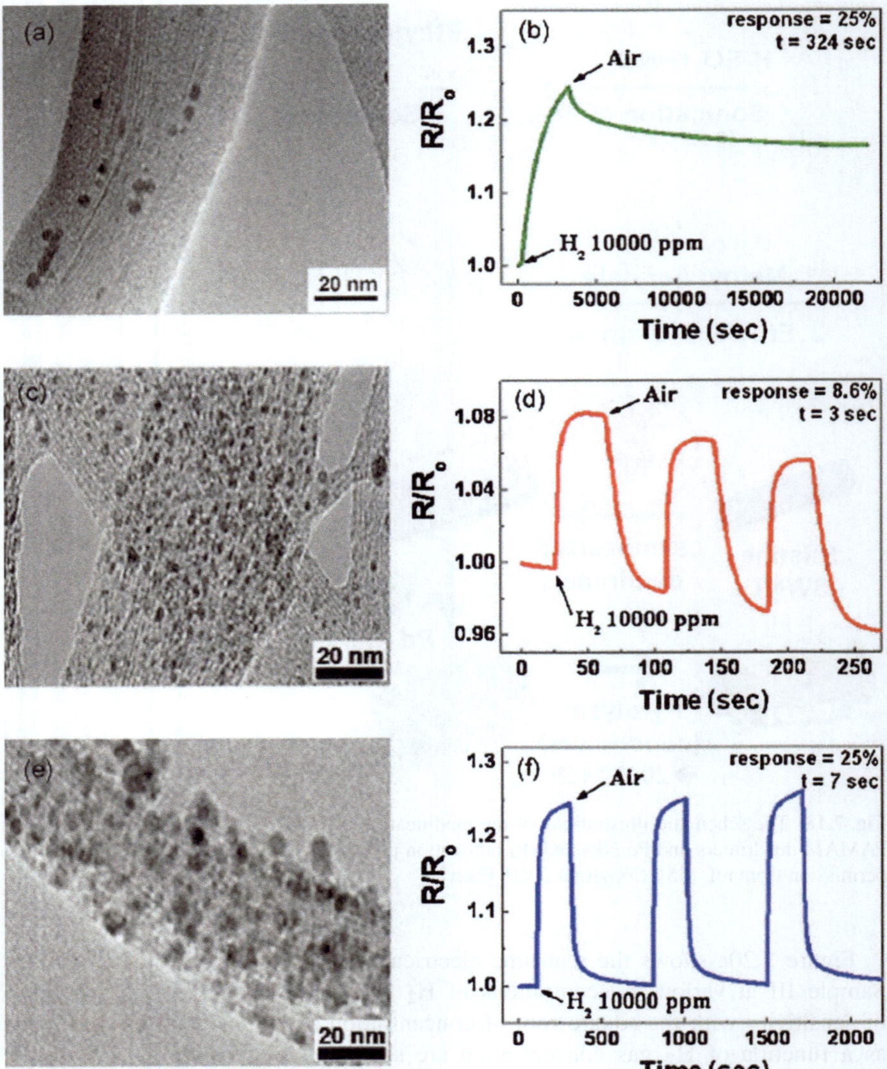

Fig. 7.19 High-resolution TEM images of (**a**) Sample I, (**c**) Sample II, and (**e**) Sample III. Representative electrical responses to 10,000 ppm H_2 gas at room temperature of (**b**) Sample I, (**d**) Sample II, and (**f**) Sample III. Reproduced with permission from ref. [30]. Copyright 2010, Elsevier

Pt nanoparticle and SWCNT. Commercially available SWCNT powder was first purified to remove catalysts and carbonaceous impurities. The purified SWCNT powder was dispersed in ethylene glycol (EG) solution with sodium dodecyl sulfate (SDS), to which $H_2PtCl_6{\cdot}H_2O$ was added. The mixture was then reduced to form

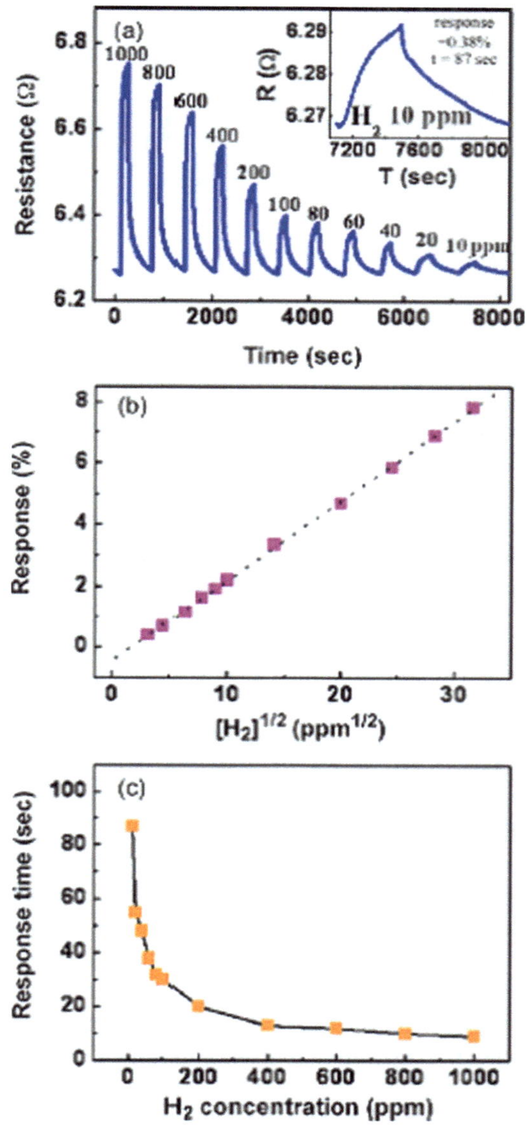

Fig. 7.20 (**a**) Real-time electrical response of Sample III to H_2 gas over a concentration range of 10–1,000 ppm at room temperature, (**b**) variation of response with the square root of H_2 concentration, and (**c**) the response time as a function of H_2 concentration. Reproduced with permission from ref. [30]. Copyright 2010, Elsevier

Pt-SWCNT hybrid nanoparticles. The hybrid nanoparticles were made into printable ink after certain necessary processes. The ink was used to deposit thin film gas sensors on Si substrates with prefabricated Au electrode arrays. The Pt-SWCNT gas sensor was sensitive to H_2 at 20 ppm.

MWCNT decorated with noble metal nanoparticles was also used to fabricate H_2 gas sensor, as reported by Zilli et al. [32]. The MWCNT powder was synthesized by floating catalyst CVD and treated by annealing and oxidation with acids. Both the

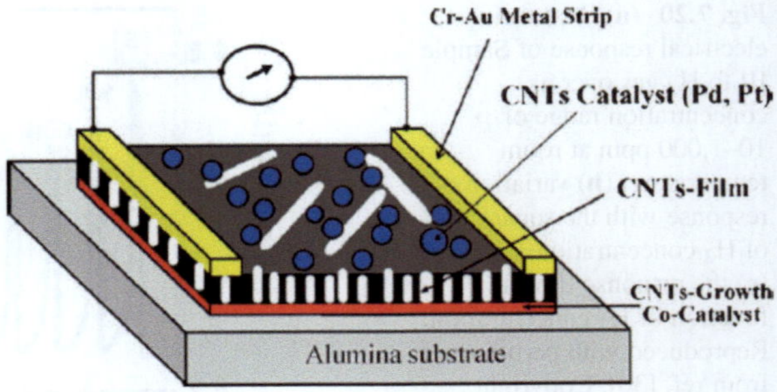

Fig. 7.21 Configuration of the gas sensor based on MWCNTs networked thin films decorated with Pt and Pd nanoclusters. Reproduced with permission from ref. [34]. Copyright 2008, Elsevier

as-synthesized and treated MWCNTs had a length of 16–20 μm and an outer average diameter of 24–50 nm. They were made into floccules-like aqueous suspensions, which were used to prepare buckypaper films through filtration. For Pd-decorated films, the treated MWCNTs were mixed with $PdCl_2$ solution and then reduced by using $NaBH_4$ before filtration. H_2 gas sensors were made with a piece of the buckypapers. The sensors exhibited a H_2 sensing limit of 70 ppm at room temperature [32]. Compared to SWCNT gas sensor, MWCNT devices are more promising, because MWCNTs are more available than SWCNTs.

CVD has been used to deposit CNT thin films directly for fabrication of gas sensors. Star et al. [33] deposited SWCNT thin films on Si substrate by using CVD at 900 °C with Fe nanoparticles as catalyst and CH_4/H_2 gas mixture as carbon source. The Si substrates had Au electrodes for electrical measurement. The gas sensor device was a field-effect transistor (FET). A number of metals, including Mg, Al, Ti, V, Cr, Mn, Fe, Co, Ni, Zn, Mo, Rh, Pd, Sn, W, Pt, Au, and Pb, were used to decorate the SWCNT thin films. The metals were incorporated by thermal or electron beam evaporation. Electrodeposition and solution chemical reduction methods were also used to prepare nanoparticles of some of the metals for decoration purpose. The gas sensors based on the SWCNT FET decorated with different metal catalysts had different responses to H_2, CH_4, CO, and H_2S. It meant that selective sensing of the gas sensors was realized by using different metal decorations. The sensors were combined to form sensor arrays. Electronic responses of all the sensors and the sensor array data could be analyzed by using pattern-recognition analysis tool, which would be useful in detecting and identifying toxic or combustible gases.

Penza et al. [34] used CVD to deposit MWCNT thin films on alumina substrates. After deposition, the MWCNT films were decorated by 5-nm-thick Pt and Pd nanoclusters, deposited by using a magnetron sputtering. Configuration of the gas sensors is shown in Fig. 7.21. The sensors had high sensitivities to NO_2, H_2S, NH_3, and CO, at sub-ppm level.

There were also reports that used nanosized oxide particles to decorate CNTs for gas sensor applications. Various oxides, such as SnO_2 [36–41], TiO_2 [42, 43], WO_3 [44], and In_2O_3 [45], have been used for such a purpose. For example, Mao et al. [36] developed an ultrafast H_2 sensor with randomly distributed SWCNTs decorated

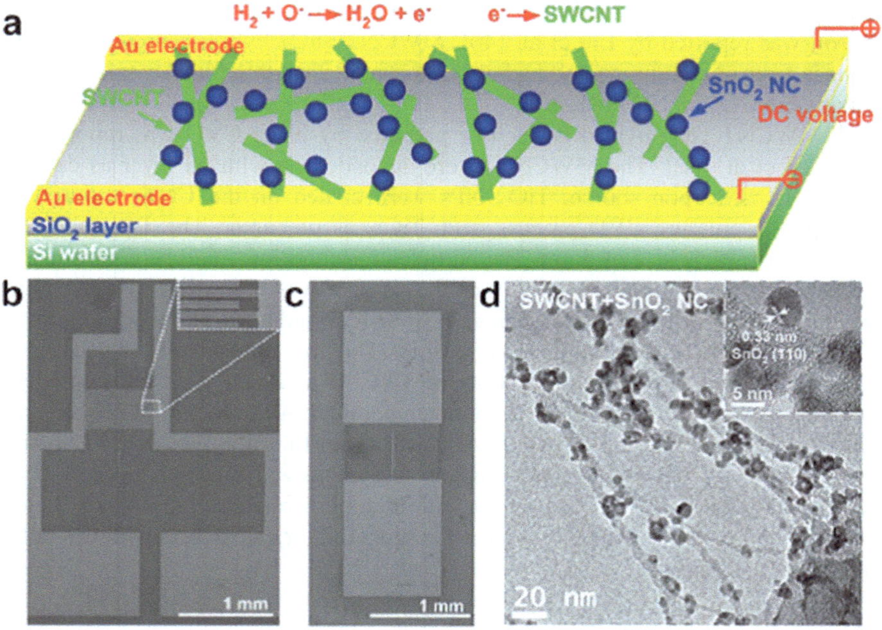

Fig. 7.22 (**a**) Schematic of the gas sensor made of SWCNTs decorated with SnO_2 NPs. (**b**) SEM image of the multiple-pair Au finger electrodes. (**c**) SEM image of the single-pair Au finger electrodes. (**d**) TEM image of the SnO_2 NPs decorated SWCNTs, with the *inset* being HR-TEM image of the hybrid. Reproduced with permission from ref. [36]. Copyright 2012, The Royal Society of Chemistry

with nanosized SnO_2 particles. Figure 7.22a shows a schematic of the gas sensor configuration. The SWCNT network acted as conducing channels. As H_2 molecules were adsorbed onto SnO_2 NPs, there was electron transfer from the gas molecules to the NPs and then to the CNT network, leading the change in conductivity of the sensor. Two types of Au electrodes (Fig. 7.22b, c) were used. The multiple-pair Au finger electrode was for randomly dispersed SWCNTs, while the single-pair finger electrode was for the directly grown SWCNT network film. As shown in Fig. 7.22d, the SnO_2 NPs had sizes of 3–6 nm, which were distributed on the surface of the SWCNTs. The SnO_2 NPs were high-quality crystals of rutile. The sensor possessed a response time of 2–3 s to 1 % H_2 at room temperature [36].

Wei et al. [37] found that the incorporation of SWCNT to SnO_2 thin film sensor enhanced its sensitivity to NO_2. The hybrid thin films were prepared by using spin-coating. Liang et al. [38] used a thin layer of SnO_2 to coat MWCNTs. Gas sensors made of the SnO_2-coated MWCNTs had lower resistance than the sensors based on SnO_2 nanobelt. The hybrid gas sensor was sensitive to NO_2 at 2 ppm. Du et al. [39] employed layer-by-layer deposition method to assemble a ternary hybrid of CNT, SnO_2, and Au, which showed high sensitivity to CO at room temperature. Another ternary hybrid, consisting of CNT, SnO_2, and Sb, used as indoor air pollutant gas sensor, was reported by Liu et al. [40]. MWCNT-SnO_2 hybrid sensors were also made in form of membrane, by shadow mask coating, and thin films by spin-coating, sensitive to CO and NO_2 [41, 46].

A MWCNT-TiO_2 hybrid gas sensor prepared by using CVD was reported by Khalilian et al. [42]. The MWCNTs were deposited on with Ni layer as catalyst and C_2H_2 as carbon source. TiO_2 NPs were coated on the CNTs through an atmospheric pressure CVD process, with $TiCl_4$ as precursor. The TiO_2 NPs had an average particle size of 5 nm. The particle concentration could be controlled by the deposition condition. Sensitivity of the sensors was optimized by controlling the surface area of the hybrid. Ueda et al. [43] reported a simple way to fabricate SWCNT-TiO_2 hybrid thin film gas sensors. The thin films were coated on Si substrates with a suspension of mixture of SWCNT and nanosized TiO_2 powders. The sensors were sensitive to NO at room temperature. A mixture of MWCNTs and nanosized WO_3 was used to prepare suspensions for thin film gas sensor sensitive to NO_2 [44]. A hybrid of CNT-In_2O_3 was synthesized through layer-by-layer deposition, leading to NH_3 gas sensor with high performances [45]. It is believed that many other oxides will be used as decorators to fabricate hybrid with CNTs for gas sensor applications.

7.3.2 Hybrids Based on Graphene

The motivation of using hybrids of graphene or graphene oxide and nanoparticles of metals and oxides for the applications of gas sensors is similar to that for CNTs. Graphene or graphene oxide films decorated with noble metal nanoparticles could be prepared by either CVD method [47–52] or dispersion method [53, 54]. An example demonstrated by Chung et al. [47] is used to show the detailed process of such gas sensors. The authors used CVD to deposit graphene films on copper (Cu) foils. The graphene films were then coated with a layer of PMMA, which were transferred onto PET substrates. After the transformation, the PMMA was removed. Pd nanoparticles were deposited on the graphene films by using thermal evaporation. Au lead wires were connected to the graphene film using Ag paste to form gas sensors. The fabrication process is illustrated in Fig. 7.23 [47]. The performances of the graphene sensors could be optimized. The sensor with 3-nm-thick Pd exhibited a response of 33 % to 1,000 ppm H_2 and sensing limit of as low as 20 ppm H_2 at room temperature.

In a separate study, Johnson et al. used Pd NPs to functionalize multilayer graphene nanoribbon (MLGN) networks [53]. Figure 7.24 shows microstructures of the graphene nanoribbons, schematic of the sensor device made with them, and Raman characterization results. The MLGN networks were synthesized by dispersing expanded flake graphite in a surfactant-water solution, followed by sonication and vacuum filtration. After filtration and drying, the filtrating substrates were dissolved, releasing the MLGN membrane, which was transferred to sensor

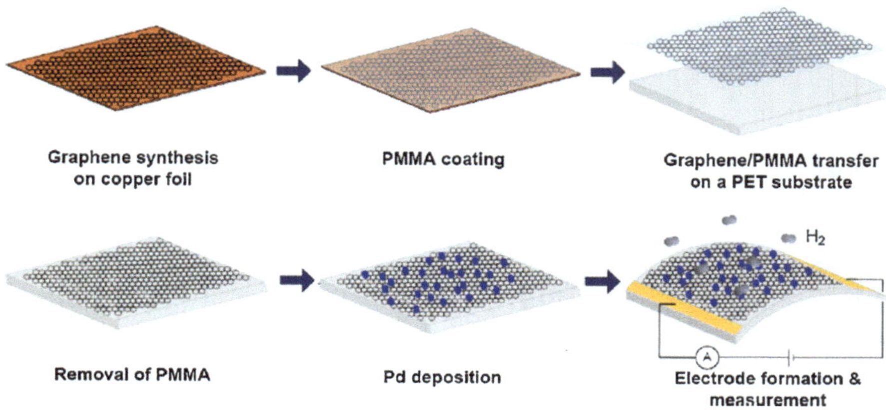

Fig. 7.23 Fabrication procedures of the graphene-Pd hybrid gas sensor. Reproduced with permission from ref. [47]. Copyright 2012, Elsevier

substrates. Inorganic alumina anodic filter membranes with a 100 nm pore size were used for the filtration experiment. The graphene nanoribbons had an average thickness of 7.5 nm, an average width of 6.6 nm, and lengths of 0.5 to a few μm. The graphene nanoribbons had very high graphitic quality, as demonstrated by the appearance of the prominent tangential G-band peak, a 2D peak, and a small defect D peak.

Sensing characteristics of the Pd-MLGN network gas sensor to H_2 at different concentrations are shown in Fig. 7.25. The Pd-MLGN sensor showed a very high sensitivity to H_2 at room temperature. For example, its sensitivity ($\Delta R/R$) was as high as 55 % for 40 ppm H_2, being gradually increasing with increasing H_2 concentration. It also exhibited a very fast response speed and short recovery time, as well as good repeatability. A similar process was used by Kumar et al. [54] to prepare graphene layer decorated by Pd-Pt binary NPs.

Oxides that have been used to decorate graphene for applications as gas sensors, as examples in this chapter, include SnO_2 [55, 56], ZnO [57, 58], Cu_2O [59], WO_3 [60], and Al_2O_3 [61]. Song et al. [55] and Zhang et al. [55] both reported hydrothermal synthesis of SnO_2-decorated graphene sheets for gas sensor applications. In the first example [55], synthesized graphene nanosheet powders were dispersed in $SnCl_2$ solution in HCl, which were then hydrothermally treated at 80 °C for 12 h. Gas sensors made of this kind of hybrid had very high sensitivity and selectivity to propanol. In the second case [56], SnO_2 NPs were synthesized with $SnCl_4$. Nanorods of SnO_2 were aligned on graphene nanosheets through two steps. This hybrid was used to fabricate gas sensor with high sensitivity to H_2S.

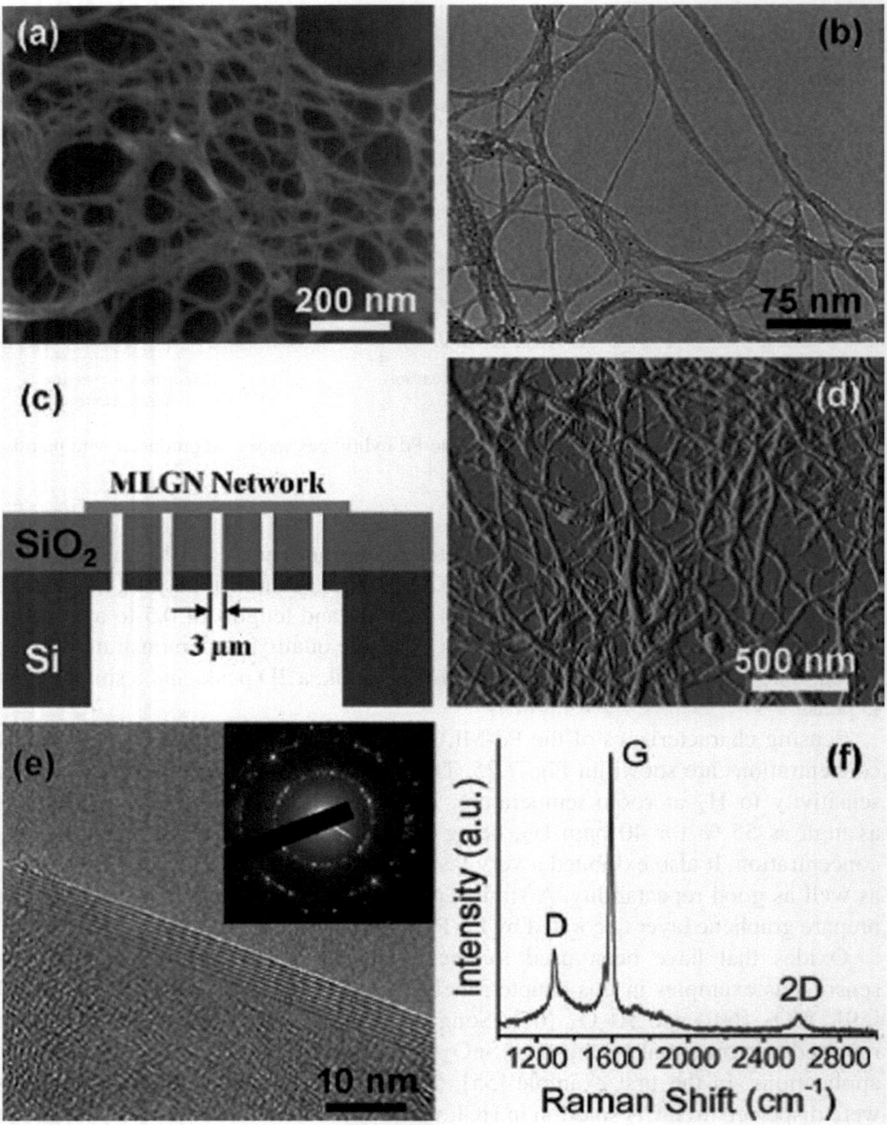

Fig. 7.24 (**a**) SEM and (**b**) TEM image of the multilayer graphene nanoribbon (MLGN) network suspended over a micromachined silicon substrate. (**c**) Cross-sectional schematic of the micromachined silicon substrate showing narrow open slits over which the MLGN network was suspended in parts (**a**) and (**b**) (not to scale). (**d**) AFM phase image of the MLGN network deposited on a Si/SiO_2 substrate. (**e**) High-resolution TEM image of the folded edge of an individual ribbon, where eight graphene layers are visible. The *inset* shows a selected area electron diffraction (SAED) pattern for the same nanoribbon. (**f**) Raman spectrum of a MLGN on a Si/SiO_2 substrate showing G, 2D, and D peaks. Reproduced with permission from ref. [53]. Copyright 2010, Wiley-VCH

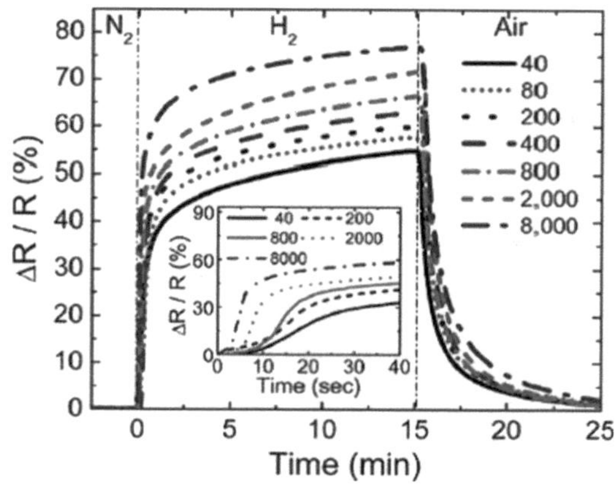

Fig. 7.25 Response ($\Delta R/R$) characteristics of the Pd-functionalized MLGN network sensor when exposed to H_2 in N_2 with concentrations ranging from 40 to 8,000 ppm. There was a monotonic increase in $\Delta R/R$ with increasing H_2 concentration. The initial resistance was that measured in N_2, and the sensor recovery was conducted in air. The *inset* shows the first 40 s of the sensor response to H_2 at different concentrations. Reproduced with permission from ref. [53]. Copyright 2010, Wiley-VCH

7.3.3 CNTs-Graphene Hybrids

Recently, a new group of hybrid consisting of CNTs and graphene emerged to be used as gas sensors [62, 63]. Since both CNTs and graphene are promising candidates for gas sensor applications, their combination would be a natural choice, which has been one of the strategies in the research of materials science and engineering.

A simple mixing method was used by Kaniyoor and Ramaprabhu [62]. They simply mixed MWCNT and graphene nanoplatelets (GNPs) suspensions. At the same time, Pt NPs were incorporated into the mixed suspensions. Figure 7.26 shows representative TEM and SEM images of the hybrid samples decorated with Pt NPs, together with EDX analysis result. The GNPs were multilayered particles, while the MWCNTs had outer diameters of 20–50 nm. The presence of the latter prevented the former from restacking. This was because the MWCNTs of various diameters and lengths interspersed in-between the stacks of the GNPs. Pt nanoparticles were dispersed on the carbon nanostructures homogeneously, due to the functionalization procedure for the nanocarbons. Average size of the Pt NPs was analyzed by using XRD, as demonstrated in Fig. 7.27, which was in the range of 8–10 nm.

The ternary hybrid powders were ultrasonically dispersed to form suspensions, which could be used to fabricate thin film gas sensors on alumina substrates. The sensors derived from this kind of hybrid showed high sensitivity to H_2 at room temperature. Sensing performances of a representative sensor device are shown

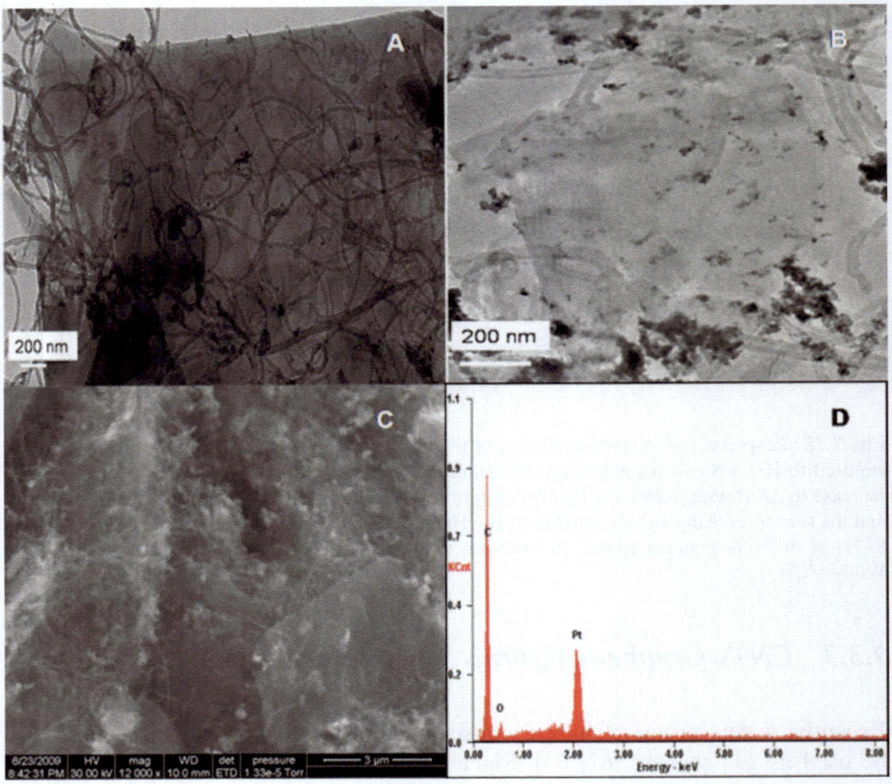

Fig. 7.26 TEM images of GNPs: MWCNTs (**a**) and 20 wt.% Pt/f-GNPs:f-MWCNTs (**b**) MWCNTs interspersed between layers of GNPs and Pt nanoparticles dispersed on the hybrid. FE-SEM (**c**) showing that the GNPs were uniformly covered with MWCNTs and Pt nanoparticles and EDX (**d**) image of the Pt/f-GNPs:f-MWCNTs, confirming a Pt loading of 20 wt.%. Reproduced with permission from ref. [62]. Copyright 2011, Elsevier

in Fig. 7.27. Both sensitivity and response time almost linearly increased with increasing concentration of H_2. In this study, the effects of various influence factors on performances of the H_2 gas sensor were systematically investigated. The results could be used as guidance to future research of this kind of sensing materials (Fig. 7.28).

Jeong et al. [63] found that the presence of a thin layer of graphene oxide was critical to ensure the sensing property of a CNT array. The graphene oxide layer was deposited on Si substrates by spin-coating with an aqueous suspension. Nanopatterned Fe catalyst was then deposited on top of the graphene oxide layer. A CVD process was used to grow CNT array on the patterned catalyst. After being detached from Si substrates, the hybrid structure was transferred onto polyimide substrates to form gas sensors. A two-terminal method was adopted to measure the

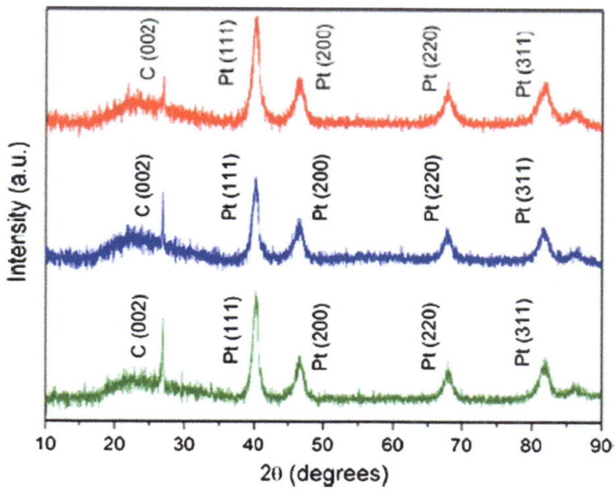

Fig. 7.27 XRD patterns of the samples: (**a**)Pt/f-GNPs: f-MWCNTs (1:3), (**b**) Pt/f-GNPs: f-MWCNTs (1:1), and (**c**) Pt/f-GNPs: f-MWCNTs (3:1). Reproduced with permission from ref. [62]. Copyright 2011, Elsevier

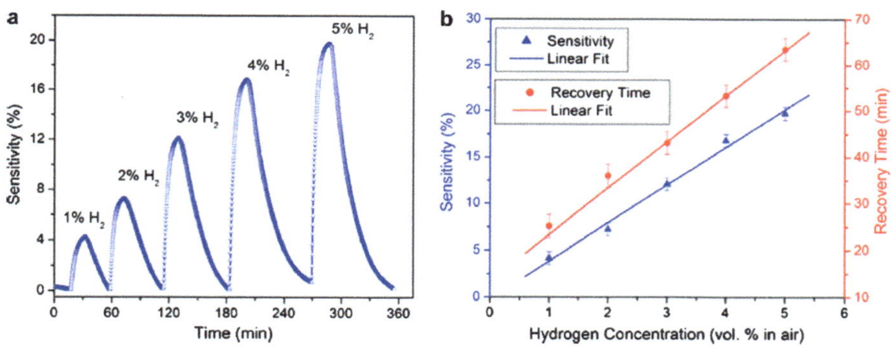

Fig. 7.28 (**a**) Hydrogen sensing performance of Nafion-solubilized Pt/f-GNPs:f-MWCNTs sensor to H_2 gas at different concentrations at 25 °C. (**b**) Sensitivity and recovery time of the sensor as a function of hydrogen concentration. Reproduced with permission from ref. [62]. Copyright 2011, Elsevier

electrical responses of the sensors. Before gas sensing experiments, the graphene oxide layer was reduced at 600 °C in a mixture of H_2 and NH_3. Schematic illustration of the CNTs/graphene (oxide) hybrid films and photograph of a sensor device are shown in Fig. 7.29.

Both the reduced graphene film and the CNTs (20 μm)/reduced graphene hybrid film supported on polyimide substrates showed linear and symmetric *I*–*V* curves, when measured by using two-terminal method. This implied that there was Ohmic contact between the films and the Au electrodes. The estimated resistances were 493 kΩ for the reduced graphene film and 331 Ω for the CNTs/reduced graphene film, respectively. Resistance of the reduced graphene films was in the range of hundreds of kiloohm. In contrast, the much lower resistance of the CNTs/reduced

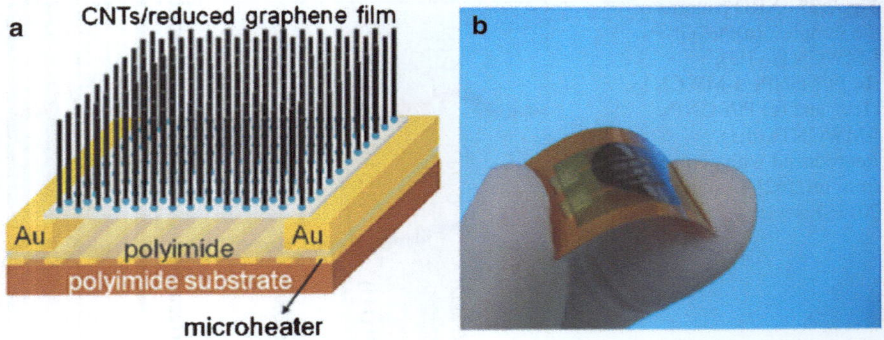

Fig. 7.29 A schematic illustration of the CNTs (20 μm)/reduced graphene sensor device (**a**) and a photograph of the sensor device (**b**). Reproduced with permission from ref. [63]. Copyright 2010, American Institute of Physics

graphene hybrid film was attributed to the highly conductive CNT arrays. Therefore, the sensor device consisting of the CNTs/reduced graphene hybrid film was similar to a parallel circuit containing a high-resistance reduced graphene layer and a low-resistance CNT array.

The hybrid gas sensor exhibited high sensitivity to NO_2. Performances of a representative device are shown in Fig. 7.30. The reduced graphene films exhibited a negligible sensing response at 10 ppm NO_2. The response curves showed an unusual feature at all NO_2 concentrations. It increased in the beginning of exposure to NO_2 and subsequently decreased after several minutes. The degree of resistance increase depended on the concentration of NO_2. The peak time for the highest resistance value was shortened as the concentration of NO_2 was increased. The abnormal behavior was attributed by the authors to a *n*–*p* transition behavior. The *n*-type characteristics of the reduced graphene films were closely related to the growth conditions, where nitrogen gas was used. Nitrogen-doped carbon-based materials generally behave as *n*-type semiconductors. Upon absorption of NO_2 molecules, a transition from *n*-type to *p*-type semiconductor was induced.

7.4 Concluding Remarks

Nanocarbon materials, including carbon nanotubes and graphene, have been shown to be good candidates for applications in thin film gas sensors. Various methods have been used to prepare nanocarbon materials for gas sensor applications. In solution casting methods, such as simple casting, dropping/drying, spin-coating, and dip-coating, suspensions of CNTs or graphene nanosheets/nanoplatelets were used to deposit thin films on substrates for sensing devices. These methods are simple and

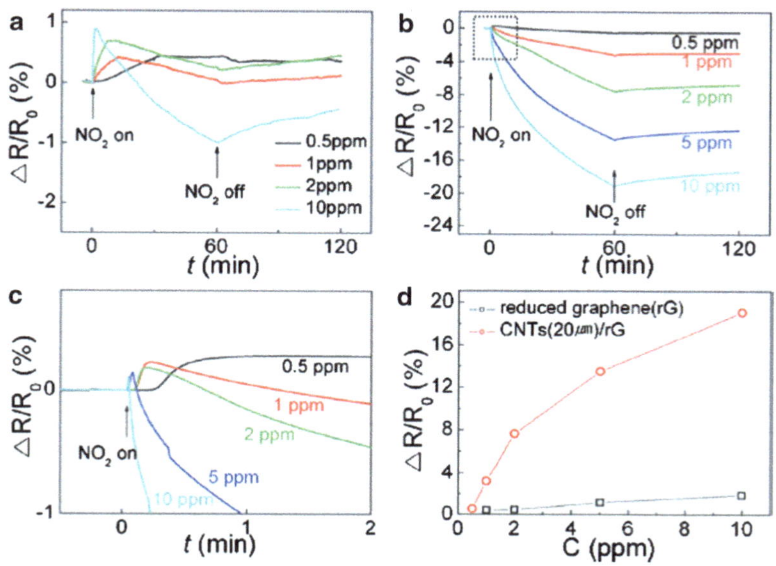

Fig. 7.30 (**a**) Resistance variation in reduced graphene films upon exposure to various concentrations of NO_2 gas. (**b**) Room-temperature NO_2 sensing behavior of CNTs (20 μm)/reduced graphene hybrid film devices at NO_2 concentrations from 0.5 to 10 ppm. (**c**) A magnification of the region of (**b**) marked with *dotted square*. (**d**) Resistance response of reduced graphene and CNTs (20 μm)/reduced graphene gas sensors measured after 60 min exposure to NO_2 at various concentrations. Reproduced with permission from ref. [63]. Copyright 2010, American Institute of Physics

cheap and can be easily scaled up. However, they also have disadvantages, such as difficulty in ensuring reproducibility. CVD methods are advantageous in controlling over homogeneity of the thin film deposited and in the high compatibility with state-of-the-art silicon technology. However, CVD requires relatively expensive facilities and is more difficult to deposit hybrid thin films.

Hybrids of nanocarbons and nanoparticles of noble metals and oxides have been paid almost equal attentions by the worldwide researchers. The advantages of using hybrids as sensing materials have been convincingly demonstrated. It is necessary to continue the research on this class of nanomaterials for gas sensor applications. More importantly, the lack of theoretical studies on these new gas sensing materials should not be neglected.

Acknowledgement One of the authors (L.B.K.) would like to acknowledge the financial support of Start-up Grant (M4080845) of the Nanyang Technological University (2012) and AcRF TIER 1 (RG44/12), MOE, Singapore.

References

1. Kauffman, D.R., Star, A.: Carbon nanotube gas and vapor sensors. Angew. Chem. Int. Ed. **47**, 6550–6570 (2008)
2. Zhang, T., Mubeen, S., Myung, N.V., Deshusses, M.A.: Recent progress in carbon nanotube-based gas sensors. Nanotechnology **19**, 332001 (2008)
3. Bondavalli, P., Legagneus, P., Pribat, D.: Carbon nanotubes based transistors as gas sensors: state of the art and critical review. Sens. Actuator B Chem. **140**, 304–318 (2009)
4. Yavari, F., Koratkar, N.: Graphene-based chemical sensors. J. Phys. Chem. Lett. **3**, 1746–1753 (2012)
5. Li, J., Lu, Y., Ye, Q., Cinke, M., Han, J., Meyyappan, M.: Carbon nanotube sensors for gas and organic vapor detection. Nano. Lett. **3**, 929–933 (2003)
6. Ndiaye, A.L., Varenne, C., Bonnet, P., Petit, E., Spinelle, L., Brunet, J., Pauly, A., Lauron, B.: Elaboration of single wall carbon nanotubes-based gas sensors: evaluating the bundling effect on the sensor performance. Thin Solid Films **520**, 4465–4469 (2012)
7. Hoa, N.D., Guy, N.V., Cho, Y., Kim, J.: Porous single-wall carbon nanotube films formed by in situ arc-discharge deposition for gas sensors application. Sens. Actuator B Chem. **135**, 656–663 (2009)
8. Wongwiriyapan, W., Honda, S., Konishi, H., Mizuta, T., Ohmori, T., Ito, T., Maekawa, T., Suzuki, K., Ishikawa, H., Murkami, T., Kisoda, K., Harima, H., Oura, K., Katayama, M.: Direct growth of single-walled carbon nanotube networks on alumina substrate: a novel route to ultrasensitive gas sensor fabrication. Jpn. J. Appl. Phys. **44**, 8227–8230 (2005)
9. Han, Z.J., Mehdipour, H., Li, X., Shen, J., Randeniya, L., Yang, H.Y., Ostrikov, K.: SWCNT networks on nanoporous silica catalyst support: morphological and connectivity control for nanoelectronic, gas-sensing and biosensing devices. ACS. Nano **6**, 5809–5819 (2012)
10. Ganzhorn, M., Vijayaraghavan, A., Dehm, S., Hennrich, F., Green, A.A., Fichtner, M., Voigt, A., Rapp, M., von Lohneysen, H., Hersam, M.C., Kappes, M.M., Krupke, R.: Hydrogen sensing with diameter- and chirality-sorted carbon nanotubes. ACS. Nano. **5**, 1670–1676 (2011)
11. Lee, H., Shaker, G., Naishadham, K., Song, X., Mckinley, M., Wagner, B., Tentzeris, M.: Carbon-nanotube loaded antenna-based ammonia gas sensor. IEEE. Trans. Microw. Theor. Tech. **59**, 2665–2673 (2011)
12. An, K.H., Jeong, S.Y., Hwang, H.R., Lee, Y.H.: Enhanced sensitivity of a gas sensor incorporating single-walled carbon nanotube-polypyrrole nanocomposites. Adv. Mater. **14**, 1005–1009 (2004)
13. Valentini, L., Armentano, I., Kenny, J.M., Dantalini, C., Lozzi, L., Santucci, S.: Sensors for sub-ppm NO_2 gas detection based on carbon nanotube thin films. Appl. Phys. Lett. **82**, 961–963 (2003)
14. Cho, W.S., Moon, S.I., Paek, K.K., Lee, Y.H., Park, J.H., Ju, B.K.: Patterned multiwall carbon nanotube films as materials of NO_2 gas sensors. Sens. Actuator B. Chem. **119**, 180–185 (2006)
15. Valentini, L., Lozzi, L., Cantalini, C., Armentano, I., Kenny, J.M., Ottaviano, L., Santucci, S.: Effects of oxygen annealing on gas sensing properties of carbon nanotube thin films. Thin Solid Films **436**, 95–100 (2003)
16. Feng, Y., Cabezas, A.L., Chen, Q., Zheng, L.R., Zhang, Z.B.: Flexible UHF resistive humidity sensors based on carbon nanotubes. IEEE. Sensor. J. **12**, 2844–2850 (2012)
17. Cava, C.E., Salvatierra, R.V., Alves, D.C.B., Ferlauto, A.S., Zarbin, A.J.G., Roman, L.S.: Self-assembled films of multi-wall carbon nanotubes used in gas sensors to increase the sensitivity limit for oxygen detection. Carbon **50**, 1953–1958 (2012)
18. Sayago, I., Santos, H., Horrillo, M.C., Aleixandre, M., Fernandez, J.M., Terrado, E., Tacchini, I., Aroz, R., Maser, W.K., Martinez, M.T., Gutierrez, J., Munoz, E.: Carbon nanotube networks as gas sensors for NO_2 detection. Talanta **77**, 758–764 (2008)
19. Su, P.G., Lee, C.T., Chou, C.Y., Cheng, K.H., Chuang, Y.S.: Fabrication of flexible NO_2 sensors by layer-by-layer self-assembly of multi-walled carbon nanotubes and their gas sensing properties. Sens. Actuator B Chem. **139**, 488–493 (2009)

20. Hsu, H.L., Jehng, J.M., Sung, Y., Wang, L.C., Yang, S.R.: The synthesis, characterization of oxidized multi-walled carbon nanotubes, and application to surface acoustic wave quartz crystal gas sensor. Mater. Chem. Phys. **109**, 148–155 (2008)
21. Chen, Y., Gao, B., Zhao, J.X., Cai, Q.H., Fu, H.G.: Si-doped graphene: an idea sensor for NO- or NO_2-detection and metal-free catalyst for N_2O-reduction. J. Mol. Model. **18**, 2043–2054 (2012)
22. Hwang, S., Lim, J., Park, H.G., Kim, W.K., Kim, D.H., Song, I.S., Kim, J.H., Lee, S., Woo, D.H., Jun, S.C.: Chemical vapor sensing properties of graphene based on geometrical evaluation. Curr. Appl. Phys. **12**, 1017–1022 (2012)
23. Wang, J., Kwak, Y., Lee, I.Y., Maeng, S., Kim, G.H.: Highly responsive hydrogen gas sensing by partially reduced graphite oxide thin films at room temperature. Carbon **50**, 4061–4067 (2012)
24. Salehi-Khojin, A., Estrada, D., Lin, K.Y., Ran, K., Hassch, R.T., Zuo, J.M., Pop, E., Masel, R.I.: Chemical sensors based on randomly stacked graphene flakes. Appl. Phys. Lett. **100**, 033111 (2012)
25. Huang, Q., Zeng, D., Tian, S., Xie, C.: Synthesis of defect graphene and its application for room temperature humidity sensing. Mater. Lett. **83**, 76–79 (2012)
26. Yavari, F., Castillo, E., Gullapalli, H., Ajayan, P.M., Koratkar, N.: High sensitivity detection of NO_2 and NH_3 in air using chemical vapor deposition grown graphene. Appl. Phys. Lett. **100**, 203120 (2012)
27. Chung, M.G., Kim, D.H., Lee, H.M., Kim, T., Choi, J.H., Seo, D.K., Yoo, J.B., Hong, S.H., Kang, J.J., Kim, Y.H.: Highly sensitive NO_2 gas sensor based on ozone treated graphene. Sens. Actuator B Chem. **166–167**, 172–176 (2012)
28. Lee, C., Ahn, J., Lee, K.B., Kim, D., Kim, J.: Graphene-based flexible NO_2 chemical sensors. Thin Solid Films **520**, 5459–5462 (2012)
29. Chen, G., Paronyan, T.M., Harutyunyan, A.R.: Sub-ppt gas detection with pristine graphene. Appl. Phys. Lett. **101**, 053119 (2012)
30. Ju, S., Lee, J.M., Jung, Y., Lee, E., Lee, W., Kim, S.J.: Highly sensitive hydrogen gas sensors using single-walled carbon nanotubes grafted with Pd nanoparticles. Sens. Actuator B Chem. **146**, 122–129 (2010)
31. Liu, R., Ding, H., Lin, J., Shen, F., Cui, Z., Zhang, T.: Fabrication of platinum-decorated single-walled carbon nanotube based hydrogen sensors by aerosol jet printing. Nanotechnology **23**, 505301 (2012)
32. Zilli, D., Bonelli, P.R., Cukierman, A.L.: Room temperature hydrogen gas sensor nanocomposite based on Pd-decorated multi-walled nanotubes thin films. Sens. Actuator B Chem. **157**, 169–176 (2011)
33. Star, A., Joshi, V., Skarupo, S., Thomas, D., Gabriel, J.C.P.: Gas sensor array based on metal-decorated carbon nanotubes. J. Phys. Chem. B **110**, 21014–21020 (2006)
34. Penza, M., Rossi, R., Alvisi, M., Cassano, G., Signore, M.A., Serra, E., Giorgi, R.: Pt- and Pd-nanoclusters functionalized carbon nanotubes networked films for sub-ppm gas sensors. Sens. Actuator B Chem. **135**, 289–297 (2008)
35. Yun, S., Kim, J.: Multi-walled carbon nanotubes – cellulose paper for a chemical vapor sensor. Sens. Actuator B Chem. **150**, 308–313 (2010)
36. Mao, S., Cui, S., Yu, K., Wen, Z., Lu, G., Chen, J.: Ultrafast hydrogen sensing through hybrids of semiconducting single-walled carbon nanotubes and tin oxide nanocrystals. Nanoscale **4**, 1275–1279 (2012)
37. Wei, B.Y., Hsu, M.C., Su, P.G., Lin, H.M., Wu, R.J., Lai, H.J.: A novel SnO_2 gas sensor doped with carbon nanotubes operating at room temperature. Sens. Actuator B Chem. **101**, 81–89 (2004)
38. Liang, Y.X., Chen, Y.J., Wang, T.H.: Low-resistance gas sensors fabricated from multiwalled carbon nanotubes coated with a thin tin oxide layer. Appl. Phys. Lett. **85**, 666–668 (2004)
39. Du, N., Zhang, H., Ma, X., Yang, D.: Homogeneous coating of Au and SnO_2 nanocrystals on carbon nanotubes via layer-by-layer assembly: a new ternary hybrid for a room-temperature CO gas sensor. Chem. Commun. **2008**, 6182–6184 (2008)

40. Liu, J., Guo, Z., Meng, F., Jia, Y., Liu, J.: A novel antimony-carbon nanotube-tin oxide thin film: carbon nanotubes as growth guide and energy buffer. Application for indoor air pollutants gas sensor. J. Phys. Chem. C **112**, 6119–6125 (2008)
41. Leghrib, R., Pavelko, R., Felten, A., Vasiliev, A., Cane, C., Gracia, I., Pireaux, J.J., Llobet, E.: Gas sensors based on multiwall carbon nanotubes decorated with tin oxide nanoclusters. Sens. Actuator B Chem. **145**, 411–416 (2010)
42. Khalilian, M., Abdi, Y., Arzi, E.: Formation of well-packed TiO_2 nanoparticles on multiwall carbon nanotubes using CVD method to fabricate high sensitive gas sensors. J. Nanopart. Res. **13**, 5257–5264 (2011)
43. Ueda, T., Takahashi, K., Mitsugi, F., Ikegami, T.: Preparation of single-walled carbon nanotube/TiO_2 hybrid atmospheric gas sensor operated at ambient temperature. Diam. Relat. Mater. **18**, 493–496 (2009)
44. Balazsi, C., Sedlackova, K., Llobet, E., Ionescu, R.: Novel hexagonal WO_3 nanopowder with metal decorated carbon nanotubes as NO_2 gas sensor. Sens. Actuator B Chem. **133**, 151–155 (2008)
45. Du, N., Zhang, H., Chen, B., Ma, X., Liu, Z., Wu, J., Yang, D.: Porous indium oxide nanotubes: layer-by-layer assembly on carbon-nanotube templates and application for room-temperature NH_3 gas sensors. Adv. Mater. **19**, 1641–1645 (2007)
46. Sharma, A., Tomar, M., Gupta, V.: Room temperature trace level detection of NO_2 gas using SnO_2 modified carbon nanotubes based sensor. J. Mater. Chem. **22**, 23608–23616 (2012)
47. Chung, M.G., Kim, D.H., Seo, D.K., Kim, T., Im, H.U., Lee, H.M., Yoo, J.B., Hong, S.H., Kang, T.J., Kim, Y.H.: Flexible hydrogen sensors using graphene with palladium nanoparticle decoration. Sens. Actuator B Chem. **169**, 387–392 (2012)
48. Chu, B.H., Nicolosi, J., Lo, C.F., Strupinski, W., Pearton, S.J., Ren, F.: Effect of coated platinum thickness on hydrogen detection sensitivity of graphene-based sensors. Electrochem. Solid-State Lett. **14**, K43–K45 (2011)
49. Gautam, M., Jayatissa, A.H.: Adsorption kinetics of ammonia sensing by graphene films decorated with platinum nanoparticles. J. Appl. Phys. **111**, 094317 (2012)
50. Shafiei, M., Spizzirri, P.G., Arsat, R., Yu, J., du Plessis, J., Dubin, S., Kaner, R.B., Kalantar-zedeh, K., Wlodarski, W.: Platinum/graphene nanosheet/SiC contacts and their application for hydrogen gas sensing. J. Phys. Chem. C **114**, 13796–13801 (2010)
51. Gutes, A., Hsia, B., Sussman, A., Mickelson, W., Zettl, A., Carraro, C., Madoudian, R.: Graphene decoration with metal nanoparticles: towards easy integration for sensing applications. Nanoscale **4**, 438–440 (2012)
52. Mao, S., Cui, S., Lu, G., Yu, K., Wen, Z., Chen, J.: Tuning gas-sensing properties of reduced graphene oxide using tin oxide nanocrystals. J. Mater. Chem. **22**, 11009–11013 (2012)
53. Johnson, J.L., Behnam, S., Pearton, S.J., Ural, A.: Hydrogen sensing using Pd-functionalized multi-layer graphene nanoribbon networks. Adv. Mater. **22**, 4877–4880 (2010)
54. Kumar, R., Varandani, D., Mehta, B.R., Singh, V.N., Wen, Z., Feng, X., Mullen, K.: Fast response and recovery of hydrogen sensing in Pd-Pt nanoparticle-graphene composite layers. Nanotechnology **22**, 275719 (2011)
55. Song, H., Zhang, L., He, C., Qu, Y., Tian, Y., Lv, Y.: Graphene sheets decorated with SnO_2 nanoparticles: in situ synthesis and highly efficient materials for cataluminescence gas sensors. J. Mater. Chem. **21**, 5972–5977 (2011)
56. Zhang, Z., Zou, R., Song, G., Yu, L., Chen, Z., Hu, J.: Highly aligned SnO_2 nanorods on graphene sheets for gas sensors. J. Mater. Chem. **21**, 17360–17365 (2011)
57. Cuong, T.B., Pham, V.H., Chung, J.S., Shin, E.W., Yoo, D.H., Hahn, S.H., Huh, J.S., Rue, G.H., Kim, E.J., Hur, S.H., Kohl, P.A.: Solution-processed ZnO-chemically converted graphene gas sensor. Mater. Lett. **64**, 2479–2482 (2010)
58. Yi, J., Lee, J.M., Park, W.I.: Vertically aligned ZnO nanorods and graphene hybrid architectures for high-sensitive flexible sensors. Sens. Actuator B Chem. **155**, 264–269 (2011)
59. Deng, S., Tjoa, V., Fan, H.M., Tan, H.R., Sayle, D.C., Olivo, M., Mhaisalkar, S., Wei, J., Sow, C.H.: Reduced graphene oxide conjugated Cu_2O nanowire mesocrystals for high-performance NO_2 gas sensor. J. Am. Chem. Soc. **134**, 4905–4917 (2012)

60. Srivastava, S., Jain, K., Singh, V.N., Singh, S., Vijayan, N., Dilawar, N., Gupta, G., Senguttuvan, T.D.: Faster response of NO_2 sensing in graphene-WO_3 nanocomposites. Nanotechnology **23**, 205501 (2012)
61. Jiang, Z., Wang, J., Meng, L., Huang, Y., Liu, L.: A highly efficient chemical sensor material for ethanol: Al_2O_3/graphene nanocomposites fabricated from graphene oxide. Chem. Commun. **47**, 6350–6352 (2011)
62. Kaniyoor, A., Ramaprabhu, S.: Hybrid carbon nanostructured ensembles as chemiresistive hydrogen gas sensors. Carbon **49**, 227–236 (2011)
63. Jeong, H.Y., Lee, D.S., Choi, H.K., Lee, D.H., Kim, J.E., Lee, J.Y., Lee, W.J., Kim, S.O., Choi, S.Y.: Flexible room-temperature NO_2 gas sensors based on carbon nanotubes/reduced graphene hybrid films. Appl. Phys. Lett. **96**, 213105 (2010)

Chapter 8
A Do-It-Yourself (DIY) Guide to Using Carbon Nanotubes for Stretchable Electronics and Sensors

Daniel J. Cohen and Michel M. Maharbiz

Abstract The primary purpose of this chapter is to assist and encourage researchers who are new to the field of carbon nanotubes and stretchable sensors in beginning to explore and prototype nanotube-enabled devices and materials. Given this, our primary focus will be on how to use "off-the-shelf" components to prepare carbon nanotubes for use in a wide variety of applications, although our specific emphases will be on 2D patterning of nanotubes and stretchable electronics. Specifically, stretchable strain gauges were selected as the case study because they are an excellent showcase for the versatility and ease of manufacturing of nanotube-based devices, while the techniques to produce them are the same techniques that would be used for many other nanotube-based devices. We will begin with an overview of the challenges of sensing large strains and the limitations of status quo devices. Armed with that knowledge, we will then discuss why carbon nanotubes are particularly well suited to address the problem and how to acquire, prepare, and post-process them into a sensor. Finally, we will present an inexpensive, simple strain gauge design based on piezocapacitive strain transduction capable of straining to a minimum of 20× farther than the best traditional metal foil gauge.

D.J. Cohen (✉)
Joint Graduate Program in Bioengineering, University of California at Berkeley and San Francisco, UC Berkeley, 656A Sutardja Dai Hall, Berkeley, CA 94720, USA
e-mail: daniel.cohen@berkeley.edu

M.M. Maharbiz
Department of Electrical Engineering, University of California, 646 Sutardja Dai Hall, Berkeley, USA
e-mail: maharbiz@eecs.berkeley.edu

S. Li et al. (eds.), *Nanoscale Sensors*, Lecture Notes in Nanoscale Science and Technology 19, DOI 10.1007/978-3-319-02772-2_8,

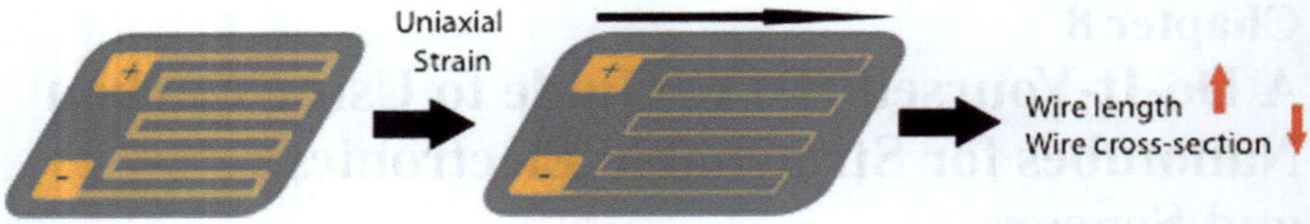

Fig. 8.1 Piezoresistance strain gauge mechanism. Representation of metallic strain gauge with a polymeric backing layer on which a fine metallic serpentine trace has been deposited. Axial strain causes the fine traces to both lengthen and contract in cross section; both effects serve to increase resistance [see (8.1)]

8.1 The Challenge of Gauging Large Strains in Soft Substrates

For decades, the fundamental tool for measuring strain has been the piezoresistive strain gauge. Its simplest form, as shown in Fig. 8.1 (left), is comprised of a rigid, polymeric substrate (such as mylar) onto which has been patterned a thin, metallic, serpentine wire, over which a second polymeric shielding layer is placed. The mechanism of operation is straightforward. First, the sensor is aligned with the direction strain to be monitored in the sample (such as a bridge stay), and then it is rigidly adhered to the substrate. The strength and rigidity of the adherent is selected so as to ensure that any strain in the underlying substrate is directly transmitted to the gauge itself. This planar, uniaxial strain serves to stretch the serpentine wire of the gauge (Fig. 8.1, right) and thereby increase the resistance of the strain gauge according to (8.1). The effect of resistance scaling with strain is known as *piezoresistance* and is the fundamental basis of nearly all modern planar strain gauges. Such a gauge is monitored and calibrated using a Wheatstone resistor bridge, and the efficiency of the gauge is measured with respect to the so-called gauge factor (GF, (8.2)), which is a measure of the percentage change in resistance of the gauge relative to the actual strain it experienced. Most metallic strain gauges have gauge factors ranging from 2 to 6 [1].

$$R_{\text{strained}} = N\rho\left(\frac{L+\Delta L}{A-\Delta A}\right) \tag{8.1}$$

$$\text{GF} = \frac{\Delta R/R}{\varepsilon} \tag{8.2}$$

Here, N is the number of the long sections in the serpentine, ρ is the resistivity of the metal, L is the length of each long section, A is the cross section, and ε is the applied strain.

While metallic strain gauges are quite successful for basic strain monitoring in rigid substrates and offer a linear output over their entire working range, they are exceptionally poor at monitoring soft substrates and large strains. In the former case, the material of the gauge is stiffer than the substrate and thereby reinforces the local region of substrate to which the gauge is adhered. In the latter case, both the yield strains (the maximum strain after which a material can elastically recover) of the metal electrodes and the polymeric backing layer are relatively low, contributing to a maximum measurable strain of typically $<5\ \%$ [1, 2]. This is a major limitation in the fields of biomechanical monitoring, medical devices, and soft robotics, where large, biologically relevant deformations are typical and the substrates themselves are often less stiff than traditional sensors. For example, many muscles experience contractile strains of at least 25 %, while bio-inspired robotics systems increasingly rely on high-strain linkages, compliant joints, and tactile interfaces [3–8]. In order to monitor such systems, both a softer sensor backing layer and a much more resilient electrode material are necessary. Further, as we continue to explore the potential of wearable electronics and mobile health, there should be a growing demand for sensors and devices that move and deform with the body.

8.2 Soft Sensor Design: Percolation and Carbon Nanotube Electrodes

The problem of a soft substrate is typically solved by using some form of silicone rubber. Biocompatible, inexpensive, highly tunable, and moldable silicone rubbers are typically $\sim$1,000$\times$ less stiff than the mylar backing found in traditional metal strain gauges and exhibit yield strains in excess of 100 % [9]. As such, silicone is one of the most common substrates used for stretchable strain gauges. The next problem is selecting and incorporating an electrode material with the silicone in order to produce a strain gauge, and the key here is finding an electrode material that is robust under strains larger than 5 %, and the key to this constraint appears to lie in the phenomenon of *percolation*.

In its simplest form, percolation is a collective property of aggregates of particles whereby above a critical density of particles (the percolation threshold), there is always at least one path from one side of the particle collective to the other side via contiguous particle connections. When using conductive particles above the percolation threshold, the resulting network ensures that there is always some contact or overlap between the particles to ensure that conductivity is maintained, even during deformation of the network, although conductive particle densities below the percolation threshold will exhibit no conductivity. The simple model for this type of behavior follows a split-function power law, as shown in (8.3):

$$\sigma = \left\{ \begin{array}{ccc} \sigma_o(\rho - \rho_c)^s & \text{if} & \rho > \rho_c \\ 0 & \text{if} & \rho < \rho_c \end{array} \right\} \tag{8.3}$$

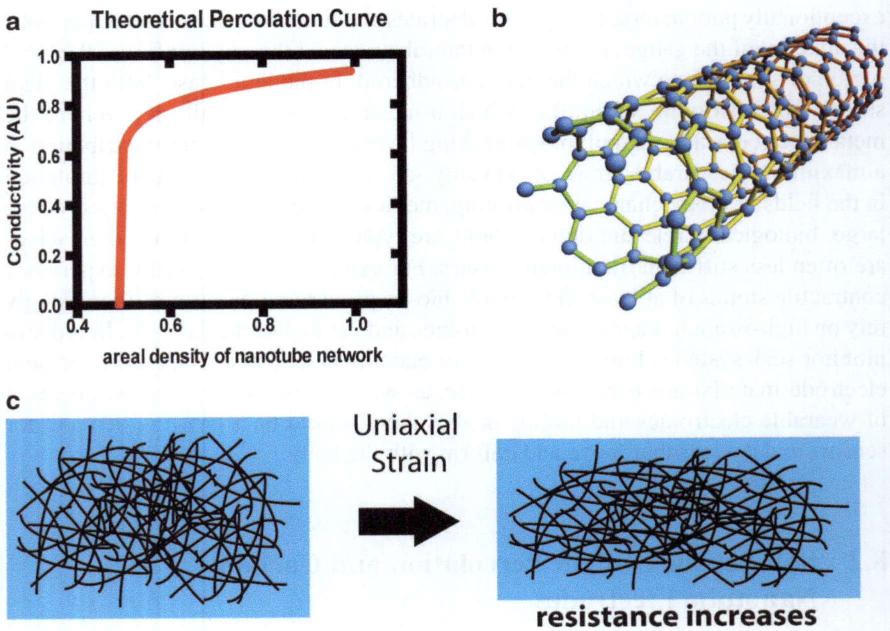

Fig. 8.2 Nanotube percolation overview. (**a**) Simulated percolation network using the power-law fit and a percolation threshold of 50 %. (**b**) Rendering of a single-walled nanotube to emphasize large aspect ratio (length/width) and the lattice structure (designed with Avogadro software). (**c**) *Left* shows a percolating network of nanotubes in a silicone matrix. It undergoes uniaxial strain that induces expansion in *x* and Poisson contraction in *y* resulting in a deformed nanotube network that still maintains conductivity while increasing the resistance

Here, σ is the conductivity of the system and σ_o is the baseline conductivity. The area or volume fraction of the particles is represented by ρ, while ρ_c is the actual percolation threshold, above which conductive networks are capable of forming, giving rise to true percolation. The exponent, *s*, is the critical exponent that controls the scaling rate. We can visualize this in Fig. 8.2a, where the plot represents the theoretical conductivity of a nanotube percolation network as a function of the space-filling fraction of nanotubes (ρ) with a percolation threshold of $\rho_c = 0.5$ [10, 11].

In practice, percolation efficiency depends strongly on both the baseline conductivity and the aspect ratio of the particles in the network, scaling directly the aspect ratio of the particles (length/width). As far as these parameters are concerned, carbon nanotubes make excellent percolation particles, possessed of both aspect ratios ranging from 100× to 1,000× and high intrinsic conductivities. As a result of the ease with which carbon nanotubes percolate, the requisite ρ_c is actually relatively low (often far lower than in the simulation in Fig. 8.2a). By comparison, traditional percolation networks made of pseudo-spherical particles (gold micro-plates and carbon black) cannot percolate as efficiently and exhibit much higher percolation thresholds than carbon nanotube networks [10, 12–19]. Carbon nanotubes' properties are derived from their structure, shown in Fig. 8.2b. The hexagonal lattice of carbon atoms affords exceptional mechanical properties and high conductivity and favors particles that are longer than they are wide (high aspect ratio). Taken together these attributes make nanotubes excellent percolating agents.

This percolation efficiency is enormously important in device design for two key reasons. First, lower percolation thresholds reduce the cost of a device by reducing the amount of conducting particles needed and also allow for transparent electrodes to be produced. A great deal of excellent work has been performed characterizing the applicability and spectral characteristics of nanotubes in transparent, stretchable electronics. While less electrically efficient than traditional transparent electrodes made from indium tin oxide (ITO), they are considerably cheaper and can be stretched, something an oxide can never do [11, 20, 21]. Focusing on this stretchability, we see the second key attribute of nanotube percolation networks: strain resistance and piezoresistivity. Fundamentally, when carbon nanotube networks are stretched, they still maintain conductivity. From a binary perspective, percolating nanotube networks can undergo enormous strains (100 %+) without losing conductivity. However, the trade-off is that stretching increases the contact resistance between adjacent nanotubes as they are pulled slightly apart from each other. This increased resistance manifests itself as piezoresistance [1, 11, 22] and is depicted in Fig. 8.2c where we see a resting network strained laterally while still maintaining local connectivity (despite the resistance change). This attribute, while ultimately quite difficult to stabilize in nanotube networks, has been the dominant mechanism employed in both traditional metallic and nanotube strain gauges. To provide context, the remainder of this section will discuss how such gauges are produced and perform, while Sect. 8.5 will explore the benefits of our alternative approach: piezocapacitance.

Typically, there are two main approaches that have been used to produce percolating carbon nanotube electrode networks—3D nanotube/polymer matrix composites and layer-based nanotube electrodes. The former methods are the earliest and a great deal of work has gone into optimizing the material properties of the resulting nanotube network. One of the first methods of this type relied on starting with a polymer matrix base and carefully dispersing carbon nanotubes into the unpolymerized base casting the resulting mixture into a desired shape and then curing the ensemble [7, 22–26]. While this method can work quite well, it is nontrivial to reliably disperse nanotubes in a viscous prepolymer base without the use of probe sonication, homogenizers, and solvents [7]. Moreover, this method may also stiffen the resulting polymer due to the high percolation threshold concentration needed to obtain decent conductivity in the network [1]. In addition to requiring extended processing, the direct mixing approach also requires many more nanotubes to obtain proper percolation in three dimensions, and this drives the cost up. An alternative process involves infiltrating silicone prepolymer into vertically aligned forests of nanotubes [22, 27]. These methods are effective and can preserve nanotube alignment while maintaining conductivity. However, they require a great deal of clean room processing and have not been demonstrated to be effective for transducing large strains.

Development of the layer-based percolation networks was driven largely by the transparent, stretchable electrode community as thin layers of nanotubes are relatively transparent and sustain stretching [11, 21, 28–33]. Here, such a layer of nanotubes is deposited on the surface of a sheet of elastomeric polymer and produces an electrode in much the same way that the deposited metal in a metallic strain gauge functions. Given this, much effort was directed at optimizing the percolation thresholds to reduce the number of nanotubes necessary for reliable conductivity while simultaneously improving transparency [11]. If transparency is not a requirement, the layering method becomes more straightforward as simpler deposition methods can be used to create the percolation network on the surface of the polymer. Typical techniques to produce nanotube layers comprise manual spreading of nanotubes, airbrushing a dispersion, silk screening, polyelectrolyte layering, and vacuum filtration of a nanotube dispersion and subsequent transfer of the filtrate [1, 28, 31, 33–36]. This latter approach is both the fastest and most versatile (used both for transparent and general nanotube electrodes), and we chose to adopt and adapt it for the methods describes below.

8.3 Nanotube Processing for Stretchable Electrodes and Patterns: Selection, Preparation, and Post-processing

For general stretchable sensors and many surface patterning applications, we have found that the layer-based method enabled by vacuum filtration and contact stamping is the simplest approach to produce electrodes or patterns of nanotubes on a planar, elastomeric surface. In brief, the method works as follows: acquire nanotubes and disperse them in water and surfactant (if necessary); vacuum filter the dispersion through a fine-pore filtration membrane; stamp transfer the filtrate layer to a silicone substrate; and encapsulate as necessary. The general process flow is presented in Fig. 8.3, and the remainder of this section will address these steps in order.

8.3.1 Acquisition

Nanotubes have now reached the point of near ubiquity where reliable nanotubes can be purchased from online retailers such as Sigma and Cheaptubes.com. Moreover, for the work described herein, inexpensive, single-walled nanotubes from Cheaptubes were sufficient to achieve reliable performance. As application-specific needs, such as number of walls, surface functionality, diameter versus length, and purity, arise, the cost may increase, but basic nanotube research need not

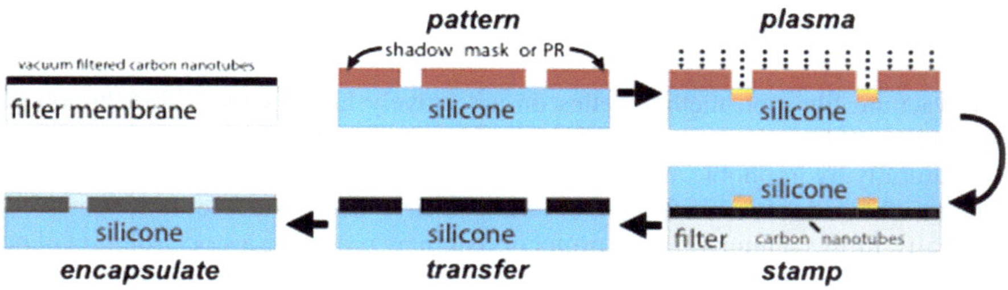

Fig. 8.3 Process flow for patterning. (*left to right*) Nanotubes are first dispersed and vacuum filtered against a filtration membrane. During this process, the silicone substrate is cut and patterned with either photoresist or a sticker shadow mask. Once masked, the silicone is plasma treated to prevent nanotubes adhering to the treated regions. Post-plasma treatment, the mask is tripped and the silicone is stamped against the filtered nanotubes, in turn resulting in the specific transfer of nanotubes only to untreated regions. Post-transfer, the nanotubes are encapsulated in a sprayed-on silicone layer. *Adapted with permission from Cohen et al., Nano Letters (2012) American Chemical Society* [37]

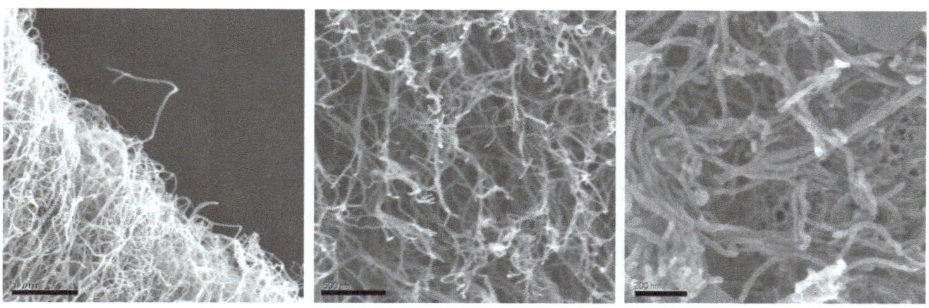

Fig. 8.4 Scanning electron microscopy sequence of prefab nanotubes. (*left*) cluster of nanotubes, scale is 1 μm; (*center*) zooming in, the interlocking network becomes clearer, scale is 500 nm; (*right*) zooming further we can see the individual tube structures (note the curvature), scale is 200 nm

be expensive. By way of example, the nanotubes used in the case study below were ordered from Cheaptubes at \$250/g for 99 % wt purity, 2 nm OD, and 3–30 μm long single-walled carbon nanotubes. As the vacuum filtration protocol requires milligrams of nanotubes, a single purchase of stock will last for quite some time. The nanotubes we selected were chosen because we found that single-walled nanotubes transferred better than multiwalled nanotubes, and they had a high aspect ratio (length vs. diameter) of ~5,000, which improves their percolation performance. Figure 8.4 depicts a magnification sequence of images captured with scanning electron microscopy of these particular single-walled nanotubes.

8.3.2 Pre-processing: Dispersion

The vacuum filtration method relies on a relatively homogeneous distribution of nanotubes in the solvent (water). The single-walled carbon nanotubes in question are naturally hydrophobic and tend to aggregate and precipitate out of the solution without the aid of a surfactant or without first being surface functionalized. Typically, most vacuum filtration protocols begin by preparing a stock solution of no more than 5 mg/L of nanotubes in deionized, distilled water. To this solution, 1 g/mL of surfactant (SDS, Sigma) is added. The surfactant is essential for dispersion of the nanotubes, although it can later become a contaminant if care is not taken with the washing steps. *Proper ventilation should be used during the preparation of the solution to prevent inhalation of the nanotubes.* Once the solution is prepared, we treated it with bath sonication for 90–180 min and then allowed it settle overnight. Once treated, the dispersion should appear uniformly dark, although still partially transparent. The stability of the solution, and the need for surfactant and sonication, depends on the type of nanotube being used, with functionalized nanotubes offering greater stability in certain cases.

8.3.3 Vacuum Filtration

The key step to this process is to prepare an apparatus capable of using a vacuum to draw the dispersion through a fine-pore filtration membrane. Such systems can be purchased from companies such as Millipore, Sigma, and Whatman. We found that optimal performance came from either 220 nm mixed cellulose ester (MCE) filtermembranes (MF Class, Millipore) or 20 nm anodic aluminum oxide filters (Anodisc, Whatman). As the latter are more expensive, fragile, and increase the filtration time significantly, the MCE filters should be tested first. The assembled filtration system is depicted in Fig. 8.5a, where the solution reservoir sits upon the filter base, and this assembly in turn couples to two vacuum traps. Some form of vacuum trap is essential to prevent the surfactant from excessively bubbling and going directly into the vacuum line. The volume of dispersed nanotube solution to use varies with the application and must be tuned in a case-specific manner. For the methods discussed here, we found 30 mL of solution to be optimal. During vacuum filtration, it is important to wash a significant amount of liquid through the filter to remove residual surfactant, and it is also essential not to let this added liquid agitate the freshly deposited nanotube layer. To handle this, we would filter the nanotubes until only a thin layer of liquid remained on the tubes and then very slowly and gently add 50 mL of fresh water down the side of the filtration reservoir to avoid damaging the nascent nanotube layer. Once the media has been exchanged, more water should be continually added until no surfactant bubbles are seen exiting the bottom of the filter (typically 500 mL of water). At this point, all liquid should be filtered out and the filter membrane should be removed and placed in a humidified

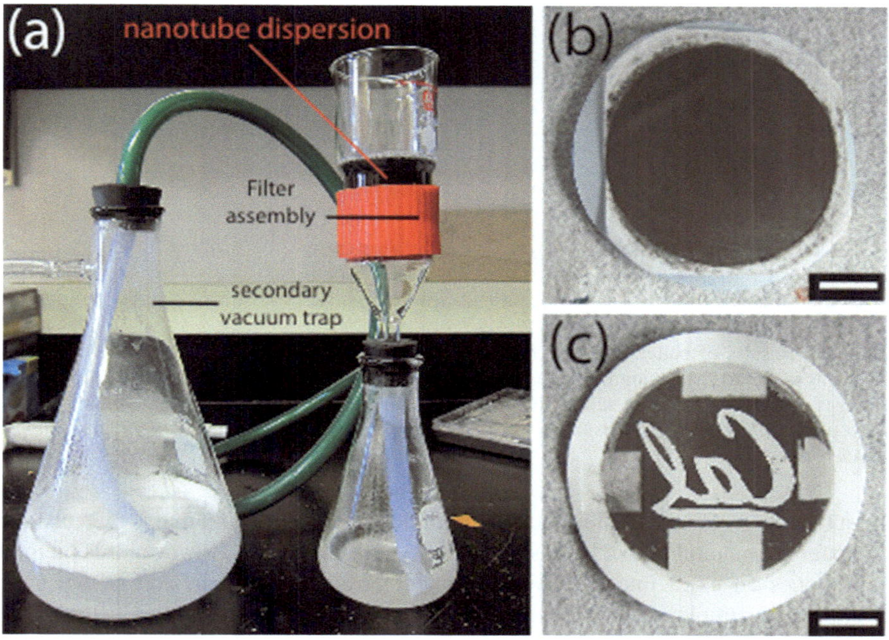

Fig. 8.5 Filtration apparatus and representative samples. (**a**) The vacuum filtration apparatus is shown using two vacuum traps, with the secondary trap useful for preventing surfactant bubbles from reaching the vacuum line. (**b**) What a filtration membrane looks like post-filtration with a gray layer of nanotubes atop it. (**c**) The negative pattern left after transferring the nanotubes to the target substrate. Scale bars are 0.75 cm

Petri dish to await the transfer step. A representative filter disc post-filtration is shown in Fig. 8.5b. In our experiments, we found that dry MCE filters tended to be less reliable during transfer than wet filters.

8.3.4 Nanotube Stamp Transfer to Silicone Substrate

Once the nanotube filtrate layer has been produced, its role becomes similar to that of ink in traditional stamp transfer with the silicone substrate for the sensor playing the role of the paper. For simple applications where the geometry of the nanotube layer on the silicone is unimportant, the silicone can be directly stamped against the wet nanotube filtrate layer, and perfect transfer should occur, resulting in a semitransparent, conductive layer of nanotubes on the surface of the silicone. While effective, this method affords no precision in the patterning process and is incompatible with techniques requiring any sort of specific geometry of patterning. To address this problem, previous studies have demonstrated that nanotubes can

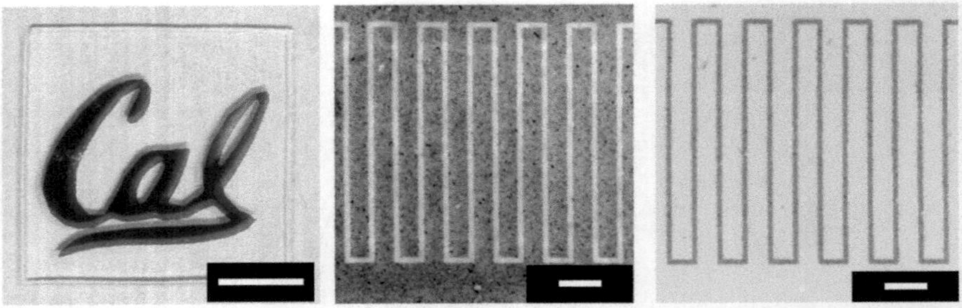

Fig. 8.6 Results of patterning. (*left to right*) "Cal" logo was produced using a sticker mask. Blurring is due to shadowing from the nanotube layer, and scale bar is 7.5 mm. The two *rightmost* images were made from the same filter—one is the negative of the other, and the scale bars are 750 μm. *Adapted with permission from Cohen et al., Nano Letters (2012) American Chemical Society* [37]

first be transferred to an intermediate, premolded polymer stamp that is then used to transfer the nanotubes to a fresh sheet of silicone [28, 38–41]. However, this method introduces an extra interface where transfer errors may occur, and so we have developed a simpler, faster method of direct transfer of high-resolution patterns.

Our method relies first on the fact that silicone can be converted between hydrophobic and hydrophilic surface properties through a brief exposure to atmospheric or oxygen plasma and second on the native hydrophobicity of single-walled carbon nanotubes. In brief, we take a silicone substrate, mask the regions intended for nanotubes, expose the silicone to atmospheric plasma for 30 s in an inexpensive plasma cleaner (Harrick Plasma), strip the mask, and then stamp the plasma-treated silicone directly against the nanotube filtrate on the filter substrate. Using this protocol, we have observed nearly perfect transfer of nanotubes only to the untreated (hydrophobic) regions of the silicone substrates, as the examples in Figs. 8.5c and 8.6 demonstrate.

We begin by using prefabricated, ~250 μm thick silicone sheeting (Bisco HT264, Stockwell Elastomerics). While we have tested the method using traditional, hand-mixed silicone (Sylgard 184, Dow Corning), we have found that the Bisco silicone has more reliable and uniform mechanical properties for strain sensing than hand-mixed silicone. Moreover, prefabricated silicone sheets are much simpler to cut into customized dies using automated machines such as a laser cutter (VLS2.3, ULS, Inc.) or an inexpensive vinyl cutter (300 μm resolution Cameo Cutter, Silhouette, Inc.). These machines were used to produce the dog bone patterns presented in the case study. With respect to masking, there are two methods—one rapid and with coarse resolution and the other lengthy but more precise.

8.3.4.1 Rapid, Coarse Patterning

The simple, rapid method relies on using sticker paper to make a custom mask that is stuck to the silicone to maintain local hydrophobicity. Stickers were cut using either a laser or vinyl cutter. While resolutions can be obtained down to 300 μm, the ease of placing the sticker by hand depends on the complexity of the pattern. Features down to 1 mm rarely pose a problem, but smaller features require empirical testing to determine if it is possible to achieve the appropriate alignment by hand. Figure 8.6a demonstrates a typical pattern from this method, while Fig. 8.5c shows the negative pattern left behind on the filter post-transfer. Overall, this method is the most versatile, as even hand-cut stickers can make effective masks.

8.3.4.2 Slow, Fine Patterning

For features down to 100 μm, we recommend photolithography. Specifically, we spin coated S1818 positive photoresist (500 rpm/5 s and 4,000 rpm/30 s) directly onto our silicone substrates, performed a 2.5 min soft bake at 115 °C, and then exposed the photoresist to 65 mJ/cm^2 I-line UV. Postexposure was performed in CD-30 for 45 s. Once masked, the samples were then exposed to plasma and immediately washed with acetone post-plasma and prior to stamping against the nanotube substrate. Features down to 100 μm are achievable using this method, although it does require more specialized equipment and time. Figure 8.6b, c presents two complimentary patterns resulting from this method.

8.3.5 Stabilizing and Encapsulating the Nanotube Layer

Stamp transfer methods require uniform pressure all across the stamp and the substrate, and we attempted to control this using two methods. Both methods begin by taking freshly plasma-treated silicone and gently reversibly adhering it to a glass slide. The slide was then placed silicone side down against the nanotube filtrate on the filter. Following this, we used a flat aluminum plate to help apply even pressure across the surface. For repeatability, we used a hydraulic press set at 0.25 MPa, although this is not strictly necessary. In the procedure described in the case study, we needed to coat both sides of a silicone sheet with different nanotube patterns, so we masked the silicone using sticker sheets on either side, plasma treated both sides simultaneously, and then performed the aforementioned stamping steps on both sides, serially, taking care not to damage the layers post-stamping. Contact with the glass stabilizer did not appear to modify the hydrophobicity, and both surfaces appeared to have transferred properly.

Once the nanotubes are safely transferred to the silicone substrate, it may be desirable to encapsulate the layer to protect it. While many methods exist to do this, we opted to use a spray-based method where Sylgard 184 silicone is mixed in a 10:10:1 ratio of hexane, silicone, and curing agent, and subsequently loaded into an airbrush. The sample to be encapsulated can be safely taped against a new glass slide to protect both the nanotube layer and any regions that need to make electrical contact post-encapsulation. This slide is then heated on a hot

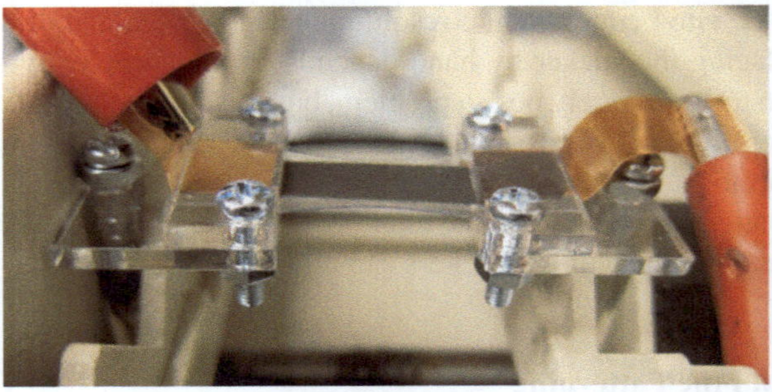

Fig. 8.7 Device mounted and fully wired. A device has been fully mounted using two custom-made acrylic clamps. Each clamp respectively presses a copper electrode sheet against the appropriate nanotube surface to make a tight connection with low contact resistance. The whole assembly has been mounted on a syringe pump for cyclic mechanical testing

plate at 110 °C. Once heated, the diluted silicone is sprayed against the exposed nanotube/silicone. Hexane is highly volatile and is almost entirely gone by the time it reaches the surface, where the high heat serves to drive off any residual hexane while simultaneously curing the silicone. This method produces a relatively uniform layer that can be controlled to be <100 μm. Alternately, electrical contacts can be made ahead of time and then the whole sample can be immersed in silicone or another elastomer.

8.3.6 Electrical Contacts

A major hurdle in any sensor with a silicone substrate is actually attaching external power/sense electrodes to the sensor since soldering is out of the question. A common method is to make conductive silicone by mixing in a percolating volume fraction of either carbon black, silver nanoparticles, or carbon nanotubes (as described earlier) to produce a percolating network that can be used like a conductive glue. While this can work, it can be difficult to get a sufficiently conductive glue. To simplify things, we opted for a clamping approach where all of our electrical contacts were made by building small plastic clamps fixed by screws that held wires (copper sheeting) in close contact with exposed regions of nanotubes after encapsulation as shown in Fig. 8.7. We recommend this approach for situations where the sensor geometry permits it. In other cases, conductive glue may be more effective.

8.4 Case Study: Poisson Capacitor for Highly Elastic Strain Sensing

Using the simplest of the techniques described above, we will now walk through how to build a robust strain sensor capable of supporting strains of at least 100 % over thousands of cycles with minimal hysteresis. The basic design is presented in Fig. 8.6 and discussed in the following section [37].

8.4.1 Piezoresistance Versus Piezocapacitance

While piezoresistance-based nanotube sensors are certainly effective, they have a number of limitations that make them less attractive from a practical perspective. The primary drawback to a piezoresistive design is that the stability of the performance (the lack of hysteresis or drift) depends entirely on the stability of the percolation network. Should the statistical average of connection routes and contact resistances throughout the network shift from cycle to cycle, then the measured strain will also drift. This problem can partially be addressed by how the nanotubes are adhered to the silicone substrate. However, these methods require more complicated and expensive processing of the nanotubes. Moreover, even the best piezoresistive sensors for high strain sensing have exhibited nonlinear performance and gauge factors [1]. While still quite usable, this does complicate the calibration process as well as reduce the device sensitivity and effective range. For these reasons, and because we sought to simplify the fabrication process, we opted for an entirely different mode of strain transduction based on capacitance.

Piezocapacitance occurs when a shift in the capacitance of a system is accompanied by a corresponding shift in strain in the system. To implement this using our platform, we needed to create two percolation electrodes sandwiching a dielectric (silicone) in order to produce a parallel plate capacitor. This sandwich design has traditionally been used to transduce strains normal to the surface (compression), but it has a mode of action that also allows it to transduce planar strains [8]. The general mode of action is presented in Fig. 8.8a, b, with an image of a completed device presented in Fig. 8.8c, d. The mechanism behind this kind of sensor relies expressly on Poisson contraction, whereby all elastic materials, when strained uniaxially, will experience strain of the opposite sign in the normal axes—if you pull on something it becomes thinner. In our device, this thinning process brings the nanotube electrodes closer together, resulting in an increase in capacitance. A simple model of the mechanics is described by equations (8.4)–(8.7), while (8.8) provides a simplified model of how parallel plate capacitance can be coupled to Poisson contraction in this kind of sensor:

$$\frac{\Delta L}{L} = \varepsilon_z = \frac{\sigma_z}{E} \tag{8.4}$$

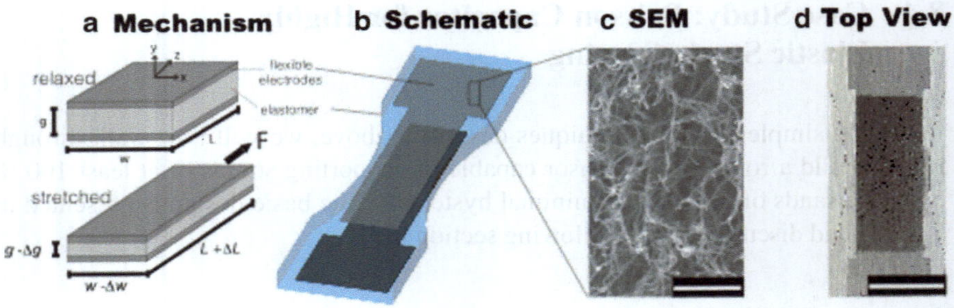

Fig. 8.8 Mechanism, design, and fabrication of a Poisson capacitor. (**a**) Poisson contraction converts planar strain to a decrease in the gap distance between the percolating electrodes and an increase in capacitance. (**b**) Schematic $L_{\text{stretch}} = L + \Delta L = L + \varepsilon_z L$ of our device geometry; all sensing is carried out in the middle region of the sensor. (**c**) SEM data demonstrating percolation of the nanotubes within the electrode; scale bar is 500 nm. (**d**) Closeup image of the sensing region of the device. *Darker* region is due to overlap of electrodes, and *striations* are due to texture of the background; scale bar is 0.75 cm. *Adapted with permission from Cohen et al., Nano Letters (2012) American Chemical Society* [37]

$$\varepsilon_x = \varepsilon_y = \nu\varepsilon_z \tag{8.5}$$

$$L_{\text{stretch}} = L + \Delta L = L + \varepsilon_z L \tag{8.6}$$

$$w_{\text{stretch}} = w - \Delta w = w - \nu\varepsilon_z w \tag{8.7}$$

Here, ν is Poisson's ratio (<0.5 for most silicone elastomers); σ is stress; ε is strain; E is the elastic modulus of silicones; and L, w, and g are the initial dimensions of the capacitor. Equation 8.4 is Hooke's law linking modulus to stress and strain. Despite the large target strains (100 %), which normally require nonlinear elasticity models, an empirical fit from mechanical testing data ($R^2 = 0.997$) indicated that this standard linear model is sufficient to predict how the device would function. Accepting linear elasticity, we can then move to (8.6) and (8.7) where we see how Poisson's ratio serves as the coupling parameter for strains in orthogonal axes. From these, we can then substitute these relations into the simple model of a parallel plate capacitor in (8.8), to arrive at an analytical expression of how capacitance should vary as a function of planar strain in the sensor:

$$\begin{aligned}\Delta C_{\text{simple}} &= e_0 e_{\text{silicone}} \left(\frac{w_{\text{stretch}} L_{\text{stretch}}}{g_{\text{stretch}}} \right) = e_0 e_{\text{silicone}} \left(\frac{w(1 - \upsilon\varepsilon_z) L (1 + \varepsilon_z)}{g_{\text{stretch}} (1 - \upsilon\varepsilon_z)} \right) \\ &= e_0 e_{\text{silicone}} (1 + \varepsilon_z) \frac{wL}{g}\end{aligned} \tag{8.8}$$

Here, e_0 is the permittivity of free space, and e_{silicone} is the relative permittivity of silicone. It is perhaps unintuitive to note that the final equation, despite resulting from Poisson contraction, does not depend on the value of Poisson's ratio as long as the material is assumed to be isotropic (a fair assumption for many silicones). Moreover, this equation also predicts that the capacitance will change linearly with strain so long as the electrode geometry scales elastically. What is particularly important about this point is that it means the performance of the device is now entirely independent of resistive properties of the electrodes so long as they maintain conductivity across the surface. In other words, even if there is resistive hysteresis and shifting in the structure of the percolation networks, so long as the networks can still conduct electricity and occupy roughly the same footprint and relative overlap across the sandwich, the sensor performance will depend only on Poisson contraction. It is this feature that enables such a simple fabrication method to produce a reliable sensor, so long as capacitance can be monitored accurately. That said, there is one final modification to the model that should be made, which is to take into account fringing fields. Equation 8.8, while simple, ultimately assumes an infinite parallel plate capacitor. Real capacitors will always experience electric fields at the borders that diverge from the simple model. However, these can be accounted for using the Palmer model, which acts as a scaling factor on all terms of the simple model and goes according to (8.9) [42]. It was this model that was ultimately used for the simulation curves presented below [variables are consistent with those in (8.8)]:

$$\Delta C_{\text{palmer}} = e_0 e_{\text{PDMS}} (1 - \varepsilon_z) \frac{wL}{g} \left(1 + \frac{g}{\pi w}\left(1 + \log \frac{2\pi w}{g}\right)\right) \times \left(1 + \frac{g}{\pi L}\left(1 + \log \frac{2\pi w}{L}\right)\right) \tag{8.9}$$

8.4.2 *Testing Apparatus*

The sensor depicted in Fig. 8.6d was affixed between two custom-made clamps that held copper sheets tightly against the unexposed patches of nanotubes on the top and bottom of the capacitive sandwich, respectively. These clamps were then attached to a syringe pump (Harvard Apparatus) which was programmed to cycle continuously between a state of no tensile strain and 100 % tensile strain (see Fig. 8.7). While slow (2 mm/s), syringe pumps make quite precise, inexpensive mechanical cycling machines. Capacitance data itself was logged using an electrical impedance spectrometer (EIS) stimulating with a 10 kHz AC signal and sensing at 3 Hz. While such systems are helpful, the same measurements can be obtained using a simple capacitive Wheatstone bridge (analogous to the bridge used for piezoresistive gauges, except with capacitors replacing the resistors) or an off-the-shelf integrated circuit capable of simple impedance analysis. Both approaches are relatively inexpensive and simple to build and can also be scaled down in size to suit robotics and medical applications where a full EIS system is impractical and unnecessary.

8.4.3 Performance Metrics

As a general baseline, the first parameter we tested was the piezoresistive performance of one of the electrodes. To do so, we built another sensor consisting of an electrode strip that went entirely across the sensor, along with clamps that attached to either side of the same electrode to produce a resistor. Using the cycling parameters described above (100 % strain, 2 mm/s), we confirmed the presence of hysteresis and nonlinearity. As shown in Fig. 8.9a, hysteresis resulted in a permanent increase in the baseline resistance of the system (measured by 2-point sensing). Further, the inset shows the presence of stress relaxation in the nanotube percolation network when the sensor was strained to 100 % and held for 10 s while monitoring the impedance. The result of these phenomena was that the gauge factor itself also shifted from 0.5 to 0.25 after very few cycles.

However, when two of these electrodes were combined to make the capacitive sandwich described earlier, the performance and stability markedly improved. All devices tested ($n = 6$) averaged a baseline capacitance of 16 pF, which implies that the fabrication method is reliable, even with variability in the percolation performance. To characterize the fundamental performance, model accuracy, and gauge factor, we looked at the dynamics of a representative stretching event to 100 % strain and back at a constant strain rate, as shown in Fig. 8.9b. Here, the empirical data from the sensor (blue circles) is represented as the relative change in capacitance to the baseline and cleanly and linearly captures the strain dynamics of the system. Further, the ascending and descending curves ought to be symmetric given the constant strains and rates, and indeed they differ by <2 %. Given this, the gauge factor (the relative change in capacitance divided by applied strain) is linear and averages out to 0.99 over the entirety of the 100 % strain range. As the theoretical optimum for a Poisson capacitor is 1, the device is performing well. The simulation curve generated by (8.9) is overlaid (red) on the data and exhibits an R^2 of 0.998 relative to the measured data.

If we compare Fig. 8.9a, b, we see that the prediction of the model that capacitance should be independent of hysteretic damage to the electrodes bears out. Despite the extreme fluctuations in electrode resistance, the capacitance data is quite steady. However, to properly test device stability, we need to look at cyclic strain data, as is shown in Fig. 8.9c. Cyclic testing was conducted using the same speed as previously used, but with a "step-and-hold" profile where the sensor was strained to 100 % strain and held for 12.5 s before ramping down the strain. This type of profile tests both strain cycling and stress relaxation effects, and it was repeated

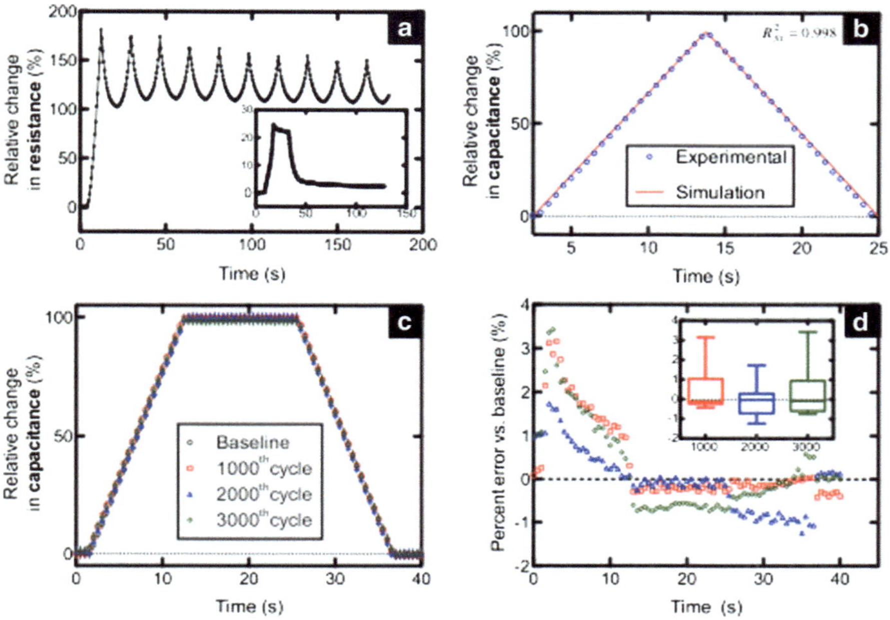

Fig. 8.9 Performance characterization of Poisson capacitor strain gauge. (**a**) Resistive performance of a single percolation electrode when undergoing 100 % cyclic strain. *Main plot* indicates significant hysteresis, while *inset* shows both relaxation during a 100 % step strain and subsequent hysteresis during recovery. (**b**) Single, 100 % strain cycle of the capacitive sensor (*blue*) overlayed (*red*) by the fit from parallel plate model with Palmer correction (*red*). (**c**) Demonstration of repeatability over 3,000 cycles of 100 % strain. After every 1,000 cycles, the sample was subjected to a 100 % step strain before being relaxed back to basal strain. (**d**) Stability plot showing how much sensor performance deviated from baseline performance over 3,000 cycles. Legend is the same as for (**c**). *Adapted with permission from Cohen et al., Nano Letters (2012) American Chemical Society* [37]

for 3,000 cycles. Figure 8.9c depicts the baseline profile and then a representative profile from each batch of 1,000 cycles throughout the trial. Taking these data, we can then calculate the deviation from the baseline over 3,000 cycles. This type of analysis is suited for identifying if there is any form of monotonic drift that would indicate underlying hysteresis. The resulting data, shown in Fig. 8.9d, indicate that while there was up to 3 % deviation from the baseline, it is not a monotonic error with respect to cycle count, and therefore it is likely due to random error rather than a systematic hysteresis. Moreover, if we normalize the error with respect to the data from the 1,000th cycle rather than the first cycle, the error drops to ~1.5 % (not shown), and this in turn indicates that there is likely some slight settling in the geometry of the electrodes that occurs over the first 1,000 cycles.

8.5 Concluding Remarks

The most unexpected result from these data is the quality of the sensor performance and patterning resolution relative to the production cost, time, and ease. In terms of the raw materials consumed for a single sensor, the cost is <\$0.50/sensor. The assembly time for a single sensor from vacuum filtration to final encapsulation is ~30 min, and this can be parallelized. While the plasma cleaner itself is ~\$5,000 new, nearly all universities have readily accessible, shared plasma cleaners, and many devices can be treated in parallel. While the strain sensor was presented as a case study, these techniques can be generalized for a variety of applications. Ultimately, we hope that the frugality, generality, and simplicity of the approach will overcome the traditional barrier to entry of nanoscience and encourage researchers to explore incorporating carbon nanotubes into their sensors, robots and assays.

References

1. Yamada, T., Hayamizu, Y., Yamamoto, Y., et al.: A stretchable carbon nanotube strain sensor for human-motion detection. Nat. Nanotechnol., 1–11 (2011). doi: 10.1038/nnano.2011.36
2. Ajovalasit, A., Zuccarello, B.: Local reinforcement effect of a strain gauge installation on low modulus materials. J. Strain Anal. Eng. Design **40**, 643–653 (2005). doi:10.1243/030932405X30894
3. Rogers, J.A., Someya, T., Huang, Y.: Materials and mechanics for stretchable electronics. Science **327**, 1603–1607 (2010). doi:10.1126/science.1182383
4. Carta, R., Jourand, P., Hermans, B., et al.: Design and implementation of advanced systems in a flexible-stretchable technology for biomedical applications. Sens. Actuators A. **156**, 79–87 (2009). doi:10.1016/j.sna.2009.03.012
5. Cutkosky, M.R., Kim, S.: Design and fabrication of multi-material structures for bioinspired robots. Phil. Trans. Math. Phys. Eng. Sci. **367**, 1799–1813 (2009). doi:10.1098/rsta.2009.0013
6. Lacour, S.P., Benmerah, S., Tarte, E., et al.: Flexible and stretchable micro-electrodes for in vitro and in vivo neural interfaces. Med. Biol. Eng. Comput. **48**, 945–954 (2010). doi:10.1007/s11517-010-0644-8
7. Sekitani, T., Nakajima, H., Maeda, H., et al.: Stretchable active-matrix organic light-emitting diode display using printable elastic conductors. Nat. Mater. **8**, 494–499 (2009). doi:10.1038/nmat2459
8. Takei, K., Takahashi, T., Ho, J.C., et al.: Nanowire active-matrix circuitry for low-voltage macroscale artificial skin. Nat. Mater. **9**, 1–6 (2010). doi:10.1038/nmat2835
9. Lee, K., Lee, S.S., Lee, J.A., et al.: Carbon nanotube film piezoresistors embedded in polymer membranes. Appl. Phys. Lett. **96**, 013511 (2010). doi:10.1063/1.3272686
10. Kyrylyuk, A.V., Hermant, M.C., Schilling, T., et al.: Controlling electrical percolation in multicomponent carbon nanotube dispersions. Nat. Nanotechnol. (2011). doi:10.1038/nnano.2011.40
11. Hu, L., Hecht, D.S., Gruner, G.: Percolation in transparent and conducting carbon nanotube networks. Nano Lett. **4**, 2513–2517 (2004)
12. Adrega, T., Lacour, S.P.: Stretchable gold conductors embedded in PDMS and patterned by photolithography: fabrication and electromechanical characterization. J. Micromech. Microeng. **20**, 055025 (2010). doi:10.1088/0960-1317/20/5/055025
13. Spitalsky, Z., Tasis, D., Papagelis, K., Galiotis, C.: Carbon nanotube–polymer composites: chemistry, processing, mechanical and electrical properties. Prog. Polym. Sci. **35**, 357–401 (2010). doi:10.1016/j.progpolymsci.2009.09.003

14. Tiwari, M.K., Yarin, A.L., Megaridis, C.M.: Electrospun fibrous nanocomposites as permeable, flexible strain sensors. J. Appl. Phys. **103**, 044305 (2008). doi:10.1063/1.2885112
15. Graz, I.M., Cotton, D.P.J., Lacour, S.P.: Extended cyclic uniaxial loading of stretchable gold thin-films on elastomeric substrates. Appl. Phys. Lett. **94**, 071902 (2009). doi:10.1063/1.3076103
16. Jha, V., Thomas, A.G., Bennett, M., Busfield, J.J.C.: Reversible electrical behavior with strain for a carbon black-filled rubber. Polymer **116**, 541–546 (2010). doi:10.1002/app
17. Cheng, C., Xiao, L., Cheung, Y., et al.: A strain gauge that uses carbon black and carbon nanotube doped silicone oil encapsulated in a PDMS microchannel. In: 2007 7th IEEE conference on nanotechnology (IEEE NANO), pp. 1199–1202 (2007). doi: 10.1109/NANO.2007.4601398
18. Lewandowski, M., Koncar, V.: A flexible strain sensor based on a conductive polymer. Sensors., 8291–8303 (2010). doi: 10.3390/s100908291
19. Byrne, M.T., Gun'ko, Y.K.: Recent advances in research on carbon nanotube-polymer composites. Adv. Mater. **22**, 1672–1688 (2010). doi:10.1002/adma.200901545
20. Hu, L., Hecht, D.S., Grüner, G.: A method of fabricating highly transparent and conductive interpenetrated carbon nanotube-parylene networks. Nanotechnology **20**, 465304 (2009). doi:10.1088/0957-4484/20/46/465304
21. Zhang, M., Fang, S., Zakhidov, A.A., et al.: Strong, transparent, multifunctional, carbon nanotube sheets. Science **309**, 1215–1219 (2005). doi:10.1126/science.1115311
22. Shin, M.K., Oh, J., Lima, M., et al.: Elastomeric conductive composites based on carbon nanotube forests. Adv. Mater. **22**, 2663–2667 (2010). doi:10.1002/adma.200904270
23. Liu, C., Choi, J.: Strain-dependent resistance of PDMS and carbon. IEEE Trans. Microw. Theor. Tech. **9**, 590–595 (2010)
24. Lü, Q., Cao, H., Song, X., et al.: Improved electrical resistance-pressure strain sensitivity of carbon nanotube network/polydimethylsiloxane composite using filtration and transfer process. Chin. Sci. Bull. **55**, 326–330 (2010). doi:10.1007/s11434-009-0562-z
25. Huang, Y.Y., Terentjev, E.M.: Tailoring the electrical properties of carbon nanotube-polymer composites. Adv. Funct. Mater. **20**, 4062–4068 (2010). doi:10.1002/adfm.201000861
26. Bokobza, L.: Multiwall carbon nanotube elastomeric composites: a review. Polymer **48**, 4907–4920 (2007). doi:10.1016/j.polymer.2007.06.046
27. Jung, Y.J., Kar, S., Talapatra, S., et al.: Aligned carbon nanotube-polymer hybrid architectures for diverse flexible electronic applications. Nano Lett. **6**, 413–418 (2006). doi:10.1021/nl052238x
28. Zhou, Y., Hu, L., Grüner, G.: A method of printing carbon nanotube thin films. Appl. Phys. Lett., 22–24 (2006). doi: 10.1063/1.2187945
29. Hu, L., Yuan, W., Brochu, P., et al.: Highly stretchable, conductive, and transparent nanotube thin films. Appl. Phys. Lett. **94**, 161108 (2009). doi:10.1063/1.3114463
30. Zhang, D., Ryu, K., Liu, X., et al.: Transparent, conductive, and flexible carbon nanotube films and their application in organic light-emitting diodes. Nano Lett. **6**, 1880–1886 (2006). doi:10.1021/nl0608543
31. Loh, K.J., Kim, J., Lynch, J.P., et al.: Multifunctional layer-by-layer carbon nanotube-polyelectrolyte thin films for strain and corrosion sensing. Smart Mater. Struct. **16**, 429–438 (2007). doi:10.1088/0964-1726/16/2/022
32. Bennett, R.D., Hart, A.J., Miller, A.C., et al.: Creating patterned carbon nanotube catalysts through the microcontact printing of block copolymer micellar thin films. Society **8273–8276** (2006)
33. Lipomi, D.J., Vosgueritchian, M., Tee, B.C.-K., et al.: Skin-like pressure and strain sensors based on transparent elastic films of carbon nanotubes. Nat. Nanotechnol., 1–6 (2011). doi: 10.1038/nnano.2011.184
34. Sangwan, V.K., Southard, A., Moore, T.L., et al.: Transfer printing approach to all-carbon nanoelectronics. Micromechanical Engineering 1–17. doi:10.1016/j.mee.2011.06.017
35. Wu, Z., Chen, Z., Du, X., et al.: Transparent, conductive carbon nanotube films. (2011). doi: 10.1126/science.1101243

36. Loh, K.J., Lynch, J.P., Shim, B.S., Kotov, N.A.: Tailoring piezoresistive sensitivity of multilayer carbon nanotube composite strain sensors. J. Intell. Mater. Syst. Struct. **19**, 747–764 (2007). doi:10.1177/1045389X07079872
37. Cohen, D.J., Mitra, D., Peterson, K., Maharbiz, M.M.: A highly elastic, capacitive strain gauge based on percolating nanotube networks. Nano Lett. **12**, 1821–1825 (2012). doi:10.1021/nl204052z
38. Pint, C.L., Xu, Y.Q., Moghazy, S., et al.: Dry contact transfer printing of aligned carbon nanotube patterns and characterization of their optical properties for diameter distribution and alignment. ACS Nano. **4**, 1131–1145 (2010)
39. Kim, B.-S., Lee, S.W., Yoon, H., et al.: Pattern transfer printing of multiwalled carbon nanotube multilayers and application in biosensors. Chem. Mater. **22**, 4791–4797 (2010). doi:10.1021/cm101401t
40. Saran, N., Parikh, K., Suh, D.-S., et al.: Fabrication and characterization of thin films of single-walled carbon nanotube bundles on flexible plastic substrates. J. Am. Chem. Soc. **126**, 4462–4463 (2004). doi:10.1021/ja037273p
41. Li, S., Yan, Y., Liu, N., et al.: Transfer printing of submicrometer patterns of aligned carbon nanotubes onto functionalized electrodes. Small **3**, 616–621 (2007). doi:10.1002/smll.200600525
42. Palmer, H.B.: Trans. Am. Inst. Electr. Eng. **56**, 363–366 (1937)

Chapter 9
Ultrasensitive In-Plane Resonant Nano-electromechanical Sensors

Faezeh Arab Hassani, Yoshishige Tsuchiya, Adrian Mihai Ionescu, and Hiroshi Mizuta

Abstract This chapter presents the design, modeling, fabrication, and characterization of in-plane resonant nano-electromechanical (NEM) sensors based on mass detection principle. The simplified NEM sensor consists of a clamped-clamped (CC) beam with two side electrodes, in-plane resonant NEM (IP-RNEM) sensor. Later, the IP-RNEM sensor is monolithically integrated with an in-plane field-effect-transistor (FET) to realize the in-plane resonant suspended gate FET (IP-RSGFET) sensor. The numerical modeling of IP-RSGFET sensor is presented using a three-dimensional finite-element-method electromechanical simulation of IP-RNEM sensor combined with a NEM-MOS hybrid circuit simulation. We investigate the impacts of the adsorbed self-assembled linker and target molecules

F. Arab Hassani (✉)
School of Materials Science, Japan Advanced Institute of Science and Technology (JAIST), 923-1292 Nomi, Japan

Department of Electrical and Computer Engineering, National University of Singapore (NUS), 4 Engineering Drive 3, 117583 Singapore
e-mail: ahfaezeh@jaist.ac.jp; elefah@nus.edu.sg

Y. Tsuchiya
School of Electronics and Computer Science, University of Southampton, SO17 1BJ Southampton, UK
e-mail: yt2@ecs.soton.ac.uk

A.M. Ionescu
Laboratory of Micro/Nanoelectronic Devices, Ecole Polytechnique Fédérale de Lausanne (EPFL), CH-1015 Lausanne, Switzerland
e-mail: adrian.ionescu@epfl.ch

H. Mizuta
School of Materials Science, Japan Advanced Institute of Science and Technology (JAIST), 923-1292 Nomi, Japan

School of Electronics and Computer Science, University of Southampton, SO17 1BJ Southampton, UK
e-mail: hm2@ecs.soton.ac.uk

S. Li et al. (eds.), *Nanoscale Sensors*, Lecture Notes in Nanoscale Science and Technology 19, DOI 10.1007/978-3-319-02772-2_9,

in various configurations on the resonance frequency of the sensor that results in the mass responsivity of 0.05 zepto g/Hz for all the different functionalization schemes, which is approximately 11 orders of magnitude smaller than that of the reported value for present quartz crystal microbalance sensors. A comparison is shown for the important parameters of the current mass detection-based sensors and the proposed NEM sensors. The effects of structural modification on the high-frequency characteristics are studied numerically for various key parameters of the sensor. Further modification and numerical analysis for the IP-RSGFET sensor are conducted to improve the DC characteristics of the sensor for the consequent RF measurement. The RF characterization of the IP-RNEM sensor is successfully conducted using a down-mixing measurement technique. The outcomes of the modeling and experimental analysis provide a realistic guideline for further improvement and characterization of these ultrasensitive sensors.

9.1 Introduction

The development of future smart sensors will be achieved by the co-integration of micro-/nanoelectrical devices and micro-/nano-mechanical devices [1], to meet challenging specifications of advanced applications. The silicon on insulator (SOI) offers the ability to provide the monolithical integration of nano-electromechanical systems (NEMS) and integrated circuits (ICs) [2, 3]. NEM sensors attract increasing interest as a new technology because they not only offer compatibility with "In-IC" integration but also provide other specific advantages such as ultraminiature-size elements, low power consumption, potential high resonant frequencies and quality factors, fast response time, high sensitivity to applied force, external damping, or additional mass [4]. The mass detection-based NEM sensors show a great progress thanking to their nanoscale dimensions and a sensitive frequency response to ultrasmall mass changes ranging from femtogram down to zeptogram using different detection schemes [5, 6]. To present the NEM sensors that can sense the molecules with higher sensitivity compare to conventional sensors, the NEM structure is integrated with a metal-oxide-semiconductor field-effect-transistor (MOSFET) in order to improve the functionalities of the sensor using the capabilities of the MOSFET [7]. In these sensors, the resonance frequency varies due to the adsorption of single viruses or molecules to the NEM structure surface.

In this chapter, we propose a mass detection-based in-plane resonant NEM (IP-RNEM) sensor that is fabricated on a heavily doped thin SOI substrate by using top-down fabrication processes. The NEM structure (i.e., the suspended beam) is integrated with an in-plane FET on the same SOI substrate in order to amplify the output signal and is named as in-plane resonant suspended gate FET (IP-RSGFET) sensor. In Sect. 9.2, the design and structure of the IP-RNEM and IP-RSGFET sensors are presented as well as analytical and numerical values for their key parameters. The modeling for the functionalization and detection processes is also presented in this section. Section 9.3 presents the simulation methods and the results obtained for the three-dimensional finite-element-method (3D FEM) modeling of the IP-RNEM sensor and the hybrid NEM-MOS circuit modeling of the IP-RSGFET sensor. We also investigate the effect of changes of gap size and biasing conditions on the output signal of the IP-RSGFET sensor. Furthermore, the 3D FEM modeling of the surface functionalization and detection processes is discussed following by the evaluation of mass responsivity and its scaling rule. A comparative study is made between the extracted numerical values of the IP-RSGFET sensor and recently proposed NEM sensors and commercial quartz crystal microbalance (QCM) biosensors. Section 9.4 presents the fabrication process for both NEM sensors following by the single-electron-microscopy (SEM) images of the fabricated devices. The DC characterization of the suspended beams and the integrated MOSFETs are conducted for the future radio-frequency (RF) characterization of the sensor. Due to the faced challenges in the RF characterization of the IP-RNEM sensor as well as the DC characteristics of the fabricated IP-RSGFET sensors, further numerical analysis and processing are done to explain the effects and improve these measured characteristics. The down-mixing technique is applied for the RF characterization of the IP-RNEM sensor as a suitable alternative technique for the sensor with current design.

9.2 In-Plane NEM Sensors: Structure and Design

The IP-RNEM sensor consists of a suspended clamped-clamped (CC) beam and two side electrodes (Fig. 9.1a), and the CC beam is excited by applying an AC voltage. When the frequency of the AC voltage source is equal to the resonance frequency of the beam, it starts resonating in-plane as shown in the 3D FEM analysis in Fig. 9.1b. The suspended CC beam is the sensing head of the NEM sensor in which the variation of the resonance frequency, $\triangle f$, is detected depending on the mass variation of the beam due to the chemical or biological species adsorption on its surface, $\triangle m_{\text{ba}}$. In order to make the surface of the CC beam sensitive to the target molecules of interest, we need to functionalize the surface by using various self-assembled linker molecules that react with the target molecules. After the functionalization process, the target molecules are introduced on the beam in a liquid or gas phase during the detection process. More details on the functionalization and detection processes are given later. The adsorbed linker and target molecules on the surface of the CC beam change the primary resonance frequency of the CC beam. In order to amplify the output of the sensor after the molecules are adsorbed onto the surface of the suspended beam, the beam is integrated monolithically with an in-plane readout MOSFET on the same SOI substrate, IP-RSGFET sensor, as shown in Fig. 9.2. In this structure, the suspended beam plays the role of the gate for the MOSFET that moves in-plane and induces a channel in the MOSFET. After the molecules are adsorbed onto the surface of the resonant beam during the functionalization and detection processes, the resonance frequency of the beam changes and consequently causes variation in the MOSFET current.

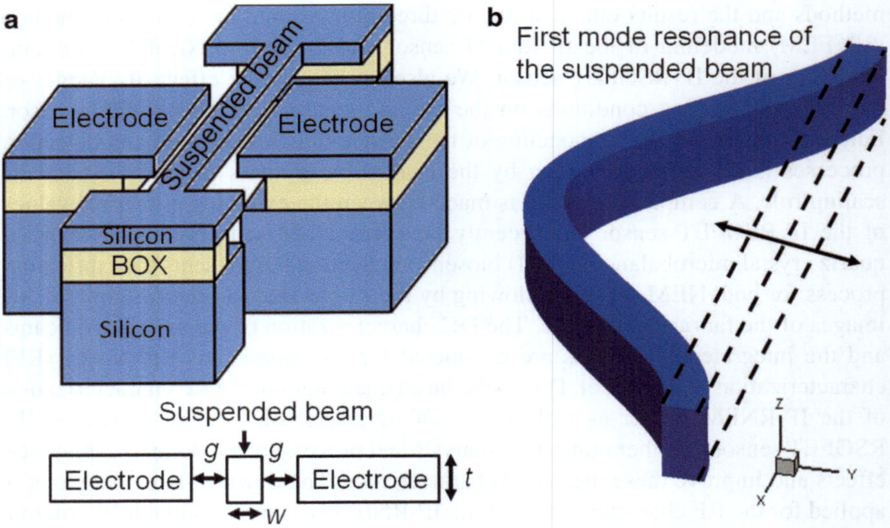

Fig. 9.1 The IP-RNEM sensor: (**a**) a suspended beam with two side electrodes and (**b**) the first resonance mode of the suspended beam

Fig. 9.2 The IP-RSGFET sensor: a suspended beam that is integrated with an in-plane MOSFET

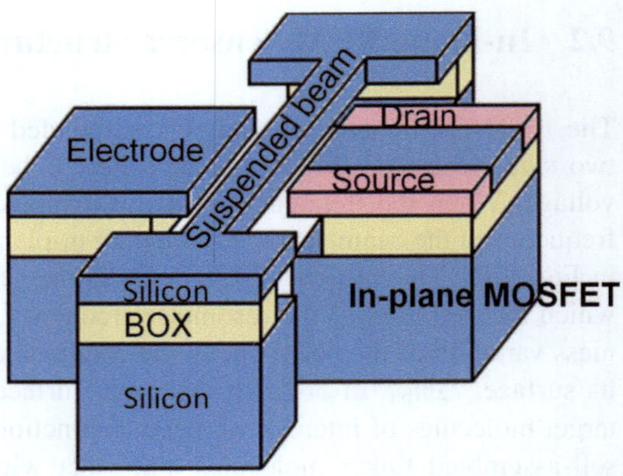

9.2.1 Key Characteristics

Some key parameters of the beam are discussed in an analytical manner as follows:

The resonance frequency of the first mode of the clamped-clamped beam is given for lateral displacement using the following equation [8–10]:

$$f_0 = 1.03\sqrt{\frac{E}{\rho}}\frac{w}{l^2} = \frac{1}{2\pi}\sqrt{\frac{k_{\mathrm{bm}}}{m_{\mathrm{b}}}} \tag{9.1}$$

where E is the Young's modulus; ρ, density; w and l, the width and length of the beam, respectively; m_b, the effective mass of the beam; and k_{bm}, the mechanical spring stiffness of the beam. m_b is given by [11]

$$m_b = 0.735\rho wlt \tag{9.2}$$

By using (9.1) and (9.2), k_{bm} is calculated as follows:

$$k_{bm} = 30.78Et\left(\frac{w}{l}\right)^3 \tag{9.3}$$

where, t is the thickness of the beam. For biasing the beam, a DC voltage was applied to the beam, V_{dc}, and an AC voltage, V_{ac}, was applied to the side electrode. More details on the applied voltages are provided later.

The induced electrostatic force due to the applied voltages causes the electrical spring stiffness, k_{be}, that is added to the mechanical spring stiffness and gives the effective spring stiffness, k_b:

$$k_b = k_{bm} + k_{be} \tag{9.4}$$

k_{be} is calculated by [12–14]

$$k_{be} = \frac{-\varepsilon_0 A V_{dc}^2}{g^3} \tag{9.5}$$

where the air permittivity, ε_0, is 8.85×10^{-12} F/m; g is the gap between the beam and electrodes; and the beam area, A, is equal to $l \times t$. As a result, the resonance frequency is calculated as follows:

$$f_r = \frac{1}{2\pi}\sqrt{\frac{k_b}{m_b}} \tag{9.6}$$

The pull-in effect that is an important issue to design the NEM/MEM structures will occur when the beam passed approximately two-thirds of the gap in the parallel-plate model. The pull-in voltage, V_p, for this model is calculated as [15]

$$V_p = 0.54\sqrt{k_b g^3/\varepsilon_0 A} \tag{9.7}$$

In order to operate the NEM sensor with a safe margin, V_{dc} was chosen to be well below V_p and assumed that $k_b \approx k_{bm}$ and consequently $f_r \approx f_0$. Energy is dissipated from the resonator in different damping mechanisms. In general, the energy in the resonator is dissipated via the ambient, $Q_{Ambient}$ (gas/liquid damping), through the anchors, Q_{Anchor} (anchor damping/loss), and its material itself, $Q_{Thermoelastic}$ (thermoelastic damping) [16]. Q_{Total} is defined as the ratio of the total energy stored in the system, W_{Total}, to the energy dissipated or lost per cycle of vibration, ΔW_{Loss} [16]:

$$Q_{\text{Total}} = 2\pi \frac{W_{\text{Total}}}{\Delta W_{\text{Loss}}} \tag{9.8}$$

Therefore, Q_{Total} consists of several individual Q-factors as follows [16]:

$$\frac{1}{Q_{\text{Total}}} = \frac{1}{Q_{\text{Ambient}}} + \frac{1}{Q_{\text{Anchor}}} + \frac{1}{Q_{\text{Thermoelastic}}} \tag{9.9}$$

Q_{Ambient} is calculated by [17]

$$Q_{\text{Ambient}} = \frac{\sqrt{k_{\text{b}} m_{\text{b}}}}{b} \tag{9.10}$$

The damping factor, b, is calculated as follows [18, 19]:

$$b = \frac{96}{\pi^4} \frac{\mu l t^3}{g^3} \tag{9.11}$$

where μ is the viscosity of the medium around the beam.

The inverse quality factor due to the thermoelastic damping, $Q^{-1}_{Thermoelastic}$, is calculated by [20]

$$Q^{-1}_{\text{Thermoelastic}} = \frac{E\alpha'^2 T}{C} \left(\frac{6}{\xi^2} - \frac{6}{\xi^3} \frac{\sinh\xi + \sin\xi}{\cosh\xi + \cos\xi} \right) \tag{9.12}$$

where α' is the thermal expansion; T is the operating temperature; and C is the heat capacity of the beam. ξ is defined by [20]

$$\xi = w\sqrt{\frac{2\pi f_0}{2\chi}} \tag{9.13}$$

where χ is the solid's thermal diffusivity.

The quality factor due to the anchor damping, Q_{Anchor}, is calculated by [21]

$$Q_{\text{Anchor}} = \frac{2 \times 2.43}{(3-\vartheta)(1+\vartheta)(\beta_n X_n)^2} \left(\frac{l}{w} \right)^3 \tag{9.14}$$

where υ is the Poisson's ratio, X_n is the shape factor, and β_n is the mode constant for a CC beam and n shows the mode number of the resonator.

An equivalent circuit model should be considered for each type of NEM structures that consists of a capacitor, an inductor, and a resistor, R_x [22]. The magnitude of the output signal of the sensor is strongly dependent to R_x at the

Table 9.1 Parameters used for calculations

Dimensions	Silicon properties	Ambient properties	Other parameters
$t = 100$ nm	$E = 130.18$ GPa	$\varepsilon_0 = 8.85 \times 10^{-12}$ F/m	$\beta_0 = 1.5056$
$w = 50$ nm	$\rho = 2{,}331$ kg/m^3	$\mu = 1.259 \times 10^{-6}$ kg/(m s)	$X_0 = -0.983$
$l = 1$ μm	$\alpha' = 2.616 \times 10^{-6}$ 1/K	$T_0 = 300$ K	
$g = 50$ nm	$C = 700$ J/(kg K)		
	$\chi = 0.86$ cm^2/s		
	$\upsilon = 0.33$		

Table 9.2 Analytical values for the IP-RNEM sensor

Key quantities	Quality factors
$f_0 = 384.95$ MHz	$Q_{\text{Ambient}} = 2{,}085.90$
$m_b = 8.56$ fg	$Q_{\text{Thermoelastic}} = 868{,}351.56$
$k_{bm} = 50.08$ N/m	$Q_{\text{Anchor}} = 4{,}998.47$
$V_p = 45.41$ V	$Q_{\text{Total}} = 1{,}469.24$
$R_x = 54.68$ kΩ	
$\eta = 1.6 \times 10^{-8}$ (V F)/m	
$\omega_c = 16.54$ GHz	
$b = 9.926 \times 10^{-12}$ (N s)/m	
$\alpha = 1.4 \times 10^{-11}$ (N s)/m	

resonance frequency especially due to the very small capacitance of the sensor. R_x can be calculated as follows:

$$R_x = \frac{\alpha}{\eta^2} \tag{9.15}$$

where α and η are calculated by [22]

$$\alpha = \frac{2\pi f_0 m_b}{Q_{\text{Total}}} \tag{9.16}$$

$$\eta = \frac{V_p \varepsilon_0 A}{g^2} \tag{9.17}$$

9.2.2 *Analytical and Numerical Values*

We assume the resonator works at the atmosphere and at the temperature of 300 K. The Q_{Total} was calculated based on three important damping sources (ambient, anchor, and thermoelastic damping) that were explained previously. The parameters used for the calculation of Q_{Total} and other key parameters such as pull-in voltage, resonant frequency, mass, and spring stiffness for the IP-RNEM sensor are given in Table 9.1, and the calculated analytical parameters are given in Table 9.2. As shown in Table 9.2, first the ambient and then the anchor dampings are smaller than the thermoelastic damping and put a limit on the Q_{Total}. By using the vacuum as the working medium and low temperature for working condition of the resonator, Q_{Ambient} and $Q_{\text{Thermoelastic}}$ will be reduced, respectively, and the anchor damping plays the dominant role among the damping sources. The anchor damping can be reduced by properly designing of the resonator so that it has reduced energy dissipation through the anchors [16, 23, 24]. Our recent modeling shows that the usage of a free-free beam for the resonator reduces Q_{Anchor} to zero after the optimization of the design. This work will be presented elsewhere.

9.2.3 Modeling of the Sensing Process

The sensing process includes two steps of functionalization and detection process. In order to detect particular biological or chemical molecules selectively, the surface of the suspended beam will be functionalized by using various self-assembled monolayer (SAM) linker molecules, for example, the silane couplers such as APTES (amino-propyltrimethoxysilane), or by using alkenes or alkyne-based molecules as shown in Fig. 9.3a. The self-assembled linker molecules were modeled simply by adding an extra surface coating layer onto the suspended beam as linker molecules are supposed to coat the surface homogeneously. In practice, the surface functionalization processes are performed either in liquid or in vapor and may result in different surface coating configurations. In this work, we studied three coating configurations: top and bottom (TB), only top (OT), and all-around (AA) coating. With the TB configuration, two surface layers with the same thickness are considered on the top and bottom of the beam. The TB configuration is shown schematically in Fig. 9.3b. This configuration may be suitable to describe the coating in liquid as it is unlikely that molecular solution flows through the nanoscale gap between the suspended beam and the electrodes as freely as above and under the beam. We also considered the OT configuration shown in Fig. 9.3c as the molecular solution does not flow under the beam when the bottom air gap is also narrow. If both the side and bottom gaps are designed wide enough, all the surface may be coated equally as shown in Fig. 9.3d. This is also a likely case for the coating in vapor as the vapor flow can go through narrow gaps more easily than the liquid flow. The total mechanical spring stiffness for the TB configuration, $k_{\text{b-TB}}$, is given by [25]

$$k_{\text{b-TB}} = 30.78(w/l)^3\left[Et + E_{\text{c}}\left(t_{\text{top}} + t_{\text{bottom}}\right)\right] \tag{9.18}$$

where E_{c} is the Young's modulus of the coating layer and t_{top} and t_{bottom}, the thickness of top and bottom layers ($t_{\text{top}} = t_{\text{bottom}}$). The calculated mass for the beam with the TB configuration is [25]

$$m_{\text{b-TB}} = m_{\text{b}} + 0.735\rho_{\text{c}}w\left(t_{\text{top}} + t_{\text{bottom}}\right)l \tag{9.19}$$

Fig. 9.3 The schematic of the (a) functionalized silicon surface with APTES, (b) top and bottom (TB), (c) only top (OT), and (d) all-around (AA) coating configurations

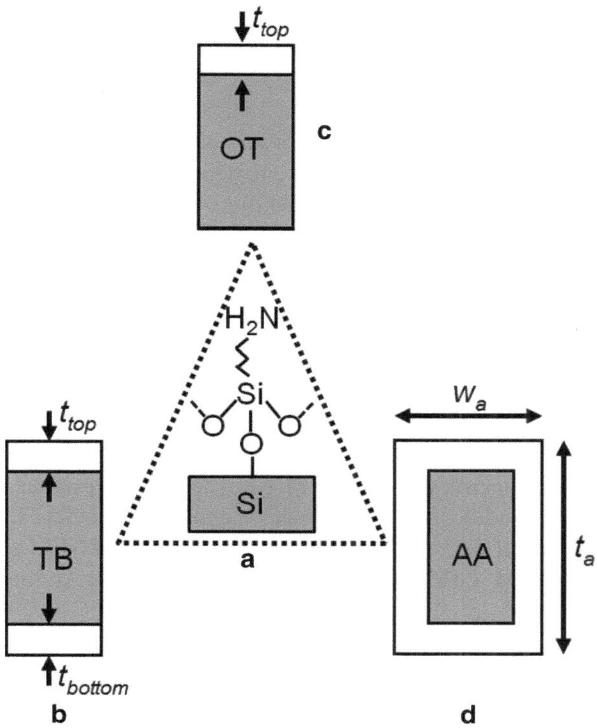

where ρ_c is the density of the coating layer. Equations (9.18) and (9.19) can be used for the OT configuration considering $t_{\text{bottom}} = 0$. $k_{\text{b-AA}}$ of the AA configuration is calculated by [25]

$$k_{\text{b-AA}} = 30.78\left\{Et(w/l)^3 + \left[\left(E_c/l^3\right)\left(t_a w_a^3 - tw^3\right)\right]\right\} \tag{9.20}$$

where t_a and w_a are the thickness and width of the beam after adding the coating layer, respectively. The calculated mass for the beam for the AA configuration is [25]

$$m_{\text{b-AA}} = m_b + 0.735\rho_c l\left(w_a t_a - wt\right) \tag{9.21}$$

Equations (9.18)–(9.21) indicate both spring stiffness and mass will increase by increasing the thickness of the coating layer. The changing rates of mass, $(\Delta m_{bl}/m_b)$, and spring stiffness, $(\Delta k_{bl}/k_b)$, depend on the configuration of the coating layer. Here $\triangle m_{bl}$ and $\triangle k_{bl}$ are changes of the mass and spring stiffness due to the linker molecules. For example for a 1-nm thick coating layer in the TB configuration, the mass of the beam changes by 1.75 %, while the spring stiffness changes by 1.07 %. According to (9.1), a larger change in the mass leads to a lower resonant frequency after adding the coating layer. Due to a 0.81 % change in the mass and a 0.51 % change in the spring stiffness for the 1 nm coating layer in the OT configuration, the resonant frequency decreases. On the other hand, for the AA configuration, a 5.60 % change in mass and a 7.90 % change in spring stiffness result in a resonant frequency larger than that for an uncoated beam. In conclusion, the coating layers on the side surfaces of the beam, which is in the direction of the beam displacement, can affect the spring stiffness more significantly than its mass. This effect is also validated using the simulation later.

9.3 Simulation Methods and Results for In-Plane NEM Sensors

CoventorWare [26] was used for simulating the NEM sensors in this chapter. This suite consists of a 3D FEM part including Designer and Analyzer and a circuit-level module, Architect. First, we conducted the 3D FEM analysis for the IP-RNEM sensor. Then, one of the electrodes in the IP-RNEM sensor was substituted with a lateral MOSFET in circuit-level modeling using Architect to realize the IP-RSGFET sensor.

9.3.1 Three-Dimensional Finite-Element-Method (3D FEM) Analysis for the IP-RNEM Sensor

The beam of IP-RNEM sensor can, in principle, resonate in plane or out of plane, but the dimensions of the beam were chosen in order to achieve the in-plane resonant behavior as its first resonance mode. The in-plane movement of the beam in the first mode is obtained by using Analyzer at the frequency of 432.77 MHz as shown in Fig. 9.1b. Harmonic analysis of the beam was conducted in the presence of a sinusoidal pressure load with the magnitude of 1 kPa. The magnitude and phase of the in-plane displacement versus frequency for the beam are shown in Fig. 9.4a, b, respectively. When the frequency of the applied pressure is equal to the resonance frequency of the beam, it resonates and shows a large peak for the displacement at the resonance frequency as shown in Fig. 9.4a. Figure 9.4b shows also a change of 180° for the displacement of the beam at the resonance frequency. The displacement versus voltage characteristics in Fig. 9.5 show the pull-in voltage of 56.95 V for the beam that shows a good compatibility to the analytical value in Table 9.2. In our modeling results, the pull-in occurred when the displacement was about half of the gap, and this is due to the fact that the displacement of the central point is significant for our clamped-clamped beam. In order to operate the NEM sensor within a safe margin, the operating voltage should be chosen well below the pull-in voltage. The derived numerical values including the resonance frequency, pull-in voltage,

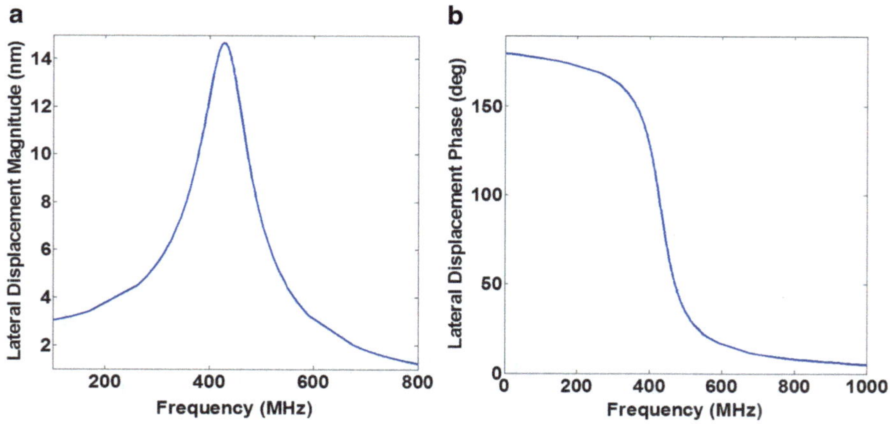

Fig. 9.4 (**a**) The magnitude and (**b**) phase of in-plane displacement of the suspended beam of the IP-RNEM sensor versus frequency in the presence of a sinusoidal pressure load

Fig. 9.5 The lateral displacement versus voltage for the calculation of the pull-in voltage

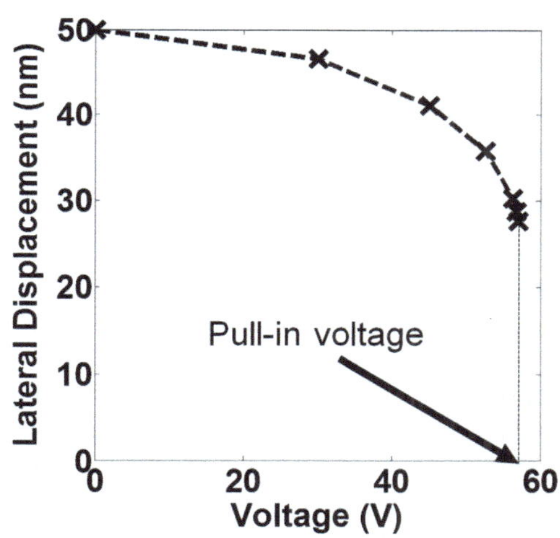

Table 9.3 Numerical values for the IP-RNEM sensor

Key quantities	Quality factors
$f_0 = 432.47$ MHz	$Q_{\text{Ambient}} = 2{,}563.08$
$b = 9.075 \times 10^{-12}$ (N s)/m	$Q_{\text{Thermoelastic}} = 680{,}632.73$
$V_{\text{p}} = 56.95$ V	$Q_{\text{Anchor}} = 3{,}153.67$
	$Q_{\text{Total}} = 1{,}411$

damping factor, and quality factors are shown in Table 9.3. A good consistency can be seen between the analytical values in Table 9.2 and the derived numerical values in Table 9.3.

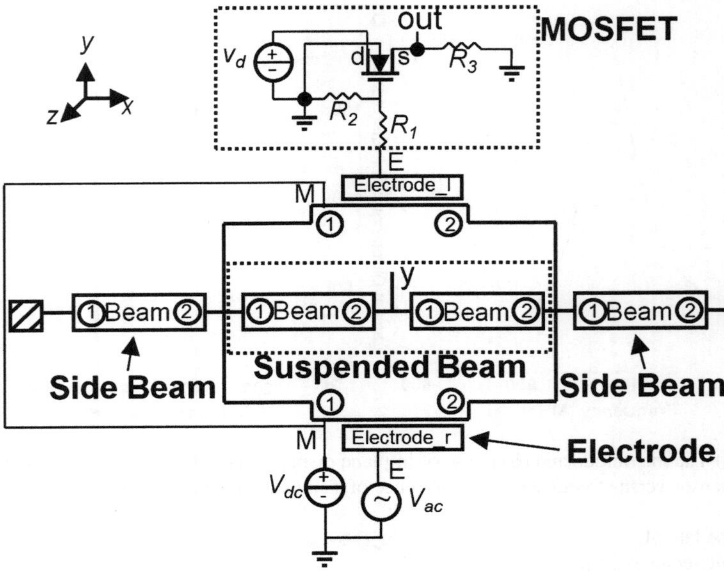

Fig. 9.6 The hybrid circuit model of the IP-RSGFET sensor in Architect

9.3.2 Hybrid Circuit Analysis for the IP-RSGFET Sensor

After conducting the 3D FEM analysis of the IP-RNEM sensor, a hybrid NEM-MOS circuit model was prepared using Architect [26], to analyze the IP-RSGFET sensor shown in Fig. 9.2. Figure 9.6 shows the hybrid circuit model for the IP-RNEM sensor, which consists of a NEM part including the electrodes and a suspended beam and a MOSFET detection part. The suspended beam is now modeled by using two central beam components with extra two side beam components. The length of the side beam components is set in order to keep the design similar to the Designer modeling in Sect. 9.2.2. Note that two central beam components are needed in order to define the node at the center and measure the displacement, *y*, as shown in Fig. 9.6. A DC voltage is applied to the beam for biasing, and an AC voltage is applied to the electrode. The output AC voltage is defined at the node shown by, out, in Fig. 9.6. The mechanical part is connected to the electrical part using resistances. R_1 was used to model the resistance of the gate, but R_2 was applied to connect the mechanical and electrical parts without receiving any error from software. The values of R_1 and R_2 were set to 10 MΩ, but later the R_2 value was reduced to the smallest value as possible, so as the resonance behavior of the sensor could be still valid for the small values of R_2. The n-type MOSFET has the 1 μm channel length same as the length of the beam.

To investigate the effects of gap changes on the resonance frequency, the gap was changed from 50 to 30 nm for three different DC voltages. The out voltage

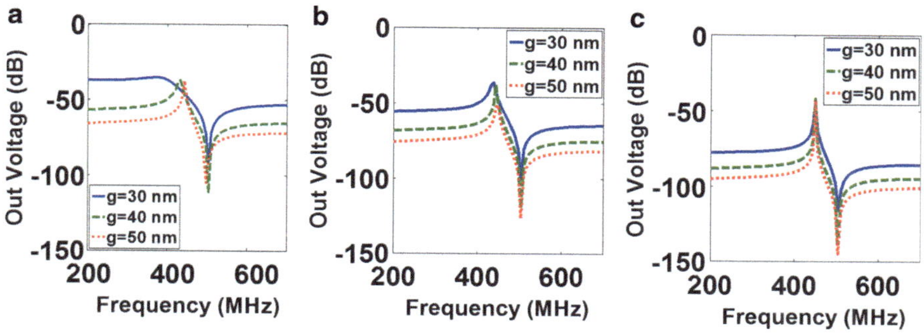

Fig. 9.7 The out voltage of the IP-RSGFET sensor versus frequency at $V_d = 10$ mV for different gaps at (**a**) $V_{dc} = 50$ V, (**b**) $V_{dc} = 30$ V, and (**c**) $V_{dc} = 10$ V

versus frequency for different gaps and voltages is shown in Fig. 9.7. The left peaks in Fig. 9.7 present the resonance frequency, while the right peaks denote the antiresonance frequency [22]. Figure 9.7a shows that at $V_{dc} = 50$ V, if the gap is decreased, the resonance frequency shifts to lower frequencies. This trend is compatible with (9.5) that shows by reducing the gap, k_{be} increases and results in smaller k_b with respect to (9.4) and consequently lower f_r with respect to (9.6). For $V_{dc} = 30$ V in Fig. 9.7b, this effect is only visible when the gap reduced from 40 to 30 nm. By reducing the applied voltage to $V_{dc} = 10$ V in Fig. 9.7c and also reducing the gap, there will not be a shift for the resonance frequency as the effect of k_{be} on k_b is smaller for smaller V_{dc}. Figure 9.7 shows that by reducing the gap in different applied DC voltages, the widening of the resonance frequency spectrum is larger which means a smaller Q-factor. This is due to the fact that only $Q_{Ambient}$ is considered as the dominant factor in these results as it was assumed that the resonator works in the atmosphere and at room temperature. By reducing the gap with respect to (9.11), the damping factor increases and results in smaller Q-factor due to (9.10).

The out voltage versus the resonance frequency for $g = 50$ nm at $V_{dc} = 10$–70 V is shown in Fig. 9.8. Figure 9.8 shows that by increasing V_{dc} from 10 to 70 V, the resonance frequency decreased due to increasing of k_{be} with respect to (9.5) and causes reduction in k_b and consequently reduction in the resonance frequency with respect to (9.4). The reduction of the resonance frequency with increasing of V_{dc} is called spring softening [13]. Reduction in k_b causes smaller $Q_{Ambient}$ with respect to (9.10) and results in wider resonance frequency spectrum as shown in Fig. 9.8.

9.3.3 Impact of Surface Modification on the Mass Responsivity

For modeling the functionalization process of the IP-RNEM sensor in different configurations (Fig. 9.3) as was described in Sect. 9.2.3, the surface coating layer

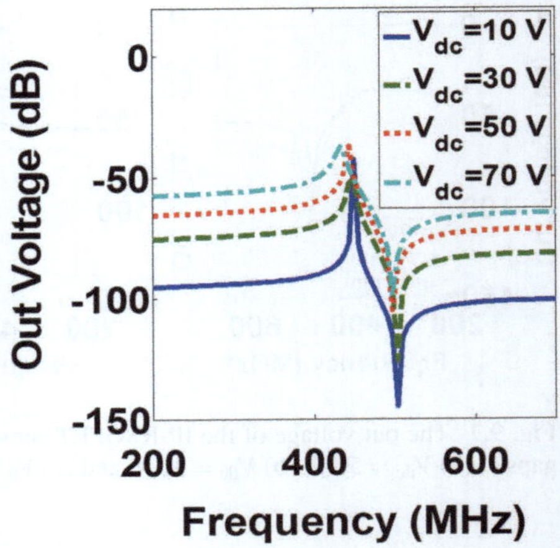

Fig. 9.8 The out voltage of the IP-RSGFET sensor versus frequency for $g = 50$ nm at different V_{dc}

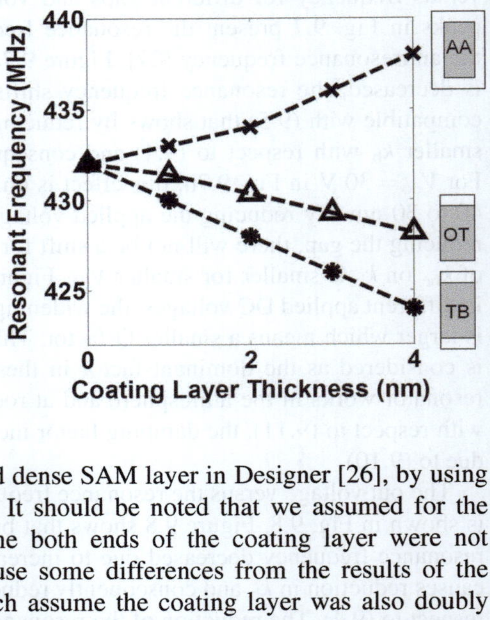

Fig. 9.9 The resonant frequency versus coating layer thickness of the TB, OT, and AA configurations for the functionalization process in Designer

was introduced as a homogenous and dense SAM layer in Designer [26], by using various insulator/polymer materials. It should be noted that we assumed for the present 3D FEM simulation that the both ends of the coating layer were not anchored. This assumption may cause some differences from the results of the analytical model in Sect. 9.2.3 which assume the coating layer was also doubly clamped. In reality, it depends on the details of the surface coating of the resonator at the clamping points whether or not the non-anchored model is more appropriate. Figure 9.9 shows the resonance frequency versus the coating layer thickness for the TB, OT, and AA configurations. In Fig. 9.9, the frequency decreases in the TB and OT configurations by increasing the coating layer thickness that shows the dominant effect of mass as was expected from the analytical model in Sect. 9.2.3. Change in the resonance frequency is lower for the OT configuration than the TB configuration due to its smaller mass value of the coating layer using (9.1). Higher resonance frequencies for the OT configuration are explained by the smaller mass of coating layer for the OT configuration than in the TB configuration. On the contrary, the

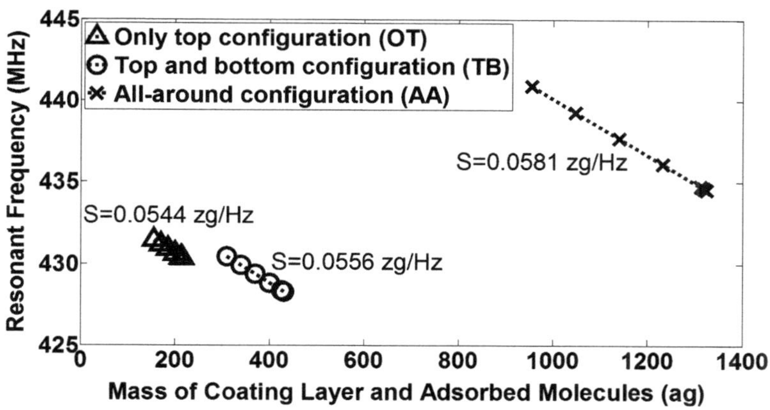

Fig. 9.10 The resonant frequency versus mass of coating layer and adsorbed molecules of the TB, OT, and AA configurations for the detection process in Designer

resonance frequency increases by increasing the coating layer thickness in Fig. 9.9 as was expected from Sect. 9.2.3 because of the dominant effect of the spring stiffness enhancement for the AA configuration.

In contrast to the functionalization process, the adsorption of target molecules to the functionalized surface is random and sparse. The sparse adsorption of target molecules will change the effective mass of the beam without much affecting the spring stiffness of the beam. In order to model the random distribution of target molecules, we simply increased only the density of the modeled coating layers with various configurations rather than introducing the random target molecules explicitly. This method enables us to study the impact of solely the mass increase due to the adsorbed target molecules on the resonant frequency and evaluate the mass responsivity of the sensor for the detection process. Figure 9.10 shows the resonant frequency calculated for the TB, OT, and AA coating configurations as a function of total mass of the surface coating layer (2 nm thick) and the adsorbed target molecules. Regardless of the different coating schemes, the increase in the mass due to the adsorbed molecules decreases the resonant frequency linearly. The inverse slope of the resonant frequency versus mass shows the mass responsivity, S. Using (9.6), S is calculated as follows [25]:

$$S \approx \Delta m_{\mathrm{ba}}/\Delta f = \Delta m_{\mathrm{ba}}^{3/2}/k_{\mathrm{b}}^{1/2} \tag{9.22}$$

where $\triangle m_{\mathrm{ba}}$ is changes of the mass due to the adsorbed molecules. Figure 9.10 shows virtually the same S values of 0.05 zepto g/Hz for all configurations regardless of modified resonant frequency with different functionalization schemes and independent of the surface area of the beam that is used for adsorption. Considering several noise processes in the operation of a resonator, Δm_{ba} is calculated as follows: [27]:

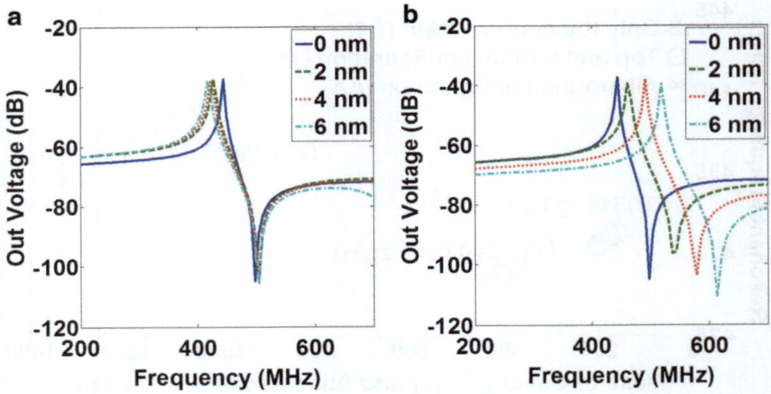

Fig. 9.11 The output voltage of the IP-RSGFET sensor versus frequency for different thicknesses of the coating layer in Architect for (**a**) the TB configuration and (**b**) the AA configuration

$$\Delta m_{\mathrm{ba}} \approx S\left(\Delta f_{\mathrm{BW}}\frac{2\pi f_0}{Q_{\mathrm{Total}}}\right)^{1/2} 10^{(-\mathrm{DR}/20)} \tag{9.23}$$

where Δf_{BW} is the maximum allowable measurement bandwidth that is f_0/Q_{Total} and DR represents the effective dynamic range intrinsic to the resonator [27]:

$$\mathrm{DR}(dB) = 10\log\left(E_{\mathrm{C}}/k_B T\right) \tag{9.24}$$

where Boltzmann constant, k_B, is 1.38×10^{-23} J/K and the maximum drive energy for the in-plane CC beam is $E_{\mathrm{C}} \approx m_{\mathrm{b}} 4\pi^2 f_0^2 0.53^2 w^2$. Using (9.22)–(9.24), Δm_{ba} is calculated equal to 11.73 atto g. By using a high-vacuum and low-temperature environment for the sensor, Δm_{ba} will reduce further.

To study the impact of adding the coating layer on the output voltage of the designed IP-RSGFET sensor, we introduced the coating layer onto the beam in our circuit model. The same assumption of non-anchored ends for the coating layer is also applied here. Figure 9.11 shows the output voltage versus frequency calculated with different coating layer thicknesses of the TB and AA configurations. The frequency spectra of the output voltage shift to low frequencies by increasing the thickness of the coating layer for the TB configuration (Fig. 9.11a), while the trend is reverse for the AA configuration (Fig. 9.11b). These results show that the effect of the mass change is dominant for the TB configuration and the effect of the spring stiffness change is dominant for the AA configuration as explained previously in this section. These results are consistent with the previous 3D simulation results in Fig. 9.9.

To investigate the scalability for the presented mass detection-based sensor, the dimensions of the beam were enlarged by the scaling factors, k, of 1, 10, 100, and 1,000. Figure 9.12 shows the calculated mass responsivity versus the scaling factor,

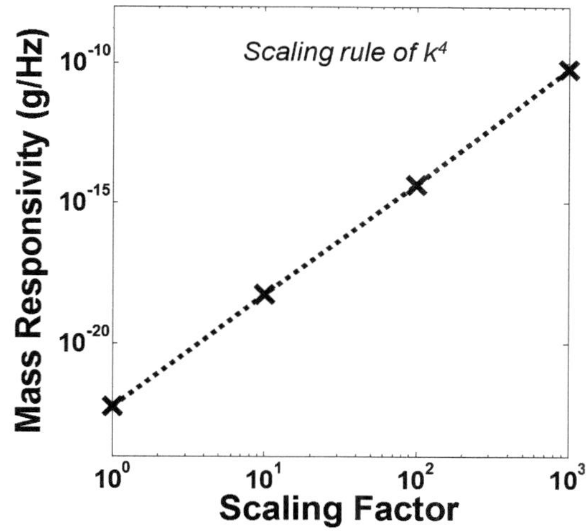

Fig. 9.12 The scaling rule for the mass responsivity of the IP-RNEM sensor

which indeed shows that the mass responsivity increases from zg/Hz to pg/Hz with the scaling rule of k^4 by increasing the scaling factor as indicated by the analytical model. This scaling rule provides a good guiding principle for the resonator design depending on our needs of mass responsivity.

9.3.4 Comparison with the Commercial and Recently Proposed Mass Detection-Based Sensors

In Table 9.4, characteristics of the IP-RNEM sensor are compared to that reported for recent commercial mass detection biosensors based on the QCM technique. The introduced NEM sensors in this paper have an extremely small sensing area (3×10^{-9} cm^2) and show a very higher sensitivity compared to the commercial QCM sensors [28–30]. The Q_{Total} and mass detection limit of IP-RNEM sensors can be improved by using a high-vacuum and low-temperature environment as discussed previously. A few of the most recent research works on mass detection-based NEM sensors using a CC beam with different materials [5, 11, 31–33] are given in Table 9.5. The IP-RNEM sensor shows higher sensitivity than that of the stated sensors in Table 9.5.

Table 9.4 Comparison of the IP-RNEM sensor and the commercial QCM biosensors

Biosensor	Mass detection limit	Q_{Total}	Mass responsivity	Sensing area	Fundamental frequency	References
Q-Sense E4	–	$\geq 10^5$	27 ng/Hz	1.54 cm^2	5 MHz	[28]
QCM Initium AFFINIX Q	100 pg	$\geq 10^5$	30 pg/Hz	0.049 cm^2	27 MHz	[29]
Mark 21 QCM sensor	56.6 µg	–	18 pg/Hz	0.1007 cm^2	50 MHz	[30]
IP-RNEM sensor	11.73 atto g	1,411	0.05 zepto g/Hz	3×10^{-9} cm^2	448.87 MHz	–

Table 9.5 Comparison of the IP-RNEM sensor and a few recent mass detection-based NEM sensors

References	Material	Type of beam	Frequency (Hz)	Q_{Total}	Temperature	Medium	Mass responsivity or detection limit
[11]	SiC	CC	32.8 M	3,000	17 K	Ultrahigh vacuum	2.53 atto g
[31]	SiC	CC	428 M	2,500	22 K	Vacuum	0.27 zepto g/Hz
[32]	SiC	CC	190 M	5,000	300 K	Ultrahigh vacuum ($<10^{-10}$ Torr)	0.86 zepto g/Hz
[5]	Carbon nanotube	CC (Fe-coated)	470 M	15	Room	Vacuum (10^{-6} Torr)	1 atto g
[33]	Silicon nano-wire	CC (metallized)	200 M	2,000	25 K	Vacuum	0.71 zepto g/Hz

9.4 In-Plane NEM Sensors: Fabrication, Characterizations, and Challenges Ahead

The IP-RNEM sensor was fabricated within four steps of lithography. First, the beam and two electrodes were patterned on the heavily doped SOI wafer of 2×10^{19} cm^{-3} that is thinned down to 50 nm, using an electron-beam (EB) lithography. Then a 200 nm SiO_2 layer was deposited on the patterned SOI layer that is followed by a photolithography step for patterning the contact holes in the deposited oxide above the side electrodes. The contact pads were then patterned in the evaporated 300 nm aluminum (Al) layer. Finally, the beam was suspended using vapor hydrofluoric acid (HF). Figure 9.13 shows the scanning electron microscope (SEM) image of the fabricated IP-RNEM sensor.

The fabrication of the IP-RSGFET sensor was also based on the SOI technology. Two types of doping strategies were used for the sensor as shown in Fig. 9.14. The thickness of SOI and BOX layers for this sensor are 40 nm and 145 nm, respectively. P-type doping of $P = 10^{15}$ cm^{-3} and $P = 10^{16}$ cm^{-3} are considered for the SOI wafers. First, the SOI wafers were implanted with different dopings of $P^+ = 10^{19}$ cm^{-3} and $N^+ = 4 \times 10^{19}$ cm^{-3} except for the channels of the MOSFETs that were protected by resist. Then the heavily doped silicon was patterned, and the beam was released using vapor HF. A 15 nm layer of thermal oxide was grown on the surface of the patterned silicon especially around the suspended beam for passivation and minimizing of the silicon surface states [34]. After that, poly silicon was deposited to fill the etched area around the beam to protect the beam from further processing steps. Contact holes above the silicon pads are opened by etching poly silicon with the etch stop layer of thermal oxide. Then oxide is deposited and contact holes are patterned in this layer. Al is deposited and patterned for contact pads and wiring afterwards. Finally, a window is opened in oxide and poly silicon above the beam for the suspension of the beam using xenon difluoride (XeF_2). The SEM view of the sensor is shown in Fig. 9.15.

For the RF characterization of the IP-RNEM sensor, Cascade SUMMIT 12000B probe station was used. The S-parameters for the sensor were measured using Agilent E8361A PNA network analyzer. Ground-signal-ground (GSG) probes with the pitch of 150 μm were used for the measurement. The characterization setup is shown in Fig. 9.16. The two-port calibration was done at the end of GSG probes. The RF characterization was done for the beam with $w = 135$ nm, $g = 80$ nm, and $l = 2{,}000$ nm that has the resonance frequency of 285 MHz based on the numerical analysis. The AC voltage with the power of −10 dBmWatt, IF bandwidth of 500 Hz, and number of point of 601 was applied. The DC voltage is applied to the beam using Agilent semiconductor device analyzer B1500. The measurement was done in the atmosphere and at room temperature.

It was not easy to distinguish the resonance peak in the transmission signal, S_{21}, from the background noise signal for this device. In order to explain this effect, the IP-RNEM sensor with the current dimensions is compared to the 14 MHz in-plane NEM resonator by Durand et al. [35]. Their resonator consists of

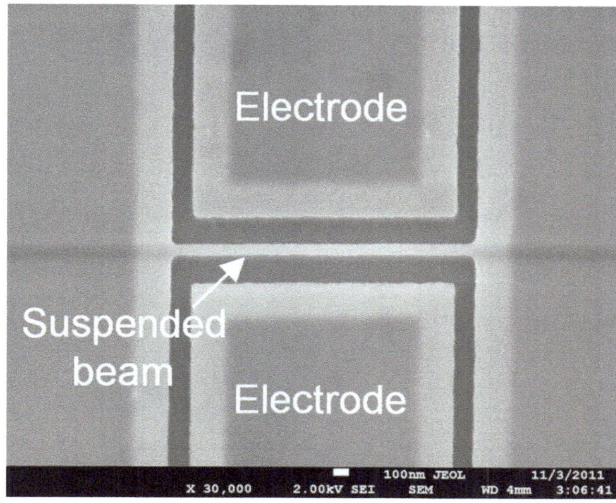

Fig. 9.13 The SEM view of the fabricated IP-RNEM sensor

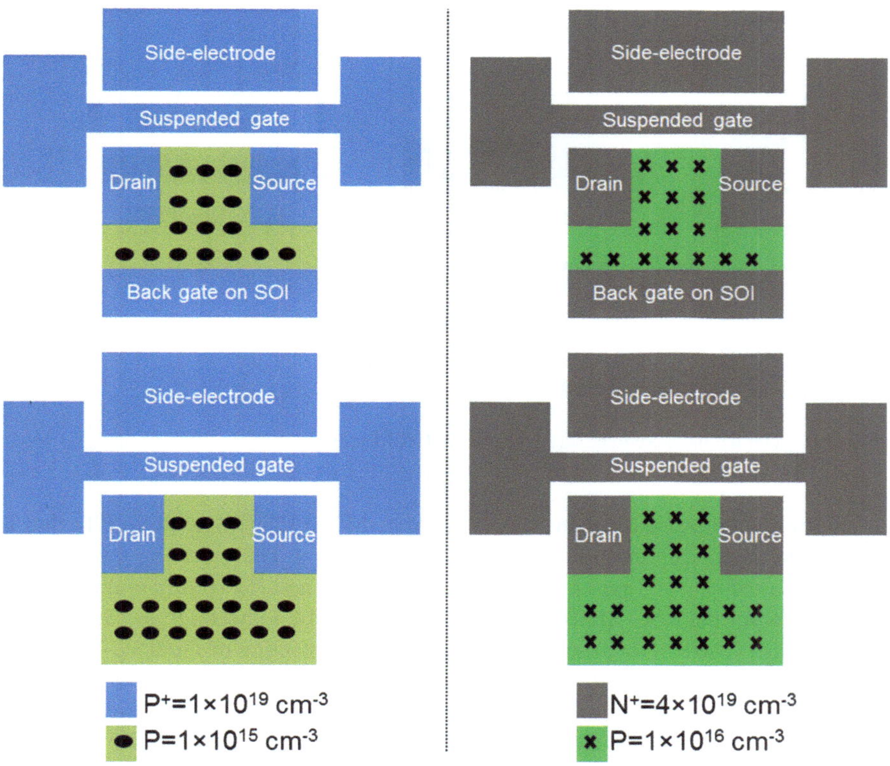

Fig. 9.14 The doping strategies for the IP-RSGFET sensor: (*left*) $P^+/P/P^+$-type and (*right*) $N^+/P/N^+$-type MOSFET with/without the back gate on SOI

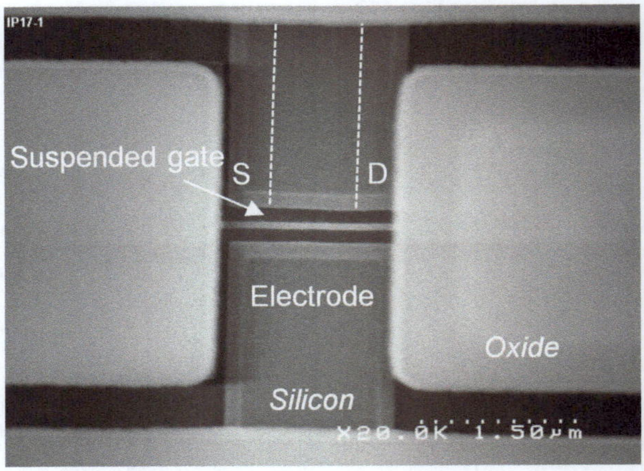

Fig. 9.15 The SEM view of the fabricated IP-RSGFET sensor

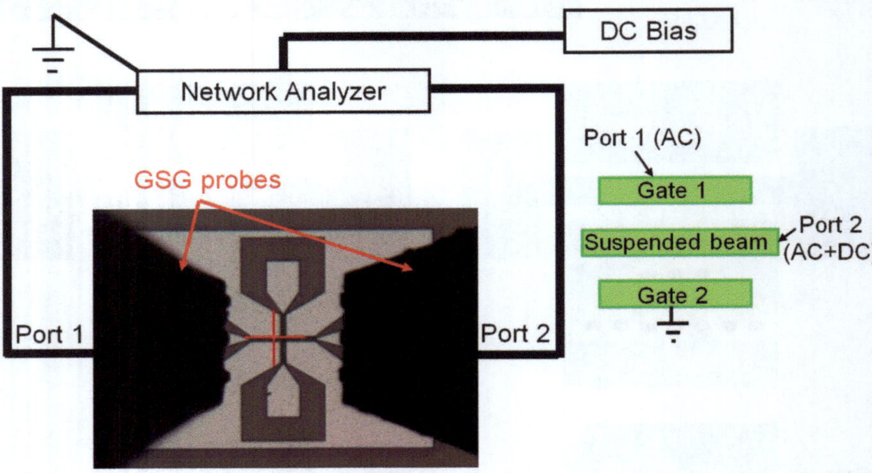

Fig. 9.16 The RF measurement setup for the IP-RNEM sensor

a vibrating gate and a resonant suspended gate MOSFET (RSG-MOSFET) that is fabricated using silicon-on-nothing (SON) technology. The vibrating gate has the dimensions of $w = 165$ nm, $t = 400$ nm, $g = 120$ nm, and $l = 10$ μm, and measured parameters for the RSG-MOSFET resonator are $f_0 = 14.43$ MHz, $R_x = 736$ kΩ, and $Q_{Total} = 700$. The electrical setup for the capacitive detection of the RSG-MOSFET resonator is similar to the setup for the IP-RNEM sensor in Fig. 9.16, and it showed 2 dB-magnitude peak for the transmission signal. The R_x for the IP-RNEM sensor that works in the atmosphere is 100 times higher than that of the RSG-MOSFET using (9.15). Due to this fact, the IP-RNEM sensor shows higher signal drop across the resistance and consequently smaller output signal than the 2 dB-magnitude transmission signal of the capacitive detection for the RSG-MOSFET.

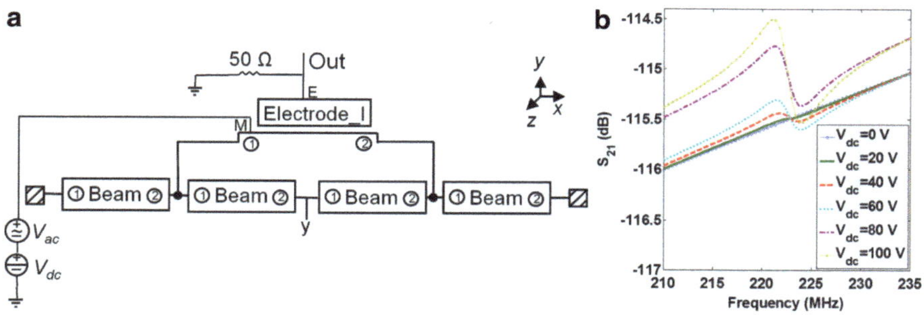

Fig. 9.17 (**a**) The circuit model for the suspended beam with one side electrode using Architect for extracting S_{21} signal and (**b**) the magnitude of the S_{21} signal for different V_{dc}

This reason explains the difficulty in distinguishing the resonance peak with very small magnitude from the background noise signal.

The circuit-level simulation of an IP-RNEM sensor with one side electrode in Fig. 9.17a was done to confirm the previous comparison between the IP-RNEM sensor and RSG-MOSFET. Figure 9.17b shows the very small magnitude of S_{21} in the presence of different applied V_{dc} that is consistent with our previous explanation. In order to reduce R_x and improve the transmission signal for the sensor, the measurement should be done in high vacuum same as the RSG-MOSFET. It is important to note that the lower working pressure reduces R_x if Q_{Total} is mainly dominated by air damping, $Q_{Ambient}$. For example Q_{Total} for the NEM sensor with $w = 135$ nm and $l = 2{,}000$ nm is dominated by anchor doping, Q_{Anchor}, and its R_x is not reduced much by using high vacuum. Using of different measurement techniques such as lock-in measurement [36–38] is a good option for the measurement of these devices that is briefly explained later in this section.

As discussed before, a MOSFET is integrated with the beam for the IP-RSGFET sensor to improve the magnitude of the transmission signal in which the output signal from the NEM structure is amplified by the intrinsic gain of MOSFET, $g_m \times r_o$, where g_m is transconductance and r_o is the output resistance of MOSFET. By optimizing the applied bias voltages in DC characteristics of the MOSFET, the maximum value for g_m is achieved that amplifies the output signal of the MOSFET as much as possible. Based on the fabrication technology, after the implantation and drive-in steps for dopants, there is an estimated lateral diffusion length of 100 nm. The voltage of 0–0.1 V was applied to drain, V_d, for the enhancement-mode n-channel MOSFET ($N^+/P/N^+$-type) with channel length, l_C, of 1,250 nm, $l = 2{,}000$ nm, $w = 135$ nm, and $g = 80$ nm as shown in Fig. 9.18a. I_d–V_d characteristics for the MOSFET for gate voltages, $V_g = 0$, 20 V, are shown in Fig. 9.18b. Figure 9.18b shows that even at $V_g = 0$ V, the channel is induced by the drain voltage. Moreover, increasing of V_g to 20 V slightly changes the drain current in inset to Fig. 9.18b that shows the very small control of gate over the channel.

I_d–V_g characteristics of the MOSFET are shown in Fig. 9.19 for V_d in the range of 200–400 mV. Figure 9.19 shows the OFF current in the order of mA and a very small

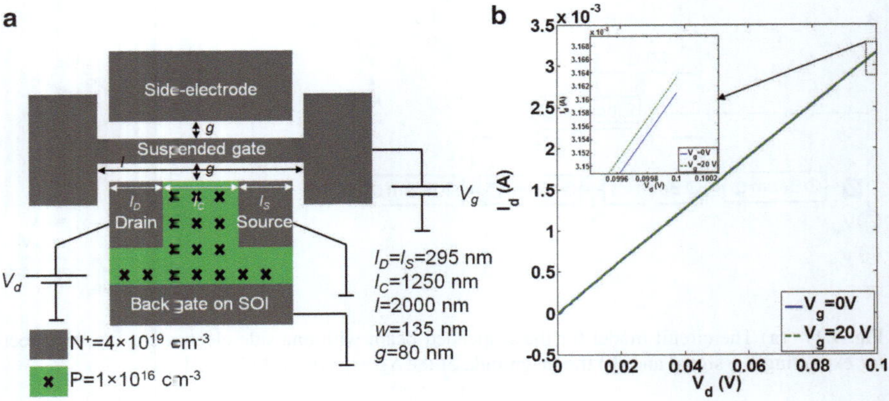

Fig. 9.18 The $N^+/P/N^+$-type in-plane MOSFET: (**a**) dimensions and the DC characterization setup and (**b**) the I_C–V_d characteristic at $V_g = 0$ and 20 V

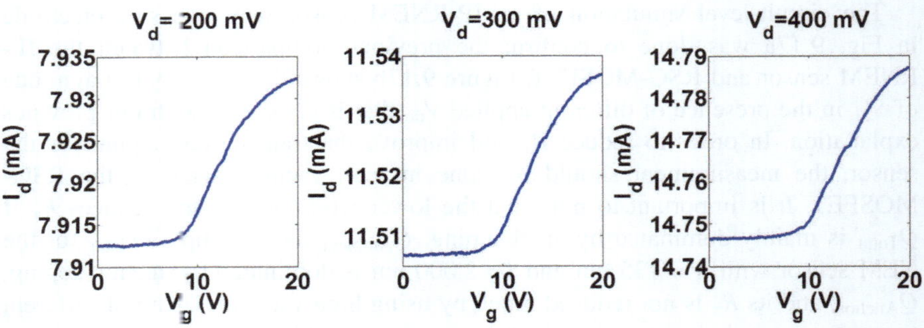

Fig. 9.19 The I_d–V_g characteristic at different V_d for the $N^+/P/N^+$-type in-plane MOSFET with $w = 135$ nm, $g = 80$ nm, and $l = 2{,}000$ nm

ON/OFF current ratio for different V_d. The current is increased by increasing V_d due to the higher impact of drain over the channel instead of the gate. The threshold voltage of $V_t \approx 9$ V is found respect to V_t extraction methods in [39] that is much larger than the analytically calculated value of $V_t = 1.75$ V for the conventional long n-channel MOSFET. The increase in the threshold voltage cannot be explained with the drain-induced-barrier-lowering (DIBL) effect which should cause the reduction in the threshold voltage [40]. The same order of OFF current has been found also for the depletion-mode p-channel MOSFET ($P^+/P/P^+$-type) with the same dimensions of the MOSFET in Fig. 9.18a.

In order to investigate the origin of high OFF current, the leakage of the MOSFET in Fig. 9.18a from source and drain to the back gate was measured. The leakage current was in the order of 10^{-4} A for both drain and source, and 1 μm distance of back gate from source and drain shows that the dopants' diffusion of source and

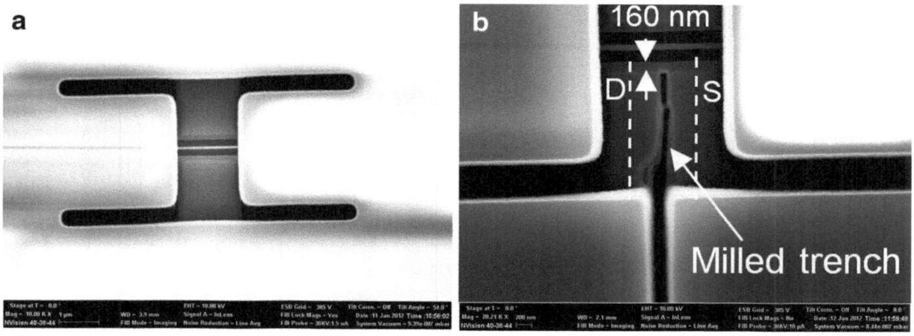

Fig. 9.20 The SEM view of the IP-RSGFET sensor with $w = 45$, $l = 1{,}750$ nm, and $l_C = 1{,}000$ nm (**a**) before and (**b**) after the FIB milling

drain towards the back gate is more than the previously stated 100 nm. The value of diffusion length that gives this level of leakage current is simulated later.

In order to isolate the source and drain as much as possible, a trench is milled between them, from the end of back gate till 160 nm away from the edge of channel using focused ion beam (FIB). This milling improved the I_d–V_g characteristics negligibly. Figure 9.20a, b shows the MOSFET with $w = 45$, $g = 200$ nm, $l = 1{,}750$ nm, and $l_C = 1{,}000$ nm before and after the FIB milling, respectively. Further improvement of the DC characteristics might be achieved considering a smaller distance than 160 nm distance between the trench and edge of channel that was not an easy task due to the alignment accuracy in the FIB milling process.

Further simulations using the device simulator ATLAS 3D [41], for the MOSFET (Fig. 9.21a), were done to investigate and confirm the reason of high threshold voltage and OFF current. The two-dimensional (2D) top view of the MOSFET is shown in Fig. 9.21b. A suspended gate is modeled by considering an air region around the beam. To make the structure more consistent with the fabricated device, a 15 nm SiO_2 layer is considered around the beam and on top of the channel as shown in the zoomed area in Fig. 9.21b. The surface mobility model, auger recombination, band gap narrowing, and Fermi-Dirac distribution are considered for the carriers in the simulation [41, 42].

In the beginning, an ideal uniform doping profile was considered for source, drain, and back gate, and the I_d–V_g characteristic of the MOSFET is shown in Fig. 9.22a. Figure 9.22a shows the OFF current in the order of 10^{-17} A and ON current in the order of μA and $V_t = 1.6$ V. Then Gaussian doping profile was considered for source, drain, and back gate, and the value for the diffusion length along with y-axis, vertical diffusion length, was varied to find the value that gives the OFF current in the order of 10^{-4} A that was previously measured. I_d–V_g characteristics of the MOSFET at $V_d = 400$ mV are shown in Fig. 9.22b with vertical diffusion length of 100 nm (solid line) and 400 nm (dotted line) while the diffusion length along x-axis, lateral diffusion length, is 50 nm for both. The solid line in

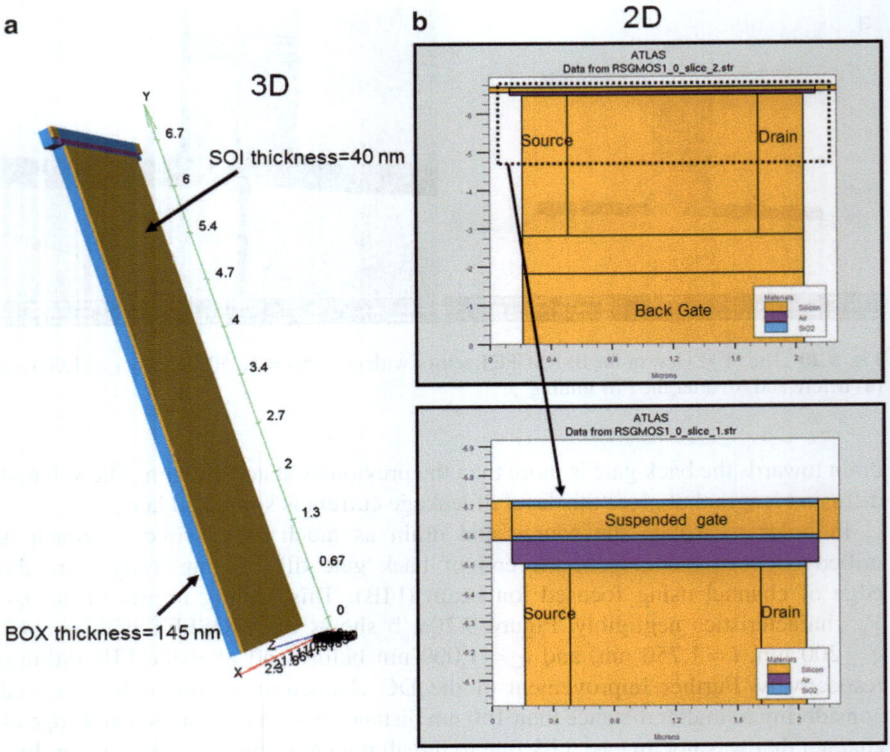

Fig. 9.21 The simulated IP-RSGFET sensor in ATLAS 3D: (**a**) the 3D and (**b**) top 2D view of the device

Fig. 9.22b shows the OFF current of nA and ON current of 8 μA and $V_t = 1.5$ V for the vertical diffusion length of 100 nm. By increasing the vertical diffusion length to 400 nm for the dotted line in Fig. 9.22b, the OFF current increases to the order of mA with the ON current in the same order that is consistent with the measurement result for $V_d = 400$ mV in Fig. 9.19. This fact also shows that how the leakage from source and drain to the back gate can affect the I_d–V_g characteristic. In Fig. 9.22b, the lateral diffusion length of 400 nm gave the leakage current in the order of 10^{-4} A from drain to the substrate and $V_t = 1.4$ V that is much smaller than the shown V_t in the measurement result for $V_d = 400$ mV in Fig. 9.19.

The lateral diffusion length is increased from 50 nm in Fig. 9.22b to 150 nm in Fig. 9.23. In Fig. 9.23 (solid line), V_t is decreased to zero volts and OFF current is increased compared to Fig. 9.22b (dotted line). Only the introduction of a negative charge ($Q_n = -1 \times 10^{12}$ cm^{-3}) in the interface of 15 nm oxide layer above the channel and channel surface causes the increase of V_t from zero for solid line to $V_t \approx 11$ V for dotted line in Fig. 9.23 that is consistent with the measurement result in Fig. 9.19. The introduced interface charge can be considered as a representative

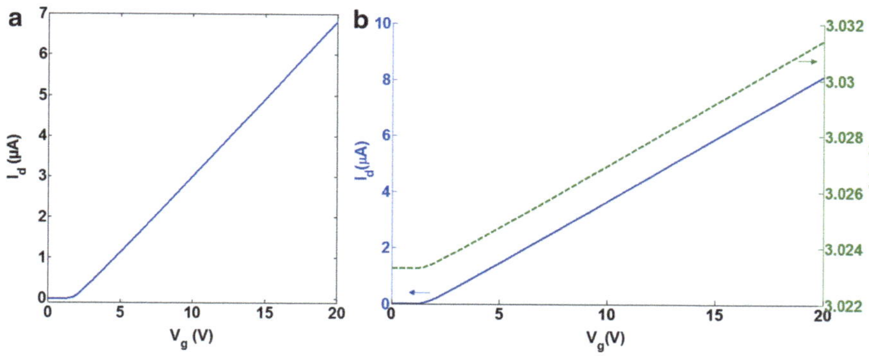

Fig. 9.22 The I_d–V_g characteristic of the N^+/P/N^+-type in-plane MOSFET in ATLAS 3D with $w = 135$ nm, $g = 80$ nm, $l = 2{,}000$ nm, and $l_C = 1{,}250$ nm at $V_d = 400$ mV considering (**a**) uniform doping for source, drain, and back gate and (**b**) Gaussian doping for source, drain, and back gate on SOI with lateral diffusion of 50 nm and vertical diffusion length of 100 nm (*solid line*) and 400 nm (*dotted line*)

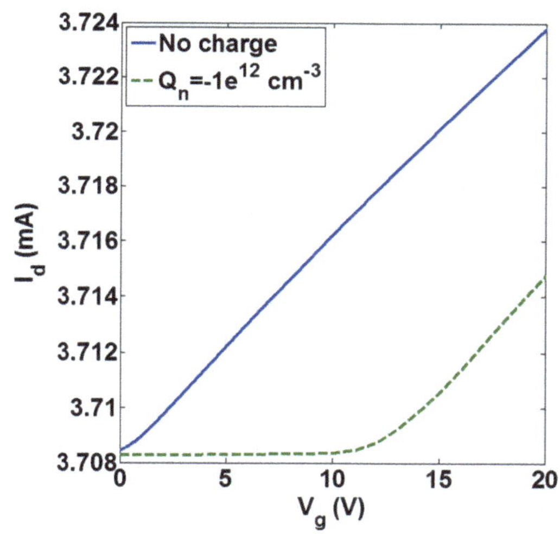

Fig. 9.23 I_d–V_g characteristics of N^+/P/N^+-type MOSFET in ATLAS 3D with $w = 135$ nm, $g = 80$ nm, $l = 2{,}000$, $l_C = 1{,}250$ nm, and lateral diffusion length of 150 nm and vertical diffusion length of 400 nm at $V_d = 400$ mV with (*dotted line*) and without Q_n (*solid line*) in the surface of channel

for the surface roughness in the channel-edge and also the possible trapped charges due to the implantation in SiO_2. The dependency of V_t to trapped charges causes the instability of DC characteristics of MOSFETs by passing time. If the same value of 400 nm for the vertical diffusion length is assumed for the lateral diffusion length, larger Q_n than -1×10^{12} cm^{-3} should be considered in the surface to have the same V_t in Fig. 9.19.

We have applied a down-mixing technique [43, 44] for the RF characterization of the IP-RNEM sensor due to the mentioned difficulties in the RF characterization of the sensor using a network analyzer. Same as the previously presented method for the RF characterization of the IP-RSGFET sensor, this current technique takes

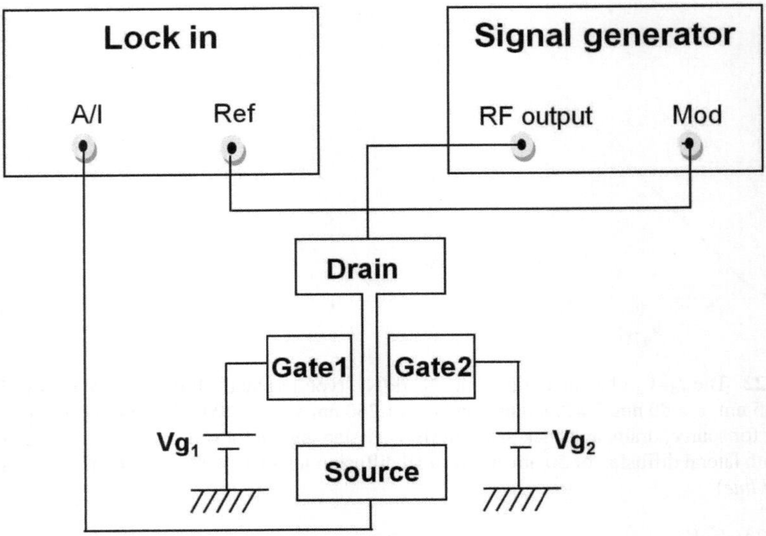

Fig. 9.24 Down-mixing measurement setup

advantage of the intrinsic gain of the integrated junction-less (JL) transistor [45] within the IP-RNEM sensor. The measurement setup is shown in Fig. 9.24. Both of the side gates were biased in order to optimize the DC bias point for the transistor. The signal generator in Fig. 9.24 was used to apply a frequency modulation signal with frequency of <100 kHz to drain. The measurements were done at high vacuum (10^{-6} mbar) and room temperature. The lock-in amplifier was used to detect the output current signal from source that is already magnified by the transconductance, g_m, of the JL transistor [43].

The DC characteristics of the sensor with $l = 1.5$ μm, $w = 45$ nm, and $t = 40$ nm are shown in Fig. 9.25a by applying same voltages to both gates. The similar DC characteristics were expected for the IP-RSGFET sensor that could not be achieved due to the high OFF current of the MOSFET. The impact of applying different voltages to gates on I_d–V_g and g_m–V_g (inset to Fig. 9.25b) characteristics are shown in Fig. 9.25b. The asymmetrical applied voltages of $V_{g1} = -20$ V and $V_{g2} = -5$ V provide higher ON current as well as high g_m.

The lock-in current versus frequency of the sensor is shown in Fig. 9.26 for various applied RF powers. The measured f_0 and Q_{Total} are less than the analytical values of $f_0 = 158.98$ MHz and $Q_{Total} = 9{,}762$ considering a 15 nm SiO_2 layer around the beam. These differences can be explained due to changes of dimensions of the suspended beam from the original designed values. Moreover, this figure shows the stability of the resonance frequency and reduction of Q_{Total} of the senor by increasing the RF power. Applying the same measurement technique for the functionalized sensor will be done in the future.

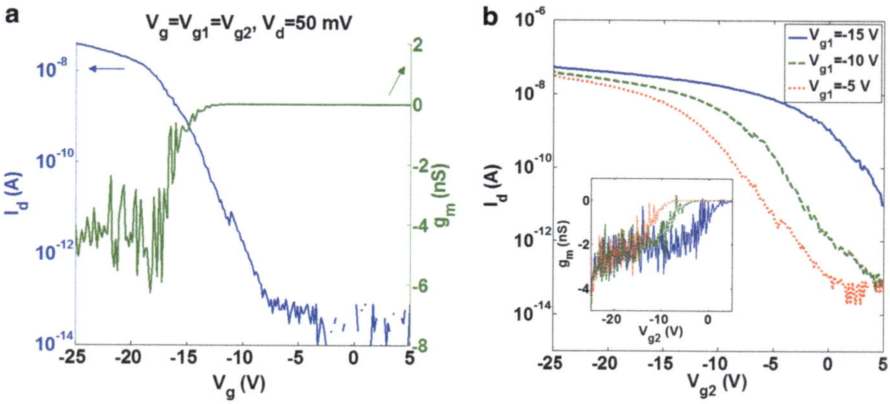

Fig. 9.25 The 1.5 μm-length NEM resonator: (**a**) DC characteristics and (**b**) the effect of various V_{g1} on DC characteristics at $V_d = 50$ mV

Fig. 9.26 The effect of various RF power on the frequency characteristics of the sensor

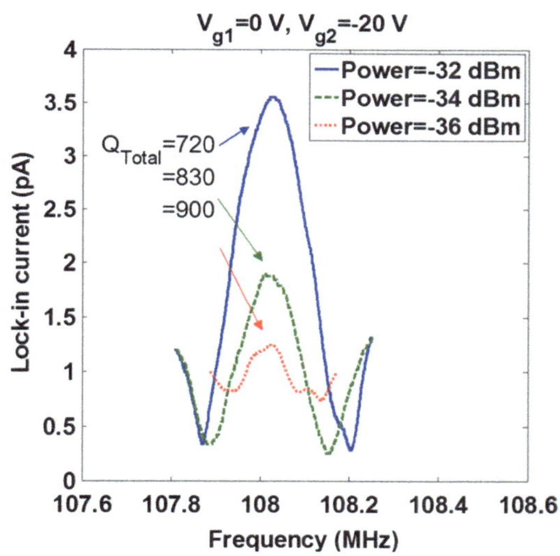

9.5 Conclusion

In this paper, we have presented the design, analysis, fabrication, and characterization of in-plane resonant NEM sensors. These sensors are based on mass detection principle and can be used as a bio/chemical sensor with a higher responsivity than that of the recent commercial QCM sensor. The proposed sensors were designed and simulated using both 3D FEM simulation and hybrid NEM-MOS circuit simulation. The surface of the suspended beam should be functionalized for adsorption of target molecules. The linker and target molecules of the sensing process have been modeled by adding extra layers to the beam in different configurations for investigating the extreme mass responsivity of 0.05 zepto g/Hz for the sensors. The scaling rule of k^4 was found for the mass responsivity of the sensor. The IP-RNEM and IP-RSGFET sensors were fabricated successfully. The RF characterization of the IP-RNEM sensor was investigated in different analytical and numerical levels to clarify the best characterization method for the sensors with current specifications. Due to the challenges of the newly proposed fabrication technology of the IP-RSGFET sensor, some post processing and simulations were done to investigate and improve the DC characteristic of this sensor. The numerical analysis shows the impact of the lateral diffusion of source/drain dopants and also the possible trapped charges in the surface of the vertical channel of the MOSFET on the DC characteristics of the sensor. Down-mixing technique was successfully applied for the RF characterization of the IP-RNEM sensors.

Acknowledgments This work is financially supported by EUFP7 project NEMSIC (hybrid nano-electromechanical/integrated circuit systems for sensing and power management applications). The authors sincerely acknowledge Dr. Cecilia Dupré and Dr. Eric Ollier (CEA-Léti), Dr. Dimitrios Tsamados (Synopsys), Mr. Sebastian T. Bartsch (EPFL), Dr. Cornel Cobianu (Honeywell Romania), and Dr. Silvia Armini and Dr. Vlamidir Cherman (IMEC-BE) for their invaluable contribution and discussion throughout the project.

References

1. Durand, C., Casset, F., Ancey, P., Judong, F., Talbot, A., Quenouillère, R., Renaud, D., Borel, S., Florin, B., Buchaillot, L.: Silicon on nothing MEMS electromechanical resonator. Microsyst. Technol. **14**, 1027–1033 (2008). doi:10.1007/s00542-007-0485-z
2. Ollier, E., Duraffourg, L., Delaye, M.T., Grange, H., Deneuville, S., Bernos, J., Dianoux, R., Marchi, F., Renaud, D., Baron, T., Andreucci, P., Robert, P.: NEMS devices for accelerometers compatible with thin SOI technology. In: 2nd IEEE int conf nano/micro engineered and molecular systems (2007)
3. Villarroya, M., Figueras, E., Montserrat, J., Verd, J., Teva, J., Abadal, G., Murano, F.P., Esteve, J., Barniol, N.: A platform for monolithic CMOS-MEMS integration on SOI wafers. J. Micromech. Microeng. **16**, 2203–2210 (2006). doi:10.1088/0960-1317/16/10/038
4. Han, X., Wu, W., Fan, J., Yan, G., Hao, Y.: Modeling of nano-resonator testing system by lumped parameter method. In: 8th int conf solid-state and integrated circuit technology (2006)
5. Yang, Y.T., Callegari, C., Feng, X.L., Ekinci, K.L., Roukes, M.L.: Zeptogram-scale nanomechanical mass sensing. Nano Lett. **6**, 583–586 (2006). doi:10.1021/nl052134m
6. Hassani, F.A., Cobianu, C., Armini, S., Petrescu, V., Merken, P., Tsamados, D., Ionescu, A.M., Tsuchiya, Y., Mizuta, H.: Design and analysis of an in-plane resonant nano-electro-mechanical sensor for sub-attogram-level molecular mass-detection. In: Extended abs int conf solid state devices and materials (2009)
7. Colinet, E., Durand, C., Duraffourg, L., Audebert, P., Dumas, G., Casset, F., Ollier, E., Ancey, P., Carpentier, J.-F., Buchaillot, L., Ionescu, A.M.: Ultra-sensitive capacitive detection based on SGMOSFET compatible with front-end CMOS process. IEEE J. Solid State Circ. **44**, 247–257 (2009). doi:10.1109/JSSC.2008.2007448
8. Clenland, A.N., Roukes, M.L.: Fabrication of high frequency nanometer scale mechanical resonators from bulk Si crystals. Appl. Phys. Lett. **69**, 2653–2655 (1996)
9. Al Khusheiny, M., Majlis, B.: Aluminum based two-port-clamped-clamped resonators. In: Proc IEEE int con semiconductor electronics (2006)

10. Rao, S.S.: Mechanical vibrations. Pearson education Inc, Pearson Prentice Hall, NJ (2004)
11. Ekinci, K.L., Huang, X.M., Roukes, M.L.: Ultrasensitive nanoelectromechanical mass detection. Appl. Phys. Lett. **84**, 4469–4471 (2004). doi:10.1063/1.1755417
12. Nathanson, H.C., Newell, W.E., Wickstrom, R.A., Davis, J.R.J.: The resonant gate transistor. IEEE Trans. Electron. Dev. **14**, 117–133 (1967)
13. Pacheco, S., Zurcher, P., Young, S., Weston, D., Dauksher, W.: RF MEMS resonator for CMOS back-end-of-line integration. Topical meeting silicon monolithic integrated circuits in RF systems (2004)
14. Bannon, F.D., Clark, J.R., Nguyen, C.T.-C.: High-Q HF microelectromechanical filters. IEEE J. Solid State Circ. **35**, 512–526 (2000)
15. Lobontiu, N., Garcia, E.: Mechanics of microelectromechanical systems. Kluwer Academic Publishers, Boston (2005)
16. Chouvion, B.: Vibration transmission and support loss in MEMS sensors. Dissertation, University of Nottingham (2010)
17. Beeby, S., Ensell, G., Kraft, M., White, N.: Inertial sensors. In: MEMS mechanical sensors, Artech House, Inc, Norwood (2004)
18. Berny, A.: Substrate effects in squeeze film damping of lateral parallel-plate sensing MEMS structures. http://www-bsac.eecs.berkeley.edu/~pister/245/project/
19. Brotz, J.: Damping in CMOS-MEMS resonators. Dissertation, Carnegie Mellon University (2004)
20. Lifshitz, R., Roukes, M.L.: Thermoelastic damping in micro- and nanomechanical systems. Phys. Rev. B. **61**, 5600–5609 (2000). doi:10.1103/PhysRevB.61.5600
21. Haoa, Z., Erbil, A., Ayazi, F.: An analytical model for support loss in micromachined beam resonators with in-plane flexural vibrations. Sens. Actuator A. **109**, 156–164 (2003). doi:10.1016/j.sna.2003.09.037
22. Abelé, N.: Design and fabrication of suspended-gate MOSFETs for MEMS resonator, switch and memory applications. Dissertation, Institute of Microelectronics and Microsystems (IMM) Electronics Laboratories (LEG) and Center of MicroNano Technology (CMI) (2007)
23. Stemme, G.: Resonant silicon sensors. J. Micromech. Microeng. **1**, 113–125 (1991). doi:10.1088/0960-1317/1/2/004
24. Hsu, W.-T., Clark, J.R., Nguyen, C.T.-C.: Q-optimized lateral free-free beam micromechanical resonators. Dig tech papers 11th int conf solid-state sensors and actuators (2001)
25. Hassani, F.A., Cobianu, C., Armini, S., Petrescu, V., Merken, P., Tsamados, D., Ionescu, A.M., Tsuchiya, Y., Mizuta, H.: Numerical analysis of zeptogram/Hz-level mass responsivity for in-plane resonant nano-electro-mechanical sensors. Microelectron. Eng. **88**, 2879–2884 (2011). doi:10.1016/j.mee.2011.03.005
26. http://www.coventor.com/
27. Ekinci, K.L., Yang, Y.T., Roukes, M.L.: Ultimate limits to inertial mass sensing based upon nanoelectromechanical systems. J. Appl. Phys. **95**, 2682–2689 (2004). doi:10.1063/1.1642738
28. http://www.qsense.com/q-sense-e4/
29. http://www.initium2000.com/en/products_qn.html
30. http://www.qcmresearch.com/#/devices/passive/
31. Feng, X.L., White, C.J., Hajimiri, A., Roukes, M.L.: A self-sustaining ultrahigh-frequency nanoelectromechanical oscillator. Nat. Nanotechnol. **3**, 342–346 (2008). doi:10.1038/nnano.2008.125
32. Peng, H.B., Chang, C.W., Aloni, S., Yuzvinsky, T.D., Zettl, A.: Ultrahigh frequency nanotube resonators. Phys. Rev. Lett. **97**, 087203 (2006). doi:10.1103/PhysRevLett.97.087203
33. Feng, X.L., He, R., Yang, P., Roukes, M.L.: Very high frequency silicon nanowire electromechanical resonators. Nano Lett. **7**, 1953–1959 (2007). doi:10.1021/nl0706695
34. Black, D.J., Mattzela, J., Ho, T., Wang, Y., Lew, K.-K., Redwing, J., Mayer, T.S.: Plasma-assisted oxidation for surface passivation of silicon nanowires. NSF EE REU PENN STATE Annu. Res. J. **2**, 121–128 (2004)

35. Durand, C., Casset, F., Renaux, P., Abelé, N., Legrand, B., Renaud, D., Ollier, E., Ancey, P., Ionescu, A.M., Buchaillot, L.: In-plane silicon-on-nothing nanometer-scale resonant suspended gate MOSFET for in-IC integration perspectives. IEEE Electron. Dev. Lett. **29**, 494–496 (2008). doi:10.1109/LED.2008.919781
36. Bartsch, S.T., Grogg, D., Lovera, A., Tsamados, D., Ayoz, S., Ionescu, A.M.: Resonant-body Fin-FETs with sub-nW power consumption. Int IEEE electron devices meeting (2010)
37. Arun, A., Campidelli, S., Filoramo, A., Derycke, V., Salet, P., Ionescu, A.M., Goffman, M.F.: SWNT array resonant gate MOS transistor. Nanotechnology **22**, 1–5 (2011). doi:10.1088/0957-4484/22/5/055204
38. Zhu, R., Wang, D., Xiang, S., Zhou, Z., Ye, X.: Piezoelectric characterization of a single zinc oxide nanowire using a nanoelectromechanical oscillator. Nanotechnology **19**, 1–5 (2008). doi:10.1088/0957-4484/19/28/285712
39. Ortiz-Conde, A., Garcia Sanchez, F.J., Liou, J.J., Cerdeira, A., Estrada, M., Yued, Y.: A review of recent MOSFET threshold voltage extraction methods. Microelectron. Reliab. **42**, 583–596 (2002). doi:10.1016/S0026-2714(02)00027-6
40. Tsividis, Y.: Operation and modeling of the MOS transistor. McGraw-Hill, Inc., New York (1987)
41. http://www.silvaco.com/
42. http://ecee.colorado.edu/~bart/book/
43. Bartsch, S.T., Rusu, A., Ionescu, A.M.: A single active nanoelectromechanical tuning fork front-end radio-frequency receiver. Nanotechnology **23**, 225501–225507 (2012). doi:10.1088/0957-4484/23/22/225501
44. Gouttenoire, V., Barois, T., Perisanu, S., Leclercq, J.L., Purcell, S.T., Vincent, P., Ayari, A.: Digital and FM demodulation of a doubly clamped single-walled carbon-nanotube oscillator: towards a nanotube cell phone. Small **6**, 1060–1065 (2010). doi:10.1002/smll.200901984
45. Bartsch, S.T., Dupré, C., Ollier, E., Ionescu, M.A.: Resonant-body silicon nanowire field effect transistor without junctions. Int. IEEE electron devices meeting (2012)

Index

S. Li et al. (eds.), *Nanoscale Sensors*, Lecture Notes in Nanoscale Science and Technology 19, DOI 10.1007/978-3-319-02772-2,

MIX
Papier aus verantwortungsvollen Quellen
Paper from responsible sources
FSC® C105338

If you have any concerns about our products,
you can contact us on
ProductSafety@springernature.com

In case Publisher is established outside the EU,
the EU authorized representative is:
Springer Nature Customer Service Center GmbH
Europaplatz 3, 69115 Heidelberg, Germany

Printed by Libri Plureos GmbH
in Hamburg, Germany